Petroleum Engineering

The Springer series in Petroleum Engineering promotes and expedites the dissemination of new research results and tutorial views in the field of exploration and production. The series contains monographs, lecture notes, and edited volumes. The subject focus is on upstream petroleum engineering, and coverage extends to all theoretical and applied aspects of the field. Material on traditional drilling and more modern methods such as fracking is of interest, as are topics including but not limited to:

- Exploration
- Formation evaluation (well logging)
- Drilling
- Economics
- Reservoir simulation
- Reservoir engineering
- Well engineering
- Artificial lift systems
- Facilities engineering

Contributions to the series can be made by submitting a proposal to the responsible Springer Editor, Charlotte Cross at charlotte.cross@springer.com or the Academic Series Editor, Dr. Gbenga Oluyemi g.f.oluyemi@rgu.ac.uk.

More information about this series at http://www.springer.com/series/15095

Hua Wang · M. Nafi Toksöz · Michael C. Fehler

Borehole Acoustic Logging—Theory and Methods

Springer

Hua Wang
School of Resources and Environment
University of Electronic Science and Technology of China
Chengdu, China

Michael C. Fehler
Earth Resources Laboratory, Department of Earth, Atmospheric and Planetary Sciences
Massachusetts Institute of Technology
Cambridge, MA, USA

M. Nafi Toksöz
Earth Resources Laboratory, Department of Earth, Atmospheric and Planetary Sciences
Massachusetts Institute of Technology
Cambridge, MA, USA

ISSN 2366-2646 ISSN 2366-2654 (electronic)
Petroleum Engineering
ISBN 978-3-030-51422-8 ISBN 978-3-030-51423-5 (eBook)
https://doi.org/10.1007/978-3-030-51423-5

This Springer imprint is published by the registered company Springer Nature Switzerland AG
The registered company address is: Gewerbestrasse 11, 6330 Cham, Switzerland

Preface

Acoustic logging relies on the analysis of seismic waves generated and recorded within a borehole. These waves provide information about the borehole, the surrounding formation, and the casing and cement of a cased hole. Since the early days, when a single source and single receiver were used, acoustic logging has advanced into a sophisticated system with a broad spectrum of applications aimed at obtaining information about the borehole and the surrounding medium. Many advanced technologies developed in recent years have improved our ability to extract additional information from acoustic logging. Among them is acoustic logging-while-drilling, in which real-time data are acquired in vertical, deviated, or horizontal wells while drilling. Another newly developed technology is borehole acoustic reflection imaging, which has the capability to delineate geological structures as far as 10–20 m away from the borehole. Other developments include high-tech sonic/ultrasonic methods, which can be employed to evaluate the condition of the cement bond in cased holes. In short, acoustic logging has grown into an extensive field with advanced applications. However, the most recent books concerning acoustic well logging were published more than 16 years ago (Paillet and Cheng 1991; Tang and Cheng 2004) and do not cover all of these newly developed technologies. To mitigate this gap in books, this book provides an in-depth review of acoustic logging using extensive numerical modeling examples of waves in and around a borehole to help visualize and understand the waveforms resulting from complicated boreholes and measurement geometries.

This book covers the principles, historical development, and applications of many acoustic logging methods, including acoustic logging-while-drilling and cased hole logging methods. State-of-the-art simulation methods, such as the discrete wavenumber integration method (DWM) and the finite difference method (FDM), are introduced to tackle the numerical challenges associated with models containing large material contrasts, such as the contrasts between borehole fluids and steel casings. In addition, waveforms and pressure snapshots are shown to help the reader understand the wavefields under various conditions. Advanced data processing methods, including velocity analyses within the time and frequency domains, are utilized to extract the velocities of different modes. Furthermore, we discuss how various formation parameters influence the waveforms recorded in the borehole and describe the principles of both existing and potential tool designs and data acquisition schemes.

Benefiting from the rapid development of information technology, the subsurface energy resource industry is moving toward data integration to increase the efficiency of decision-making through the use of advanced big data and artificial intelligence technologies, such as machine/deep learning. However, wellbore failure may happen if evaluations of risk and infrastructure are made using data mining methods without a complete understanding of the physics of borehole measurements. Hence, the uncertainty of the data must be properly understood to implement reliable supervised machine learning methods. Processed results from borehole acoustic logging will constitute part of the input data used for data integration. Therefore, to successfully employ modern techniques for data assimilation and analysis, one must fully understand the complexity of wave mode propagation, how such propagation is influenced by the well and the materials placed within the well (i.e., the cement, casing, and drill strings), and ultimately how waves penetrate into and are influenced by geological formations.

This book will benefit geophysicists (including borehole geophysicists and seismologists), petrophysicists, and petroleum engineers who are interested in formation evaluation and cementation conditions. In addition, this work will be of interest to researchers in the acoustic sciences and to fourth-year undergraduate and postgraduate students in the areas of geophysics and acoustical physics.

This book greatly benefits from the research and knowledge generated over four decades at the Earth Resources Laboratory (ERL) of the Massachusetts Institute of Technology (MIT) under its acoustic logging program. We thank the sponsors, researchers, staff, and graduates of ERL for providing the knowledge base from which this book has benefited. Hua Wang feels honored to have worked at MIT ERL for nearly 6 years. Interacting with ERL members over this period has helped Hua Wang to improve both his English and technical skills. We thank Dr. Douglas Miller, Dr. Daniel Burns, and Dr. Aimé Fournier at MIT, whose experience in data processing and mathematics stimulated our ideas on borehole acoustic data. We have greatly benefited through our close work with them and through the many resulting discussions.

Hua Wang thanks his Ph.D. thesis advisor at the China University of Petroleum (Beijing), Prof. Guo Tao, currently a Professor at Khalifa University of Science and Technology in Abu Dhabi, and Professor Lizhi Xiao at the China University of Petroleum (Beijing), for their enduring help during his research. Dr. Wei Li at the Beijing Research Center of Saudi Aramco provided some materials shown in Chap. 3. Dr. Chao Li at the Institute of Acoustics, Chinese Academy of Sciences, Dr. Meng Li at Xi'an Shiyou University, and Dr. Junxiao Li at Petroliam National Berhad (Petronas) provided and processed some of the data shown in Chap. 7. Mr. Ioan Alexandru Merciu and his colleagues at Equinor read a draft of the book and provided helpful suggestions. We also thank other former colleagues who aided and supported us during the writing of this book, including Dr. Xuefeng Shang at the Shell Exploration and Production International Company and Dr. Xinding Fang at the Southern University of Science and Technology, and we express our gratitude for the support from our many friends, including Dr. Xiao He at the Institute of Acoustics, Chinese Academy of Sciences, and Mr. Aihua Tao at China Oilfield Services Limited.

Charlotte Johnson spent countless hours editing early versions of this book and contributed greatly to its readability, and Anna Shaughnessy helped edit Chap. 6. We could not have completed the book without their dedicated efforts and help. Thank you both!

Special and deep thanks are owed to Hua Wang's wife Xichen Xu, his son, Albert (Mingyue) Wang, and his daughter, Annie (Chuoyue) Wang. The long-distance separation between Cambridge and Beijing and later between Chengdu and Beijing made it difficult for them while this book was being written. As such, we deeply appreciate their support.

Finally, we are grateful for the financial support provided by the National Science Foundation of China (No. 41404100 and 41974150), the China Postdoctoral Science Foundation (No. 2013M530106), the International Postdoctoral Exchange Fellowship Program managed by the office of the China Postdoctoral Council, the Founding Members Consortium of ERL at MIT, and MIT Energy Initiative Seed Fund Award (No. 015728-00149).

Chengdu, China
Hua Wang
Professor

Cambridge, MA, USA
M. Nafi Toksöz
Professor

Cambridge, MA, USA
Michael C. Fehler
Senior Scientist

Contents

Chapter 1
Introduction

Well logging is extensively employed within boreholes to obtain high-resolution physical measurements and determine the properties of the subsurface. Accordingly, well logging constitutes one of the key technologies in the oil and gas industry, but it also has applications in hydrology, environmental geoscience and geotechnical research. The information acquired through well logging complements surface measurements by providing high-resolution subsurface data; these borehole measurements provide detailed geological, petrophysical, and fluid properties under in situ conditions with continuous depth coverage.

The most widely used logging technology is wireline logging where the logging tool, equipped to make measurements, is lowered into the borehole (Fig. 1.1). The wireline (cable) provides a connection to the surface for both power and data transmission.

Another approach is the relatively new logging-while-drilling technique, which involves the placement of sensors on the drill collar and the transmission of information to the surface by either mud pulse telemetry (MPT) or electromagnetic telemetry (EMT) (Jarrot et al. 2018). Data may also be stored in the downhole memory and brought to the surface when the drill pipe is removed from the hole.

Borehole logging was pioneered by Conrad and Marcel Schlumberger in 1927 (Johnson 1962; Allaud and Martin 1978; Luthi 2000). The first measurement was a resistivity log (Fig. 1.2) made in a borehole in Pechelbronn, France, on September 5, 1927 (Tixier and Martin 1962). The measurement was made with an arrangement of four-electrodes at fixed intervals, and the measured resistivity was plotted on graph paper as a function of depth. An oil-bearing sandstone layer was identified by the resistivity measurements. The success of these direct resistivity measurements launched the use of logging as a standard method that could replace more expensive coring, particularly as the resistivity profiles from different wells provided a new tool to correlate beds among various wells. Hence, the success of the Schlumberger brothers marked the birth of well logging.

Today, borehole logging encompasses a wide variety of measurements using every branch of physics (Schlumberger 1987; Hilchie 1978; Tao 2006). Borehole

H. Wang et al., *Borehole Acoustic Logging—Theory and Methods*, Petroleum Engineering, https://doi.org/10.1007/978-3-030-51423-5_1

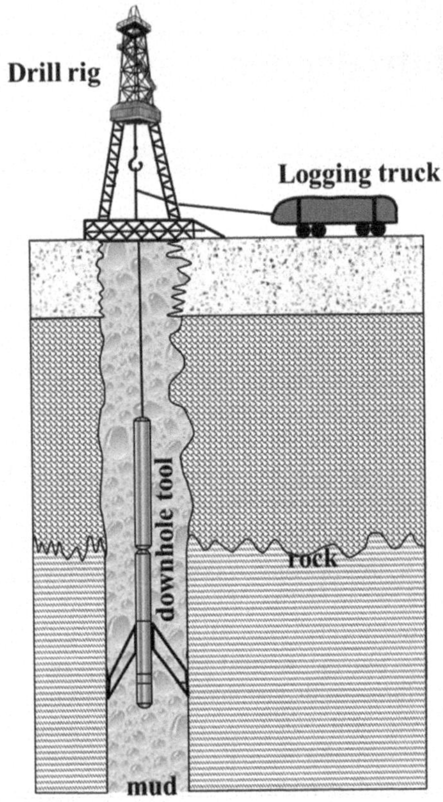

Fig. 1.1 Schematic illustration of wireline logging

logging is an interdisciplinary field that utilizes a wide array of physical measurements to generate detailed characterizations of the subsurface geology, lithology, petrophysics, fluids and saturation. Borehole measurements provide electrical (Doll 1949; Boucher et al. 1951), acoustic (Paillet and Cheng 1986; Tang and Cheng 2004), nuclear (Hilchie 1978), magnetic (Coates et al. 2000), and gamma and X-ray logs as well as information regarding the temperature and pressure (Asquith and Krygowski 2006; Ellis and Singer 2007; Lebourg et al. 1956; Chin et al. 2015).

However, seismic waves remain the primary means for imaging the Earth's interior, studying earthquakes, and exploring for oil/gas resources. Nevertheless, acoustic logging complements surface seismic acquisitions by providing direct measurement in a third dimension, i.e., the depth direction.

In this book, we cover the narrow but important field of seismic and acoustic logging. Acoustic logging plays a major role in determining of formation properties, the conditions of the borehole, casing, and cement, and assessing the well safety. We focus a great deal of attention on the use of numerical simulation to investigate the wavefields in various logging scenarios. The goal is to give the reader graphical insight into the influences of various formations and borehole geometries on the wavefields in acoustic logging.

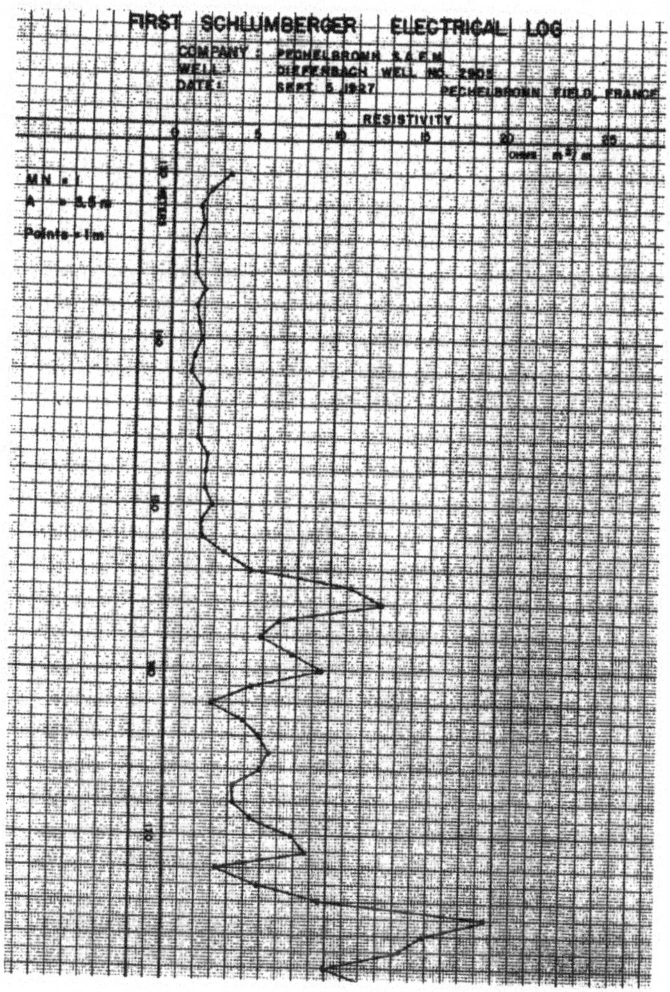

Fig. 1.2 The first resistivity log made at Pechelbronn, France, acquired on September 5, 1927 (from Fig. 1 in Johnson 1962)

The selected references given at the end of this chapter provide an extensive and comprehensive background set of references.

1.1 Evolution of Borehole Acoustic Logging

Acoustic logging was introduced in the 1930s to measure the compressional (P)-wave velocity of a formation (Schlumberger 1987). In the early years, hydrophones were lowered into the borehole by logging cables to measure the seismic travel times from sources at the surface; eventually, the source and receivers were combined into one tool system. Acoustic logging has since been employed commercially to determine the velocity of a formation by recording the travel times of first arrivals within the borehole (Summers and Broding 1952; Vogel 1952).

The evolution of acoustic logging tools is illustrated in Fig. 1.3. The simplest configuration consists of one transmitter and one receiver (Fig. 1.3 (1)). However, tools with more receivers and/or transmitters were developed to reduce the effects of tool tilt and borehole radius changes. Initially, the travel times of the first arrivals were used to determine the formation velocity; then, tools with an array of receivers (Fig. 1.3 (2) and (3)) were used to improve the accuracy and resolution of the velocity measurements. Eventually, logging tools with an acoustic isolator (ASO in Fig. 1.3

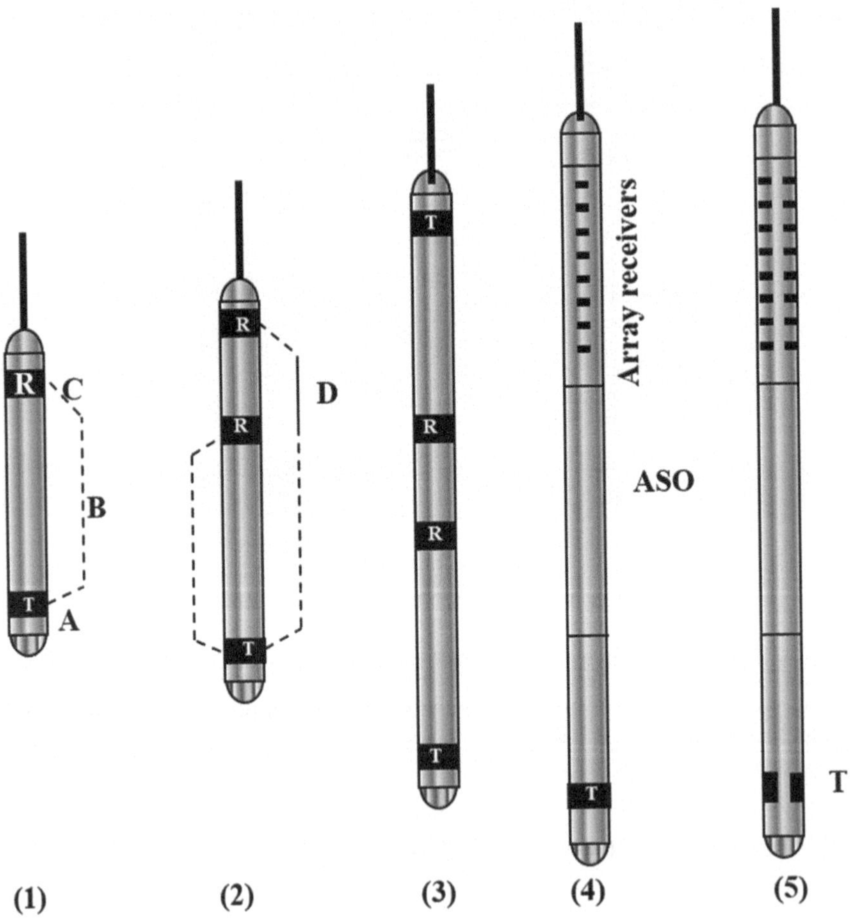

Fig. 1.3 Configurations for different wireline tools (T, R, and ASO represent the source, receivers and acoustic isolator, respectively)

(4)) situated between the transmitter and receivers were developed to eliminate waves propagating inside the tool (e.g., Chen et al. 1998). Grooves were eventually emplaced along the tool body between the source and receivers in conjunction with absorbing materials to scatter and attenuate the waves within the tool (Cowles et al. 1994; Su et al. 2015). Ultimately, the installation of sources and receivers along different azimuths (Fig. 1.3 (5)) enhanced the capability to acquire azimuthal measurements.

A major advance in acoustic logging was full waveform recording, which was enabled by developments in digital data acquisition, transmission, and recording. Figure 1.4 shows a field example of a full waveform data set collected from a borehole within a layered sedimentary section consisting of carbonates, evaporates, and shales (Arditty et al. 1981; Toksöz et al. 1984a). Different formations (marked in the figure) affect the characteristics of the full waveforms in different ways. The changes in the arrival times and waveforms define the lithological boundaries. P, S, pseudo-Rayleigh (pR), and Stoneley (ST) phases are marked in the figure. Only a small variation in the ST wave arrival time with depth is observed, whereas the changes in the arrival times of the P- and S-waves in different lithologies are clear. Even a minor change in the P-wave velocity between depths of 4900 and 5400 ft is accentuated in all the waveforms recorded in that depth interval. At depths from 5800 to 6000 ft, the S-waves and the long train of pR waves are missing; this depth interval corresponds to a case where the formation S-wave velocity is less than the borehole fluid velocity (e.g., a slow formation). Consequently, a dispersive leaky P-waves and ST waves are the only two phases on the wave train in this depth range.

1.1.1 Data Processing Methods

From array waveforms, the velocities of P, S, and ST waves can be determined using one of many advanced data processing methods. Figure 1.5 shows an example of full waveform recordings made with a monopole tool. The waveforms clearly change with depth, and the differences between the upper and the lower sets of waveforms are caused by differences in the formation lithology. The velocities of different modes are obtained from the array waveforms using the velocity-time semblance method (see Chap. 3).

With the demonstrated need for more detailed information about the subsurface, acoustic logging capabilities have improved using advances in sensor and data handling capabilities.

1.1.2 Borehole Acoustic Logging Tools

Borehole acoustic logging tools, briefly discussed at the beginning of Sect. 1.1, have been further developed to acquire improved measurements of P- and S-wave

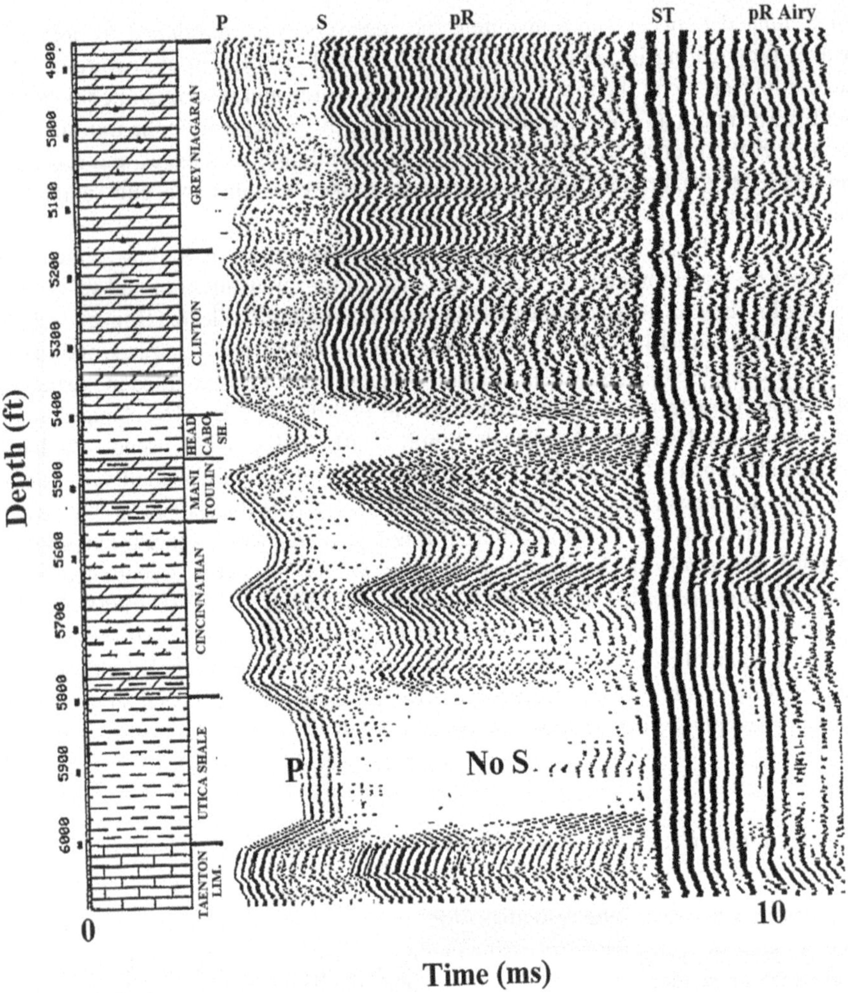

Fig. 1.4 Field data example of a full waveform monopole measurement (an iso-offset section) in a formation with alternating sections of carbonates, evaporates and shales (from Fig. 1 in Toksöz et al. 1984a). The data were logged with E.V. A. (Arditty et al. 1981), the ELF Aquitanie full waveform acoustic logging tool. The source-receiver distance is 6.75 m

velocities, elastic anisotropy and other formation properties. The frequency band of measurements has been expanded from a few hundred Hz to a few tens of kHz. In addition to monopole sources, dipole and quadrupole source and receiver geometries have been developed to determine the shear (S)-wave velocities in slow formations. Cross-dipole measurements (Fig. 1.6) are commonly used to determine the formation anisotropy. In addition, multiazimuth array tools have been employed for peripheral imaging, geosteering, and fracture mapping. Moreover, advanced sensor and data

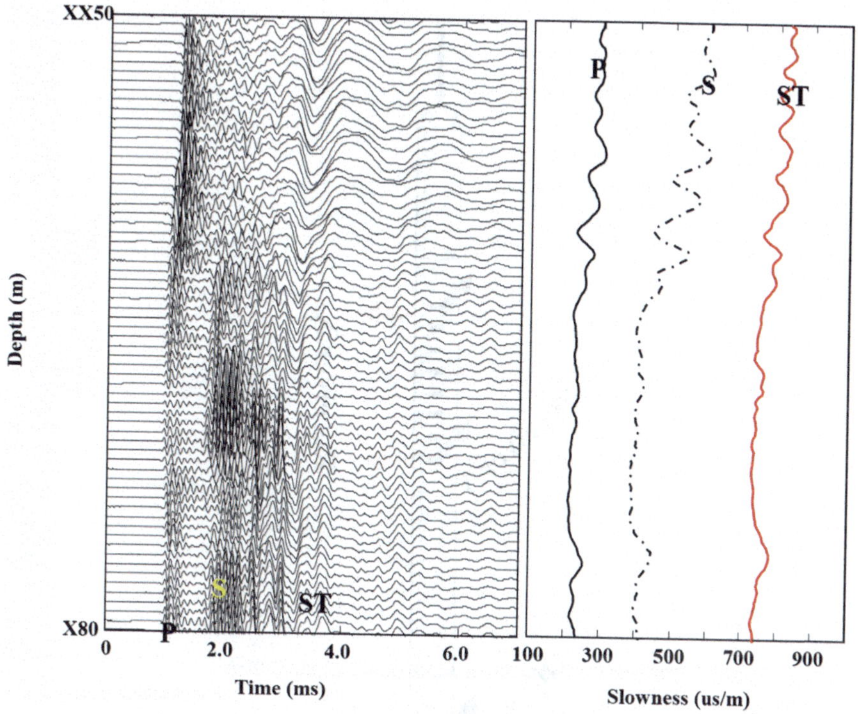

Fig. 1.5 Full waveform recording made using a monopole source and the resulting velocity profile. **a** Traces at the receivers closest to the source versus depth. There are two traces per foot. Each trace has 660 samples with a time interval of 12 μs. The source-receiver distance is 2.74 m, and the receiver interval is 0.15 m. **b** Velocities of P, S, and ST waves determined from the array waveforms using the velocity-time semblance method (see Chap. 3)

transmission technologies, improved theoretical modeling, and data analysis and interpretation capabilities have all contributed to these advances.

There are three types of sources: monopole, dipole, and quadrupole (Fig. 1.7). A monopole source is radially symmetric; hence, a tool with a monopole source can be used to measure the P- and S-wave velocities in fast formations. A dipole source can be used to generate shear motions and flexural waves to measure the S-wave velocities in both fast and slow formations (White 1967; Zemanek et al. 1984; Tang and Cheng 2004). The S-wave velocity can also be measured using a quadrupole source (see Fig. 1.7). For ultrasonic tools (where the source frequency exceeds 100 kHz), the sources are more like a piston (focused or planar), which gives the source a strong directivity.

In acoustic logging, two other tool properties are important for data acquisition and processing. As shown in Fig. 1.8, the distance between the source and the nearest receiver, referred to as the transmitter-receiver spacing (T-R), must be sufficiently large to acquire refracted P- and S-waves. In addition, the distance between two adjacent receivers is known as the interval. Large T-R allows different borehole

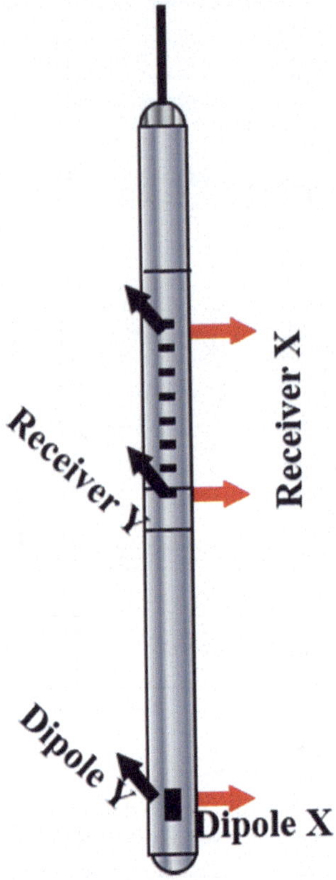

Fig. 1.6 Schematic of a cross-dipole tool. The arrows designate the polarization directions of the source and receivers

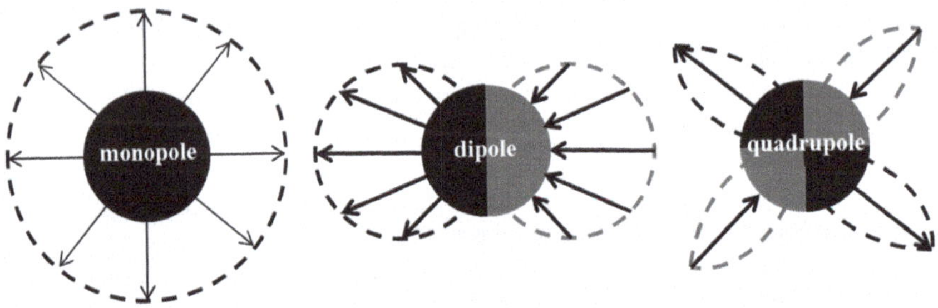

Fig. 1.7 Radiation patterns of monopole and multipole sources

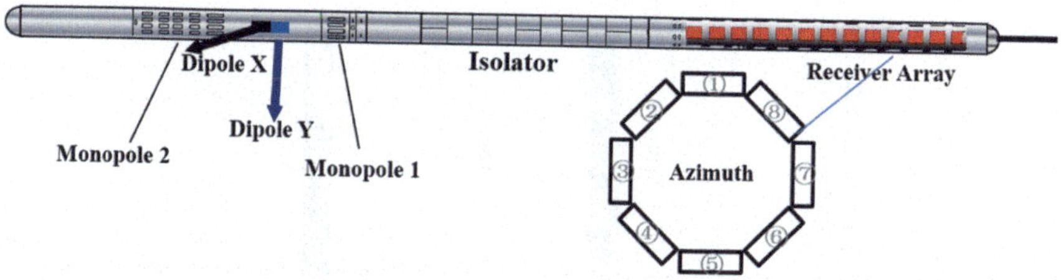

Fig. 1.8 Schematic of a multipole array acoustic logging tool. There are two monopoles: Monopole 1 is a near-monopole source and Monopole 2 is a far-monopole source. Two dipole sources, X and Y, are used for cross-dipole measurements

modes propagating at different velocities to be separated with greater clarity, but attenuation over a large propagation distance lowers the signal-to-noise ratio. In contrast, short intervals increase the spatial resolution along the borehole but can reduce the accuracy when the array length is decreased. In a typical acoustic wireline tool, the T-R spacing is approximately 3–4 m and the interval is 0.15 m.

The continuing evolution of sonic logging tools has been aimed toward broadening the source frequency band. As a result, the source frequency has shifted from the 10–30 kHz frequency range to a lower frequency range (1–10 kHz). This was made possible by improvements in transducer technology and by the need for lower-frequency waves to measure S-wave velocity in slow (low-velocity) formations (Pistre et al. 2005). Other developments have been made to increase the numbers of sources and receivers on the same tool. Figure 1.8 shows a schematic diagram of an advanced tool, on which the two monopole sources make it possible to make measurements with different source-receiver separations and different source frequencies. Receiver arrays with 8–13 nodes (with each node housing as many as 8 azimuthally distributed sensors) have been developed. Two dipole sources are used for dipole and cross-dipole measurements.

The discussion so far has been about logging in open (uncased) boreholes. However, in field surveys, many wells are cased and cemented to ensure the borehole integrity and safety and to enable proper production from the selected strata. The introduction of casing and cement creates a cylindrically layered geometry within the borehole that raises two important issues: the measurement of the formation properties behind the casing and the evaluation of the cement integrity, generally referred to as cement bond logging. The cement fills the annulus between the casing and formation to form a seal around the casing. The quality of the seal is important both for production and for safety. An overall evaluation of the cement bond quality can be accomplished using sonic logging in the 10–30 kHz frequency range. For detailed evaluations of the cement bond quality and for imaging the material behind the casing, ultrasonic frequencies (larger than 100 kHz) are used.

Two techniques are utilized to evaluate cased holes: pulse-echo and pitch-catch techniques. These methods are discussed in detail in Chap. 4. Pulse-echoes measure reflections from different interfaces. Cement bond quality is determined using estimates of the acoustic impedance contrasts between the casing, cement and fluid

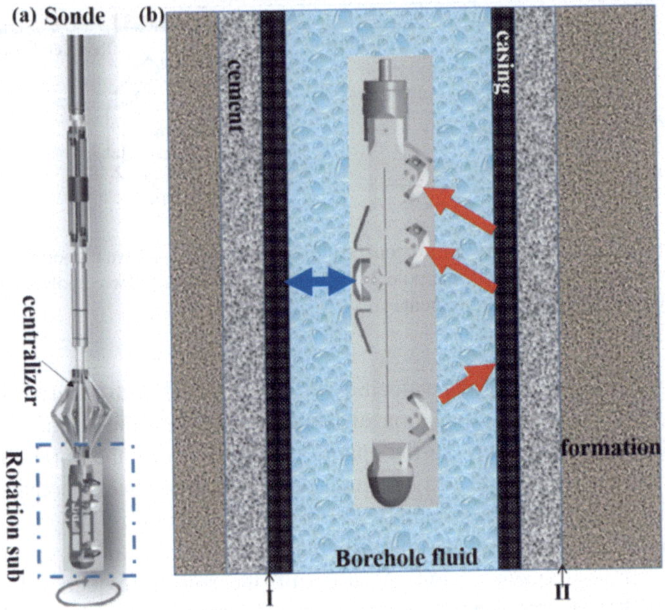

Fig. 1.9 Schematic diagram of an ultrasonic cement bond evaluation tool combining the pulse-echo and pitch-catch elements. **a** Tool configuration. **b** Measurements in a cased and cemented borehole. The blue arrow indicates a pulse-echo measurement, while red arrows indicate pitch-catch measurements

obtained from the various reflection amplitudes. In contrast, the pitch-catch method measures the shear attenuation of the material behind the casing (Zeroug and Froelich 2003), and this attenuation is interpreted to infer the cement bond quality. The combination of these two techniques provides detailed images of the cement interfaces. Figure 1.9 shows the configuration of the combined pulse-echo and pitch-catch approach, and Fig. 1.10 shows the pulse-echo and pitch-catch wavefields in a cased hole.

The introduction of Acoustic Logging-While-Drilling (ALWD) in the 1990s was an important development in the ability to measure formation properties in real time while drilling. Unlike wireline logging, ALWD data acquisition and recording occurs in real-time while drilling when relatively little fluid has invaded the formation. The ALWD evolution followed a path similar to that of wireline acoustic logging. The concept of ALWD was introduced in the 1950s. Initially, only the first arrival time was measured. Then, the interval transit time was recorded by employing one transmitter and two receivers, followed by tools with azimuthal sources and array receivers to obtain the P- and S-wave velocities in fast and slow formations.

One challenge in the acquisition of ALWD measurements is the influence of the massive collar located on the drill string on the wavefields and the resulting measurements of the P- and S-wave velocities. Another problem is the noise in the

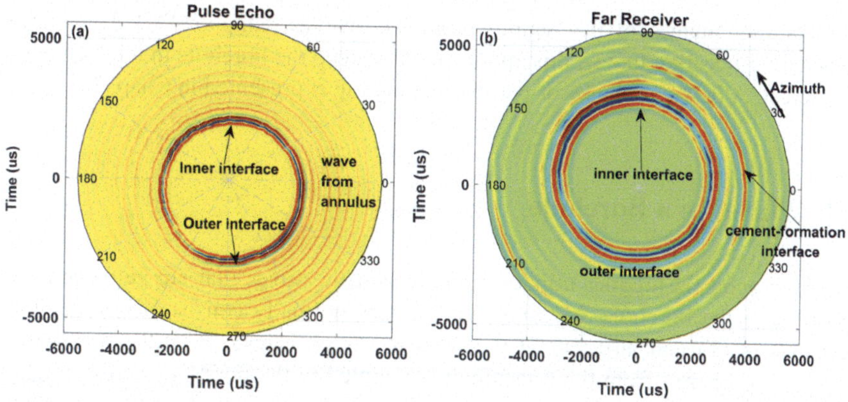

Fig. 1.10 Examples of pulse-echo (**a**) and pitch-catch (**b**) measurements in a cased-hole. See Chap. 4 for more details

borehole due to drilling activity. These issues are discussed in Chap. 5, which is devoted to ALWD.

With increasing numbers of sources and receivers, the use of larger arrays having longer source-receiver separations, and enhanced abilities to tackle larger data volumes, the utilization of borehole acoustic logging has been extended to imaging beyond boreholes. Figure 1.11 shows an example of the use of borehole acoustic logging data to map fractures tens of meters from the borehole wall. This technique obtains peripheral images around the borehole to distances of tens of meters.

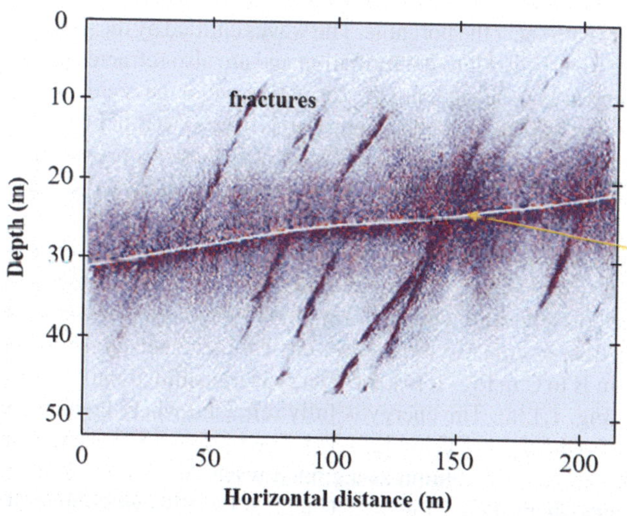

Fig. 1.11 Example of borehole reflection imaging for the mapping of fractures in a horizontal well (modified from Fig. 5 in Hirabayashi et al. (2016))

Obtaining images of the regions around boreholes using ALWD data makes geosteering feasible; that is, this approach helps keep the borehole in the formation of interest. A detailed review of peripheral imaging is presented in Chap. 7.

1.2 Waves in a Borehole

Borehole acoustic logging utilizes seismic/acoustic waves that are generated and recorded in the borehole. Measurement devices normally consist of one or more transmitters (T) and an array of receivers at some distance from the transmitter (Fig. 1.12). Energy from the transmitter is coupled to the borehole wall through a fluid; this energy (i.e., a wave) is refracted along the borehole wall when the incident angle is larger than the critical value given by Snell's law. Several refracted waves are generated that can be interpreted to determine the P- and S-wave velocities of the formation surrounding the borehole.

A fluid-filled borehole represents a very efficient waveguide. A portion of the energy from the acoustic source is often trapped in the borehole and subsequently propagates as guided waves. Figure 1.13 shows two snapshots of waves propagating in a borehole and the surrounding formation. These snapshots were output from numerical simulations of wave propagation calculated using a 3-D finite difference code (see Appendix B). The first example (Fig. 1.13a) is for a borehole in a fast formation, in which the formation S-wave velocity is greater than the borehole fluid velocity (Vs > Vf). Direct, transmitted, refracted, reflected and guided waves can be seen. The result is a long wave train, an example of which is shown in Fig. 1.12.

We illustrate the case of the fast formation with a simplified diagram (Fig. 1.14) that shows the ray paths near the borehole. The waves emitted by the source (at $x = 0$ m and $z = 0$ m) are transmitted into the formation and are also refracted along the borehole wall. These waves penetrate into the formation when the incident angle (θ_{in}) is less than the P-wave critical incidence angle (θ_{cp}) (from O to B in Fig. 1.14). Beyond the critical angle, the incident waves are refracted as P-waves propagating with the formation P-wave velocity along the borehole wall. The refracted P wave radiates energy into the borehole fluid. The borehole sonic logging tool receives this energy as primary (or head) waves. Part of the P-wave energy is converted into S-wave energy at the borehole wall and hence propagates as an S-wave. The refracted S-wave propagates along the borehole wall with the formation S-wave velocity when the incident angle is greater than or equal to the S-wave critical incidence angle (θ_{cs}). Between θ_{cp} and θ_{cs} (from B to C in Fig. 1.14), the energy is transmitted into the formation as an S-wave (see Fig. 1.13a). The energy is fully refracted when the incident angle is greater than θ_{cs} (to the right of C in Fig. 1.14). The refracted S-wave radiates energy back into the borehole fluid column as a guided wave that is related to **pR** and **ST** waves (Paillet and Cheng 1986; Tang and Cheng 2004; Ellis and Singer 2007).

In the case of a slow formation, where the S-wave velocity is less than the borehole fluid velocity (Vs < Vf) (Cheng et al. 1982; Schmitt and Bouchon 1985), there are no refracted S waves. Instead, only P and **ST** waves exist, as shown in Fig. 1.13b.

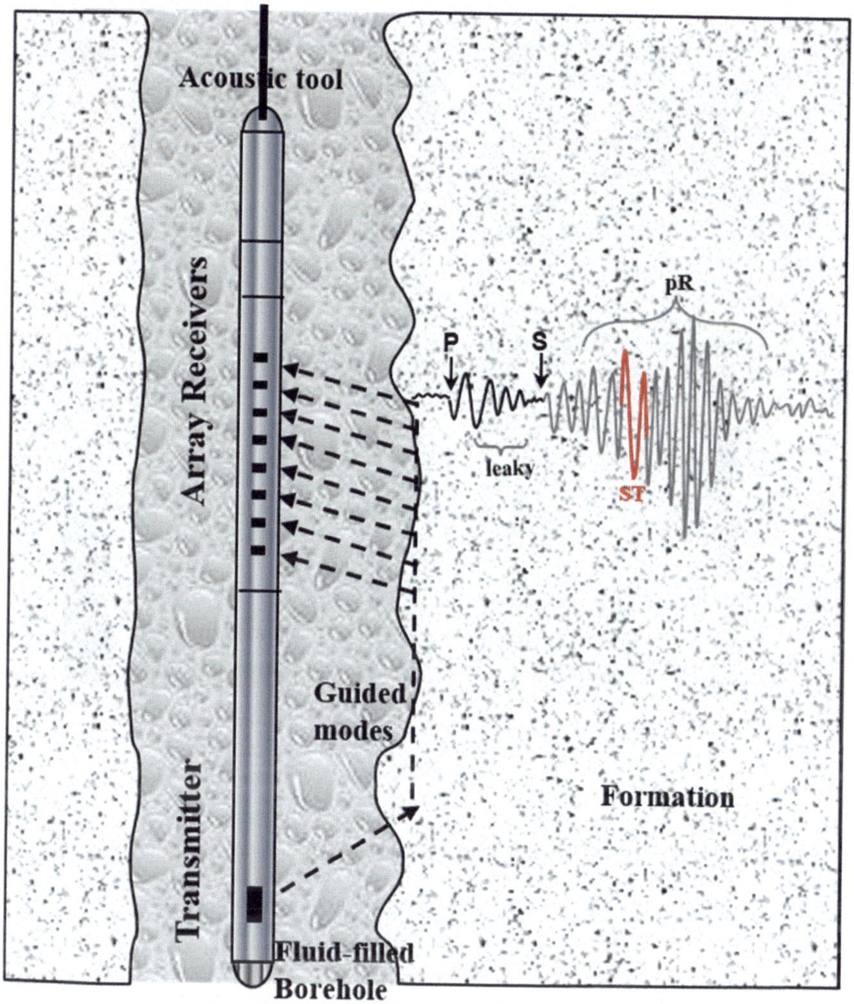

Fig. 1.12 Schematic illustration of acoustic logging with a source (transmitter) and an array of receivers. A typical received waveform shows formation P, S, pseudo-Rayleigh (pR), and Stoneley (ST) waves. The dashed lines show the wave propagation ray paths

In a fast formation, the received wave modes (for a monopole tool within the sonic frequency range) include two body waves: namely, compressional (marked as **P** in Fig. 1.13a) and shear (marked as **S** in Fig. 1.13a) waves, and two guided waves, namely, **pR** and **ST** waves. However, S and pR waves do not appear in a slow formation (Fig. 1.13b). In this case, **leaky P** waves become the first arrivals. A leaky mode is a type of elastic energy propagating in the borehole; part of the energy is reflected into the borehole from the borehole wall, while the rest radiates (or 'leaks') into the formation. In this case, ST waves become the only type of guided wave, and

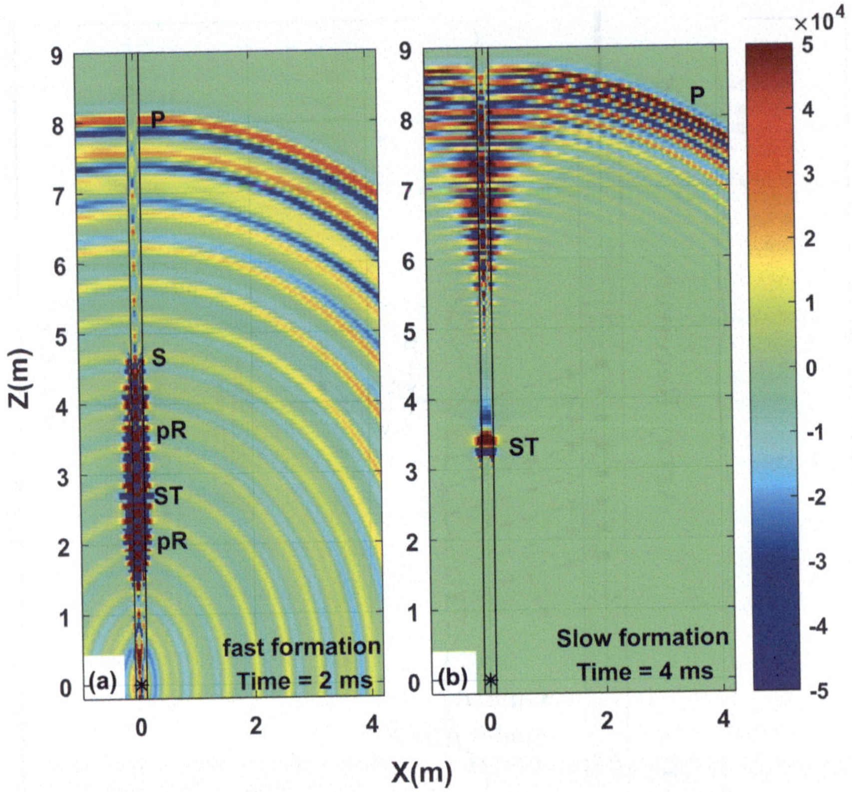

Fig. 1.13 Snapshot of the wavefield in the borehole and surrounding formation calculated by a 3-D finite difference code (see Appendix B). **a** Wavefield at 2 ms in a fast formation. **b** Wavefield at 4 ms in a slow formation. The 10 kHz monopole source located at x = 0 m and z = 0 m, is marked with a star. The borehole radius is 0.1 m, and the borehole wall is delineated by black lines

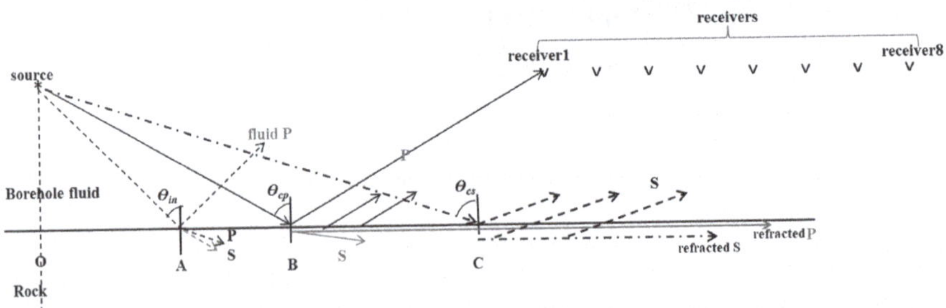

Fig. 1.14 Schematic illustration of incident, reflected, refracted and transmitted waves near the wellbore wall in a fast formation

the dispersion of ST waves depends on the formation S-wave velocity (Paillet and Cheng 1986).

1.3 Coverage of This Book

This book covers seismic/acoustic wave propagation in and around a borehole and the corresponding use of acoustic logging. Knowledge about the dependence of waveforms on seismic velocities, source types, and frequencies is important for the interpretation of the data and the design of logging tools and for deciding which measurements best suit a user's needs.

Chapter 2 covers the fundamental aspects of seismic/acoustic waves in and around boreholes. This chapter presents numerous examples of waveforms and wavefield snapshots generated by finite difference simulation that illustrate the characteristics of waveforms for wells in formations having different properties. The examples were collected using various sources and source frequencies.

Chapter 3 describes the most commonly used methods for determining the P- and S-wave velocities from an array of waveforms obtained during logging. The commonly used velocity-time semblance analysis method and other new spectral methods are described.

Frequently, acoustic logging data are obtained in cased holes to determine the casing and cement conditions and to assess the cement bond integrity. These measurements are conducted with sonic and ultrasonic tools. Models generated by 3-D finite difference calculations (shown in Chap. 4) demonstrate the differences in the wavefields and waveforms obtained under both good and poor cement conditions.

ALWD is a relatively new technology that complements wireline logging. A major difficulty for both tool design and data analysis is coping with the effects of a massive drill collar in the borehole. The modeling of ALWD waveforms, covered in Chap. 5, provides insights for both tool design and data analysis.

Chapter 6 covers the modeling of eccentered (i.e., off-centered) tools in a borehole. Eccentricity affects both the modal compositions of waveforms and the amplitudes of different phases due to loss of radial symmetry. Velocity measurements may also be impacted because of interference by extra modes.

The final chapter (Chap. 7) covers peripheral imaging around boreholes for the imaging of faults, fractures, and bed boundaries located at distances of several tens of meters from a borehole. Surface seismic survey data processing methods are applied to borehole imaging data. Geo-steering, especially in horizontal drilling, is another application of peripheral imaging.

The primary goal of this book is to provide an in-depth review of seismic/acoustic wave propagation in and around boreholes using analytical and numerical calculations. The results of 2- and 3-D finite difference simulations provide a visual display of wave propagation characteristics for a broad spectrum of logging examples. These simulated results provide a library of examples that will benefit the acoustic logging community, including tool designers, analysts, interpreters, and researchers.

Suggestions for Further Reading

Our book complements other books, such as those of White (1983), Paillet and Cheng (1991), Tang and Cheng (2004), and Chin (2014), in that each of these works emphasizes different aspects of acoustic logging and its application. We list a selected set of references for further reading.

Aki, K. and Richards, P. G., 2002, Quantitative seismology, Second Edition. University Science Books.

Chin, W. C., 2014, Wave propagation in drilling, well logging and reservoir environments. Wiley.

Ellis, D. V., and Singer, J. M., 2007, Well logging for earth scientists. Springer.

Paillet F., and Cheng C. H., 1991, Acoustic waves in boreholes. CRC Press.

Schlumberger, 1987, Log interpretation principles/applications. Schlumberger educational services.

Tao G., 2006, Logging information processing and application. China University of Petroleum Press.

Tang X. M., and Cheng C. H., 2004, Quantitative Borehole Acoustic Methods. Elsevier.

White, J. E., 1983, Underground sound: Application of seismic waves. Elsevier Science Publishers B. V.

Chapter 2
Wave Propagation in an Open Borehole

This chapter covers theoretical/numerical modeling of wave propagation in an open (uncased) borehole. It starts with the analytical formulation of seismic wave propagation and proceeds to modeling for boreholes in formations with different velocities and for different source types (monopole, dipole, and quadrupole). Numerical calculations provide an extensive suite of examples that illustrate the sensitivity of the acoustic logs to formation properties, borehole radii, and type and frequency of the sources. Field data examples are also included to illustrate the waveforms in different formations.

2.1 Waves in and Around an Open Borehole

2.1.1 Waves in the Isotropic Solid Medium

The displacement in an isotropic solid is: $\mathbf{u} = u\,\boldsymbol{r} + v\,\boldsymbol{\theta} + w\,\boldsymbol{z}$, where u, v, and w are the displacements in radial, azimuthal, and axial directions and $\boldsymbol{r}$, $\boldsymbol{\theta}$, and $\boldsymbol{z}$ are the corresponding unit vectors. The elastic wave equation in a solid medium (without source) is (Aki and Richards 1980),

$$(\lambda + \mu)\nabla(\nabla \cdot \mathbf{u}) + \mu\nabla^2\mathbf{u} = \rho\frac{\partial^2\mathbf{u}}{\partial t^2}, \tag{2.1}$$

where t is time and ρ is density. λ and μ are Lamé constants. $\mathbf{u}$ can be expressed in terms of displacement potentials ϕ and $\boldsymbol{\psi}$ (Lay and Wallace 1995): $\mathbf{u} = \nabla\phi + \nabla \times \boldsymbol{\psi}$. ϕ is the scalar potential for P-waves. $\boldsymbol{\psi}$ is the potential vector. Potential $\boldsymbol{\psi} = \chi\mathbf{z} + \nabla \times (\Gamma\mathbf{z})$, where χ and Γ are the potentials for SH and SV waves. Equation 2.1 can be expressed by three differential equations for P, SH, and SV waves,

H. Wang et al., *Borehole Acoustic Logging—Theory and Methods*, Petroleum Engineering, https://doi.org/10.1007/978-3-030-51423-5_2

$$\text{P-wave} \quad (\lambda + 2\mu)\nabla^2\phi = \rho\frac{\partial^2\phi}{\partial t^2}, \tag{2.2a}$$

$$\text{SH wave} \quad \mu\nabla^2\chi = \rho\frac{\partial^2\chi}{\partial t^2}, \tag{2.2b}$$

$$\text{SV wave} \quad \mu\nabla^2[\nabla \times \Gamma] = \rho\frac{\partial^2[\nabla \times \Gamma]}{\partial t^2}. \tag{2.2c}$$

P and S velocities are $V_P = \sqrt{\frac{\lambda+2\mu}{\rho}}$ and $V_S = \sqrt{\frac{\mu}{\rho}}$. In a cylindrical coordinate system, the Laplace operator is $\nabla^2 = \frac{\partial^2}{\partial r^2} + \frac{1}{r}\frac{\partial}{\partial r} + \frac{1}{r^2}\frac{\partial^2}{\partial \theta^2} + \frac{\partial^2}{\partial z^2}$.

Equations 2.2a, 2.2b, 2.2c in the frequency domain are as follows:

$$\begin{cases} \nabla^2\Phi + k_p^2\Phi = 0 \\ \nabla^2\bar{\chi} + k_s^2\bar{\chi} = 0, \\ \nabla^2\bar{\Gamma} + k_s^2\bar{\Gamma} = 0 \end{cases} \tag{2.3}$$

where Φ, $\bar{\chi}$, and $\bar{\Gamma}$ are the terms ϕ, χ, and Γ in the frequency domain, respectively. $k_p = \frac{\omega}{v_p}$ and $k_s = \frac{\omega}{v_s}$ are the wavenumbers of P- and S-waves. k_z is the axial wavenumber and the complex P and S radial wavenumbers are $p = \sqrt{k_z^2 - k_p^2}$ and $s = \sqrt{k_z^2 - k_s^2}$.

The general solutions, at each frequency, of Eq. 2.3 can be obtained by the method of separation of variables (Miller 1977),

$$\begin{cases} \Phi(r,\theta,z) = (A_n K_n(pr) + A_n' I_n(pr))(l_1 \sin n\theta + l_2 \cos n\theta)(c_1 e^{ik_z z} + c_2 e^{-ik_z z}) \\ \bar{\chi}(r,\theta,z) = (B_n K_n(pr) + B_n' I_n(pr))(l_1 \sin n\theta + l_2 \cos n\theta)(c_1 e^{ik_z z} + c_2 e^{-ik_z z}), \\ \bar{\Gamma}(r,\theta,z) = (C_n K_n(pr) + C_n' I_n(pr))(l_1 \sin n\theta + l_2 \cos n\theta)(c_1 e^{ik_z z} + c_2 e^{-ik_z z}) \end{cases} \tag{2.4}$$

where $I_n(x)$ and $K_n(x)$ ($n = 0, 1, 2, \ldots$) are the first and second kinds of modified Bessel functions (Lebedev 1972), corresponding to incoming and outgoing waves.

Assuming k_z to be negative in the propagation direction, the coefficients of the term $\exp(-ik_z z)$ are zeros. The cosine terms are used for P and SV waves, and sine terms are used for the SH wave. Then Eq. 2.4 can be expressed as,

$$\begin{cases} \Phi(r,\theta,z) = l\sum_{k_z}\sum_{n}\left[A_n K_n(pr) + A_n' I_n(pr)\right]\cos(n\theta)\exp(ik_z z), \\ \bar{\chi}(r,\theta,z) = l\sum_{k_z}\sum_{n}\left[B_n K_n(pr) + B_n' I_n(pr)\right]\sin(n\theta)\exp(ik_z z), \\ \bar{\Gamma}(r,\theta,z) = l\sum_{k_z}\sum_{n}\left[C_n K_n(pr) + C_n' I_n(pr)\right]\cos(n\theta)\exp(ik_z z). \end{cases} \tag{2.5}$$

For the open borehole in an infinite solid medium, there is no incoming wave from outside the borehole and coefficients of the term $I_n(pr)$ are zeros.

$$\begin{cases} \Phi(r,\theta,z) = l\sum_{k_z}\sum_{n} A_n K_n(pr)\cos(n\theta)\exp(ik_z z), \\ \overline{\chi}(r,\theta,z) = l\sum_{k_z}\sum_{n} B_n K_n(pr)\sin(n\theta)\exp(ik_z z), \\ \overline{\Gamma}(r,\theta,z) = l\sum_{k_z}\sum_{n} C_n K_n(pr)\cos(n\theta)\exp(ik_z z). \end{cases} \tag{2.6}$$

However, for other cases such as cased holes and boreholes near reflecting interfaces, the "incoming" terms need to be considered.

2.1.2 Waves in a Fluid Medium

In a fluid medium with density ρ_f and velocity V_f, inside the borehole, there is only the scalar potential ϕ and wave equation with a source is as follows (Fig. 2.1):

$$\nabla^2\phi_f = \frac{\rho_f}{\lambda}\frac{\partial^2\phi_f}{\partial t^2} + \delta(r-r_0)\delta(\theta-\theta_0)\delta(z-z_0)f(t), \tag{2.7}$$

where ϕ_f is the displacement potential in the fluid and $f(t)$ is the source time function. r_0, θ_0, and z_0 are the source coordinates along radial, azimuthal, and axial directions in the borehole. δ is a Dirichlet function (Dunham 2005).

In the frequency domain, the equation is as follows:

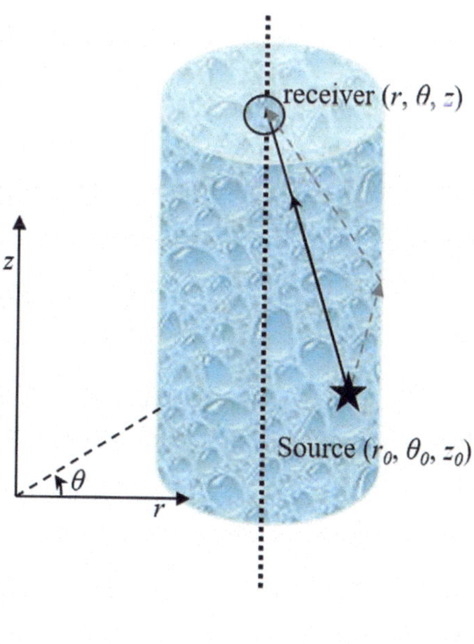

Fig. 2.1 Diagram of point source and receiver in the borehole

$$\nabla^2 \Phi_f + k_f^2 \Phi_f = \delta(r - r_0)\delta(\theta - \theta_0)\delta(z - z_0) F(\omega), \tag{2.8}$$

where Φ_f is the Fourier transform of ϕ_f and $k_f = \frac{\omega}{v_f}$ is the P-wave wavenumber in the fluid. Following the general solutions in Eq. 2.4, a solution at each frequency of Eq. 2.8 can be expressed as (Aki and Richards 1980),

$$\Phi_f(r, \theta, z) = \sum_{n=0}^{\infty} (D_n K_n(fr) + D_n' I_n(fr)) \cos n(\theta - \theta_0) e^{ik_z(z-z_0)} F(\omega), \tag{2.9}$$

where f is the wavenumber in the radial direction of P-wave in the fluid, $f = \sqrt{k_z^2 - k_f^2}$.

The displacement potential at the location (r, θ, z) of a point source is (Kurkjian and Chang 1986),

$$\phi_d(t, r, \theta, z) = \frac{f(t)e^{ik_f Z}}{Z} = \frac{f(t)e^{ik_f \sqrt{r^2 + r_0^2 - 2rr_0 \cos(\theta - \theta_0) + (z - z_0)^2}}}{\sqrt{r^2 + r_0^2 - 2rr_0 \cos(\theta - \theta_0) + (z - z_0)^2}}. \tag{2.10}$$

In the frequency domain, the expression is as follows:

$$\Phi_d(\omega, r, \theta, z) = \frac{F(\omega)e^{ik_f Z}}{Z} = \frac{F(\omega)e^{ik_f \sqrt{r^2 + r_0^2 - 2rr_0 \cos(\theta - \theta_0) + (z - z_0)^2}}}{\sqrt{r^2 + r_0^2 - 2rr_0 \cos(\theta - \theta_0) + (z - z_0)^2}}. \tag{2.11}$$

This spherical formulation can be converted into a cylindrical formulation. Ignoring the source term $F(\omega)$, the cylindrical formulation is as follows:

$$\Phi_d(r, \theta, z) = \sum_{n=0}^{\infty} \varepsilon_n \begin{Bmatrix} \frac{1}{\pi} I_n(fr_0) K_n(fr), r > r_0 \\ \frac{1}{\pi} I_n(fr) K_n(fr_0), r < r_0 \end{Bmatrix} * \cos(n(\theta - \theta_0)) e^{ik_z(z-z_0)}. \tag{2.12}$$

where $\varepsilon_n = 1$ when n is 0. $\varepsilon_n = 2$ for other cases.

From Eq. 2.9, incoming waves can be expressed,

$$\Phi_{sca}^f(r, \theta, z) = \sum_{n=0}^{\infty} D_n' I_n(fr) \cos n(\theta - \theta_0) e^{ik_z(z-z_0)}. \tag{2.13}$$

The total wavefield in the borehole is the sum of the direct (outgoing) and scattered (incoming) waves from borehole wall,

$$\Phi_{tot} = \Phi_d + \Phi_{sca}^f = P(\omega, k_z) e^{ik_z(z-z_0)}, \tag{2.14}$$

where $P(\omega, k_z) = \sum_{n=0}^{\infty} \begin{Bmatrix} \varepsilon_n \frac{1}{\pi} I_n(fr_0) K_n(fr) + D_n' I_n(fr), r > r_0 \\ \varepsilon_n \frac{1}{\pi} I_n(fr) K_n(fr_0) + D_n' I_n(fr), r < r_0 \end{Bmatrix} \cos[n(\theta - \theta_0)].$

2.1.3 *Wavefieds in the Borehole and Formation*

The coupling of the wavefields in the borehole fluid to a solid formation requires boundary conditions. The value of D_n' in Eq. 2.14 and coefficients in Eq. 2.6 can be determined from the boundary conditions.

The boundary conditions between the fluid and a solid are

$$\text{Continuity of radial displacement: } u_r^f = u_r,$$
$$\text{Continuity of normal stress: } -P_f = \sigma_{rr},$$
$$\text{Zero shear stress: } \sigma_{rz} = 0, \sigma_{r\theta} = 0,$$

where u_r^f is radial displacement in the fluid and u_r is radial displacement in the formation. P_f is pressure in the fluid and σ_{rr} is the normal stress in the formation. σ_{rz} and $\sigma_{r\theta}$ are shear stresses.

There are four equations for the fluid-solid interface. The expression for the open borehole case is as follows:

$$[m_{ij}]_{4\times4} \cdot [a_i]_{4\times1} = [b_i]_{4\times1}, \tag{2.15}$$

where $[a_i]_{4\times1} = [\, D_n \;\; A_n \;\; B_n \;\; C_n \,]^T$ and $[b_i]_{4\times1}$ is the source term. $[m_{ij}]_{4\times4}$ is given in Appendix A.

The relationship between velocities of different modes and frequencies can be determined by taking the determinant of $[m_{ij}]_{4\times4}$ as zero. This approach is used to calculate the modal dispersion which are shown throughout the book.

Waveforms can be obtained from the following equation,

$$P(r, z, \theta, t) = \int_{-\infty}^{+\infty} \mathrm{F}(\omega) e^{-i\omega t} d\omega \int_{-\infty}^{+\infty} P(\omega, k_z) e^{ik_z(z-z_0)} dk_z, \tag{2.16}$$

where wavenumber k_z is a complex variable. A few singularities exist in the complex k_z plane.

2.1.3.1 Waveform Calculations

The solution of Eq. 2.16 faces a challenge because of singularities along the complex wavenumber integration path. There are two methods of performing the integration.

First is the complex contour or branch-cut integrations (Tsang 1978; Paillet and Cheng 1991). This method transfers the integration path to go around the singularities (poles) in the complex wavenumber plane and then adds the contribution of poles (i.e., residues). Then the last integral in Eq. 2.16 becomes,

$$\begin{aligned}\int_{-\infty}^{+\infty} P(\omega, k_z)e^{ik_z(z-z_0)}dk_z &= 2\pi i \sum_{i} Res\left[P(\omega, k_z)e^{ik_z(z-z_0)}\right]_{k_l} \\ &+ \sum_{j=p,s,f} \int_{bl_j} P(\omega, k_z)e^{ik_z(z-z_0)}dk_z\end{aligned}, \quad (2.17)$$

where "Res" is the residue of $P(\omega, k_z)e^{ik_z(z-z_0)}$ for wavenumber k_l at the lth pole. The second term on the right is the contribution from the jth branch-line integral.

The integration involves the residues of the P- and S-wave branch points lying on the lower Riemann sheet. It is helpful to calculate the contribution of each pole individually because each singularity corresponds to a mode in the borehole.

The second method is the discrete wavenumber integration or real axis integration method (Bouchon and Aki 1977; Tsang 1978; Schmitt and Bouchon 1985). This method yields the full waveform by directly integrating Eq. 2.16 along the real axis of k_z. The singularities are avoided by introducing a small but finite imaginary part to the wavenumber (or frequency) to move the path into the complex plane, and it is no longer on the real axis. The introduction of the small imaginary wavenumber results in some attenuation. This can be remedied by multiplying waveforms with an appropriate exponential function.

2.1.3.2 Sources in a Borehole: Multipole Representation

The waves, generated by a point source located at (r_0, θ_0, z_0), include all modes, such as monopole (n = 0), dipole (n = 1), quadrupole (n = 2), hexapole (n = 3), and higher modes (e.g. Byun and Toksöz 2006; Chen et al. 2010). Placing receivers over a range of azimuthal angles can be used to obtain individual modes. With a point source at A and four receivers distributed at A ($\theta = 0°$), B ($\theta = 90°$), C ($\theta = 180°$), and D ($\theta = 270°$) (as shown in Fig. 2.2), one can obtain solutions for the monopole, dipole and quadrupole modes by summing (adding) and subtracting the appropriate contributions.

$$\Phi_A(\theta = 0) = M + D + Q + \varepsilon, \quad (2.18a)$$

$$\Phi_B(\theta = \pi/2) = M - Q + \varepsilon, \quad (2.18b)$$

$$\Phi_C(\theta = \pi) = M - D + Q + \varepsilon, \quad (2.18c)$$

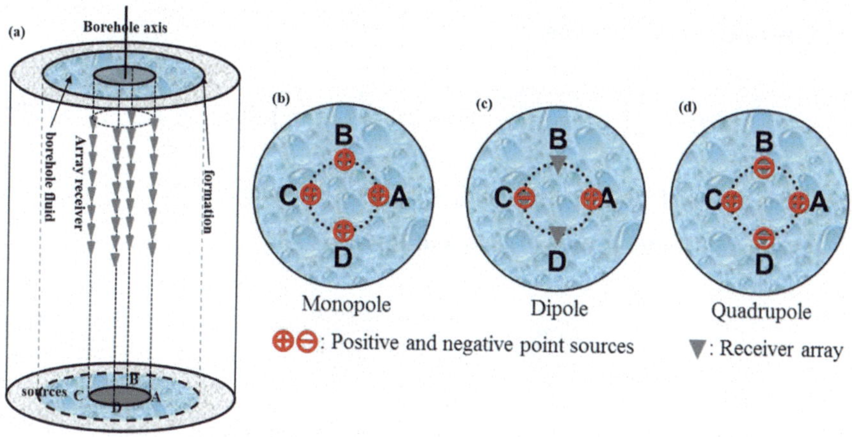

Fig. 2.2 Source and receiver locations in a borehole

$$\Phi_D(\theta = 3\pi/2) = M - Q + \varepsilon, \tag{2.18d}$$

where Φ is potential and its subscripts A, B, C, and D are the receiver locations. θ is the azimuth angle of the receiver relative to receiver A. M, D and Q are the total potentials of monopole, dipole, and quadrupole modes. ε is the sum of higher-order potentials. By adding and subtracting the potentials, given in Eqs. 2.18a, 2.18b, 2.18c, 2.18d the potentials for the monopole, dipole, and quadrupole waves can be obtained,

$$\begin{cases} M = \Phi_A + \Phi_B + \Phi_C + \Phi_D \\ D = \Phi_A - \Phi_C \\ Q = \Phi_A - \Phi_B + \Phi_C - \Phi_D \end{cases}. \tag{2.19}$$

Higher modes, such as hexapole (n = 3) and octupole (n = 4), can also be calculated but require more azimuthal receivers. From Eq. 2.14, we determine the amplitude of the point source as a function of r_0. The wavefield outside the source sphere increases when r_0 increases. For the simulation of a multipole source in wireline logging, we use $r_0 = 0.02$ m, the typical radius of an acoustic logging tool. However, in the acoustic logging-while-drilling case, r_0 is the outer radius of the drill collar (approximately 0.09 m). Examples and more interpretations can be found in Fig. 6.5 and Sect. 6.1.

2.1.3.3 Source Time Functions

There are various source time functions used for synthetic waveform calculations. The most used source functions are,

(1) **Cosine envelope function**

$$f(t) = \frac{1}{2}\left(1 - \cos\frac{2\pi}{T}t\right)\cos 2\pi f_0\left(t - \frac{T}{2}\right), 0 \le t \le T, \quad (2.20)$$

where T is the width of the pulse and f_0 is the center frequency.

(2) **Tsang and Rader source (Tsang and Rader 1979)**

$$F(\omega) = \frac{8\alpha\omega_0(\alpha + i\omega)}{\left[(\alpha + i\omega)^2 + \omega_0^2\right]^2}, \quad (2.21a)$$

where ω is the frequency. α and ω_0 give the width and center frequency of the spectrum. The function in the time domain is as follows:

$$f(t) = 4\alpha t e^{-\alpha t}\sin(\omega_0 t). \quad (2.21b)$$

(3) **Ricker source (Aki and Richards 1980)**

$$\text{Time domain,} \quad f(t) = A_0\left(1 - a^2T^2\right)\exp\left(-a^2T^2/2\right), \quad (2.22a)$$

$$\text{Frequency domain,} \quad F(\omega) = f^2/f_0^2\exp\left(-f^2/f_0^2 - i2\pi f t_s\right), \quad (2.22b)$$

where $a = 1.414\pi f_0$. A_0 is the amplitude of the source. $T = t - t_s$ and $t_s = 1.5/f_0$ (time delay). For most numerical simulations in this book, the Ricker source is used.

(4) **Gaussian sources**

The Gaussian sources consist of different types (Kelly et al. 1976; Stephen et al. 1985).

The first time function is

$$f(t) = -2\alpha T e^{-\alpha T^2}. \quad (2.23)$$

$$\text{The frequency domain, } F(\omega) = \omega e^{\left(-\frac{\omega^2}{4\alpha} + i\omega t_s\right)}\sqrt{\frac{\pi}{\alpha}}. \quad (2.24)$$

The first and second order derivatives are,

$$f'(t) = -2\alpha\left(1 - 2\alpha T^2\right)e^{-\alpha T^2}, \quad (2.25a)$$

$$f''(t) = 4\alpha^2\left(3T - 2\alpha T^3\right)e^{-\alpha T^2}, \quad (2.25b)$$

where α is the attenuation or the width parameter which equals to $f_0^2/0.1512$, and $T = t - t_s$. t_s is a given time delay.

(5) **Sweep frequency source (or linear chirp)**

$$\text{Time domain, } f(t) = \sin\left(\varphi_0 + 2\pi\left(f_0 + \kappa t^2/2\right)\right). \quad (2.26)$$

This source is useful to generate a sweep of wide frequency range. The sweep frequency goes from f_0 to f_1 in time T. Parameter κ is the rate of frequency change, $\kappa = (f_1 - f_0)/T$.

The spectra of different source functions are shown in Fig. 2.3. The start and final frequencies in the sweep frequency source are 0 and 30 kHz. For other sources, we set the center frequency at 10 kHz. The width of the pulse in the cosine envelope function is 2 ms (milliseconds). Different source functions produce waves with different frequency spectra.

In the discussions so far, the solution given by Eq. 2.16 is for a point source. To obtain solutions for other source configurations, such as a ring or a line, $F(\omega)$ in Eq. 2.16 is replaced by $F(\omega)U(k_z)$, where $U(k_z)$ is the space distribution function of the source. For example, a distribution function of a line source with a height h is as follows:

$$U(k_z) = 2z_0 \frac{\sin(k_z h)}{k_z h}. \quad (2.27)$$

Numerical methods for calculating synthetic waveforms include discrete wavenumber (Schmitt and Bouchon 1985), finite difference (Cheng 1994), finite element (Matuszyk and Torres-Verdin 2011; Wang et al. 2013b), and boundary element (Charara et al. 2011) methods. In this book, we use the discrete wavenumber

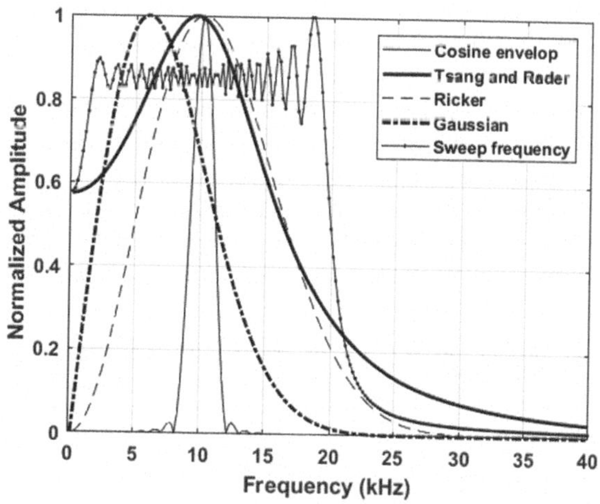

Fig. 2.3 Spectra of different source functions

and finite difference methods. The finite difference method is described in Appendix B.

2.1.3.4 Attenuation

The intrinsic or anelastic attenuation of the medium is generally expressed by the quality factor Q. In wave propagation, attenuation can be incorporated using a complex velocity (Aki and Richards 1980),

$$V(\omega) = V_0\left[1 + \frac{1}{\pi Q}\log\left(\frac{\omega}{\omega_0}\right) - \frac{i}{2Q}\right], \tag{2.28}$$

where V_0 is the velocity at a given angular frequency ω_0. Q is the quality factor. Actual attenuation is proportional to 1/Q (see Toksöz and Johnston 1981). The higher the Q, the lower the attenuation. The influence of Q on waveforms is investigated in Fig. 2.5. Attenuation is not included in the synthetic waveform calculations in this book unless specifically stated.

2.2 Wavefields in Fast Formations

A fluid filled borehole in a solid medium forms a strong waveguide. A wave generated by a source in the fluid propagates inside the fluid and in the solid formation around the borehole. Different modes are generated. Mode types depend on the properties of the borehole fluid, surrounding formation, source type, and source frequency. To demonstrate the effects of different parameters on waveforms, we use a standard set of models for the calculations. For the formation velocities, we chose four models (F1, F2, S1, and S2) that represent a suite of geological formations. Formations F1 and F2 are fast formations, which indicates that the S-wave velocity is greater than the velocity of P-wave in the borehole fluid. The model parameters are listed in Table 2.1. The borehole radius is 10 cm unless it is specifically stated. Borehole fluid (i.e., mud) velocity and density are chosen to be 1500 m/s and 1000kg/m^3, respectively.

Table 2.1 Elastic parameters of media in the open hole model

Media	Vp (m/s)	Vs (m/s)	Density (kg/m^3)	Radius (mm)
Fluid	1500	0	1000	100
F1	4500	2650	2400	–
F2	3000	1800	2000	–
S1	2300	1000	2000	–
S2	2000	500	2000	–

2.2.1 Monopole Source

2.2.1.1 Modes

The first case is a monopole source in a borehole in fast (Vs > Vf) formations. A monopole source in a fluid-filled borehole in a fast formation generates four different wave types propagating in and around the borehole: P, S, pR (pseudo Rayleigh), and ST (Stoneley). Figure 2.4a shows an example of the synthetic full waveforms in formation F1. Although advanced sonic logging tools are designed by including structure detail in numerical simulations, in this chapter the tool is simplified as a point source because the tool effect may be equivalent to a smaller borehole radius (Cheng and Toksöz 1982). The P-wave, the first arrival, has the smallest amplitude

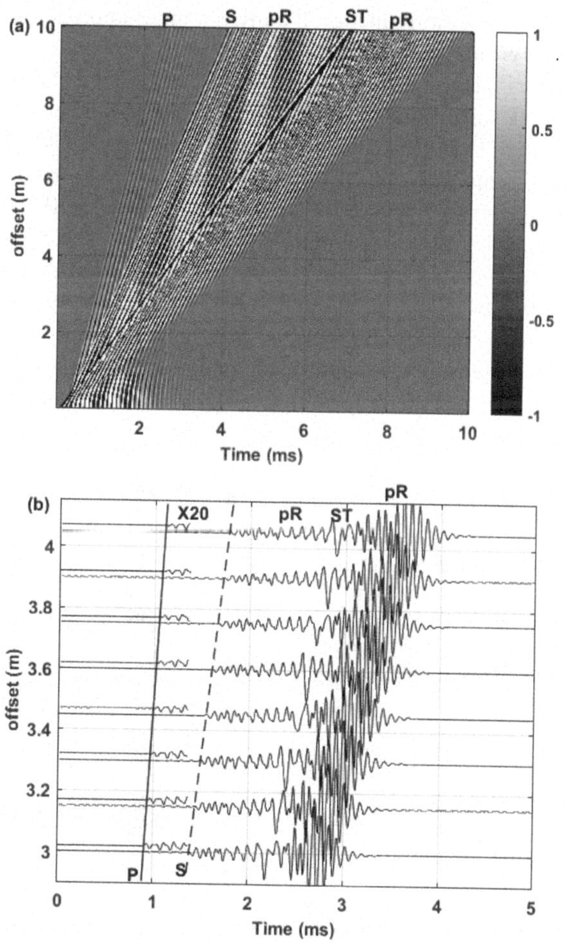

Fig. 2.4 Monopole waveforms in a fast formation (F1), a 10-kHz source frequency. The pR wave behind the ST wave is an Airy phase. **a** Dense receivers (receiver spacing of 0.01 m). **b** Waveforms at 8 receivers. P-wave is amplified 20 times

and the amplitude decreases with the increasing source-receiver distance. The S-wave refracted along the borehole is the second arrival in the waveform. The ST wave has a low frequency and it is not easy to attenuate with the source-receiver distance. The pR waves, following the S arrival, are long and dispersed waves with a wide range of phase and group velocities. A detailed display of the waveforms for an array receiver with 3 m T-R (source-receiver distance) and 0.15 m receiver spacing is shown in Fig. 2.4b. Here attenuation is not considered in the calculation. To illustrate attenuation on the waveforms, Fig. 2.5 shows the waveforms in formation F1 with different attenuation values. The Q values for fluid, P, and S velocities are listed on each trace. A larger Q indicates a smaller attenuation, which produces a lager amplitude.

Figure 2.6 shows the dispersion curves of the three pR modes and ST. The thick curves are the phase velocities and thin curves are the group velocities. pR consists of multiple modes. Only the group velocity of the fundamental pR mode is plotted. At the lowest frequency, the pR velocities approach the formation S-wave velocity. The group velocity (thin curve) slope is steep and covers a wide velocity range. An Airy Phase, marked with an ellipse, has a velocity slower than the fluid wave velocity or ST wave velocity. The pR behind the ST wave in Fig. 2.4 is the Airy phase. The duration of pR waves is long and dominates the wave train. The pR waves have a low frequency cutoff (or low frequency limit). Each pR mode has its own cutoff frequency. pR modes do not exist below the cutoff frequency. Velocity at the cutoff frequency is the formation S-wave velocity. The ST wave exists at all frequencies, with no cutoff frequency, and exhibits a slight dispersion. The ST wave velocity increases as frequency increases to approximately 0.95 V_f at high frequencies. The group velocity of ST wave is slightly larger than its phase velocity.

The entire wavefield inside and around the borehole is calculated by the finite difference code (Appendix B). The total wavefield snapshots are shown in Fig. 2.7.

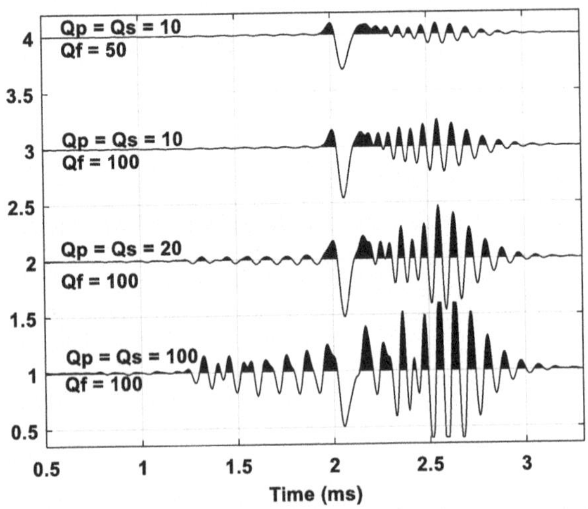

Fig. 2.5 Monopole waveforms with different Q values in a fast formation (F1), a 10-kHz source frequency. Q values are listed on the plot. The source-receiver distance is 3 m. Borehole radius is 0.1 m

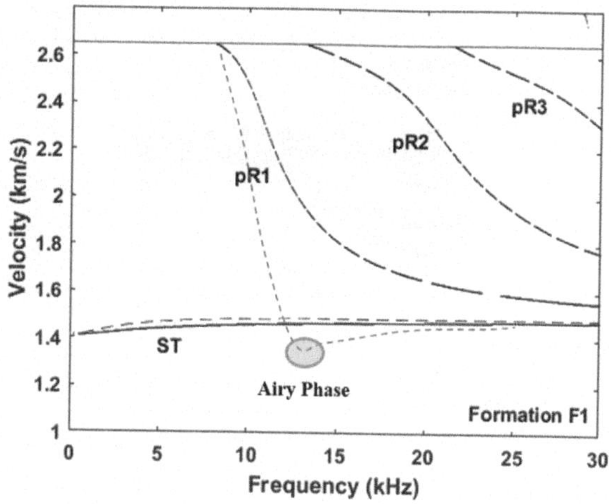

Fig. 2.6 Phase (thick curves) and group (dashed thin curves) velocity dispersion curves in a fast formation (formation F1 in Table 2.1). The dispersion curves for the three modes of pR waves are shown. Thin curves are the group velocity of the fundamental pR mode and ST mode

The P-wave in the formation propagates as a head wave (marked as P in Fig. 2.7a), as does the S-wave. The maximum amplitude of the P-wave inside the borehole is at the borehole center (Fig. 2.7b). The pR waves propagate as dispersed waves guided by the borehole. The maximum amplitude is located at the borehole center, and the absolute value of the amplitude outside the borehole decreases exponentially away from the borehole wall (Fig. 2.7c). The ST mode arrives in the middle of the dispersed pR waves (Fig. 2.7a). It has the slowest phase velocity. Its amplitude decreases with distance away from the borehole wall, both inside and outside of the borehole. The maximum absolute value of the amplitude is at the borehole wall.

The waves in a borehole in formation F2 with slightly lower velocities, are displayed in Fig. 2.8. The ST stands out with the largest amplitude. The pR waves are not as strongly dispersed as in the F1 formation case, and the Airy Phase is no longer the last arrival. This illustrates the sensitivity of dispersed waves to formation velocities.

2.2.1.2 A Note on Stoneley Waves

Stoneley (ST) waves, sometime referred to as tube waves, are the most common waves in the borehole. There are mainly four mechanisms for generating ST waves (Cheng and Toksöz 1982; Li et al. 2017). The most efficient source for ST waves is a monopole source in the borehole. Other sources are: surface waves crossing the well head and scattered into the borehole (Hardage 1981); and body waves impinging on the heterogeneous borehole, especially at "borehole washouts" or fractures (Huang and Hunter 1980; Peng and Toksöz 1992; Bakku et al. 2013).

A schematic diagram of the particle motion of the ST wave is shown in Fig. 2.9 (Cheng and Toksöz 1982). There is no radial displacement at the borehole center and the particle motion is rectilinear along the borehole axis. More radial displacement

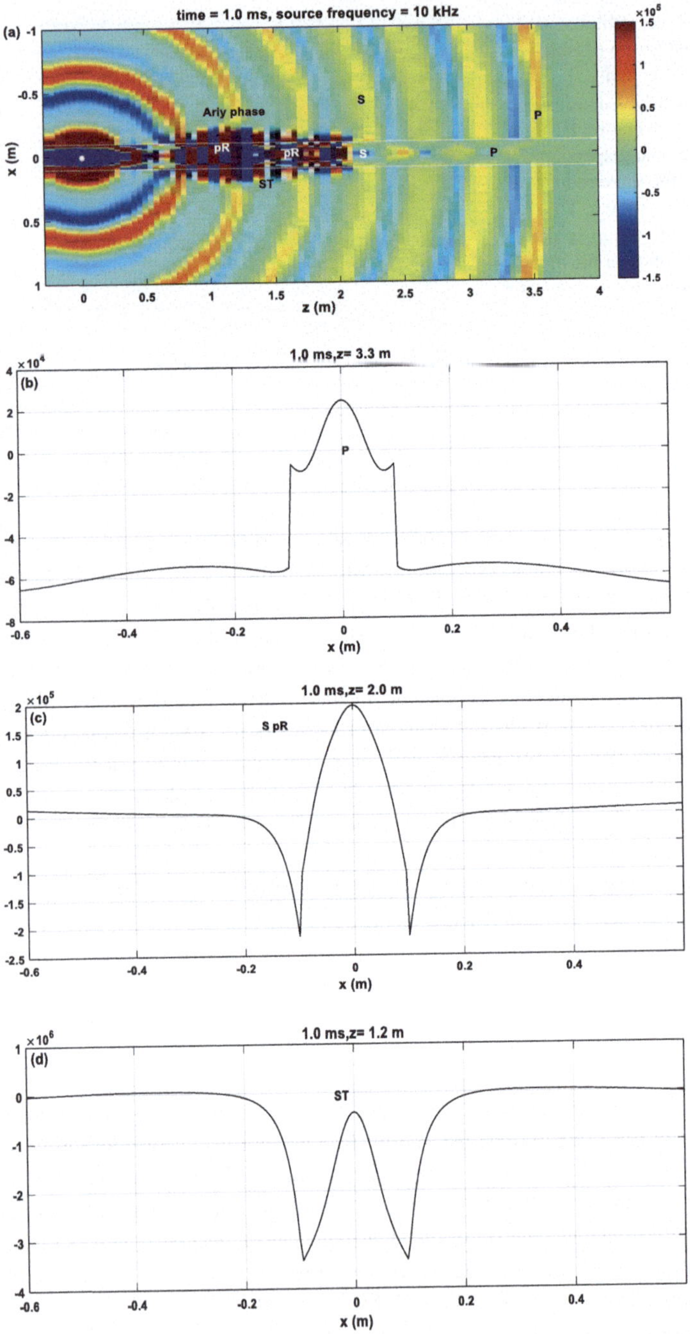

Fig. 2.7 Snapshot and amplitudes of different waves as a function of radius. **a** Snapshot at 1 ms for the monopole source (10 kHz) in formation F1. **b–d** Amplitude of waves as a function of radial position at a fixed source-receiver distance for **b** P, **c** S and pR, and **d** ST. Borehole radius is 100 mm

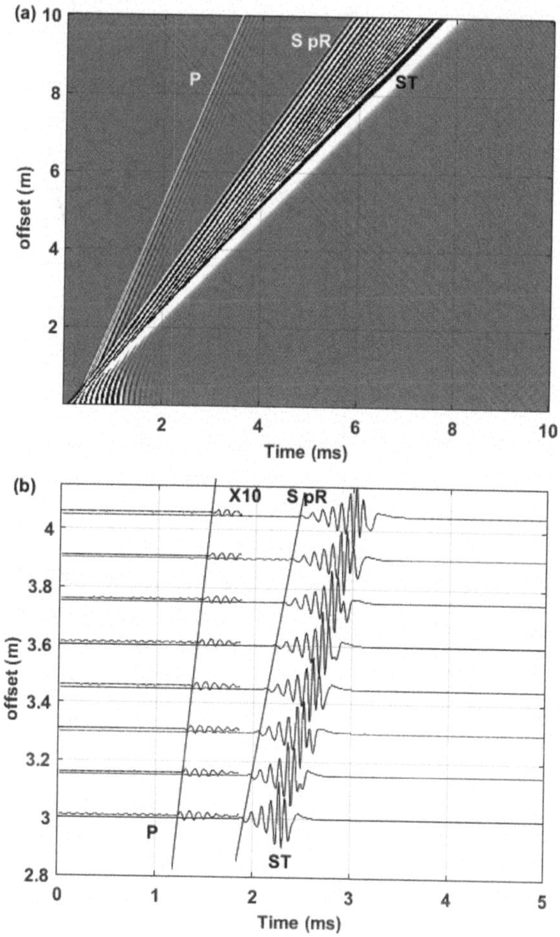

Fig. 2.8 Monopole wavefield in a fast formation (F2), a 10-kHz source frequency. **a** Dense receivers (receiver space of 0.01 m). **b** Selected 8 receivers showing details of the waveforms. P-wave is amplified 10 times

appears at the borehole wall. For a typical borehole, ellipticity depends on frequency and formation S-wave velocity.

2.2.1.3 Factors that Influence the Different Modes

Borehole Radius

pR waves are the guided waves related to the reflections from the borehole wall. The borehole radius, and formation velocities, are the primary factors that control pR waves. The cutoff frequency of the fundamental pR mode decreases when borehole radius increases, as shown in Fig. 2.10a. This produces an approximate equivalence between the cutoff frequency and borehole radius (Cheng and Toksöz 1980, 1981;

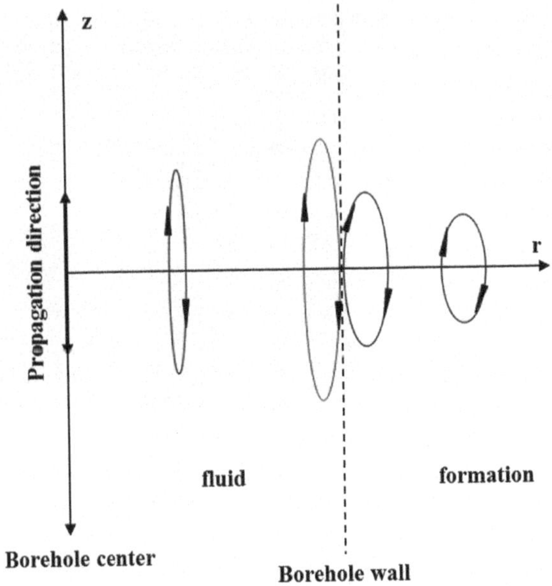

Fig. 2.9 Schematic diagram of the particle motion of the ST wave in the borehole fluid and formation (Modified from Fig. 6 in Cheng and Toksöz (1982))

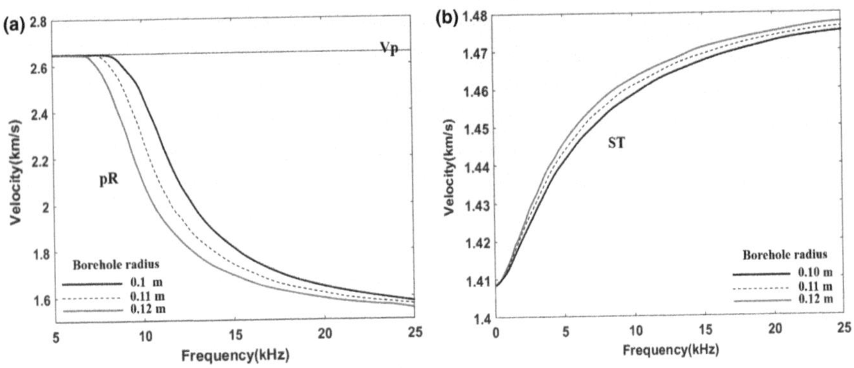

Fig. 2.10 Phase velocity dispersion curves for various borehole radii in formation F1. **a** Fundamental pR wave. **b** ST wave. The formation properties are given in Table 2.1

Paillet 1981). The effect of borehole radius on the ST mode is very small. The ST wave velocity slightly increases when borehole radius increases (as shown in Fig. 2.10b).

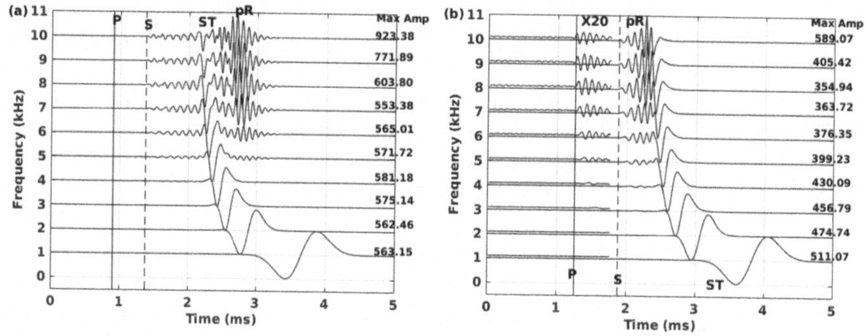

Fig. 2.11 Waveforms (3 m offset) at different source frequencies. Each waveform is normalized by their maximum amplitude (listed on each plot). **a** Formation F1. **b** Formation F2. Formation properties are given in Table 2.1

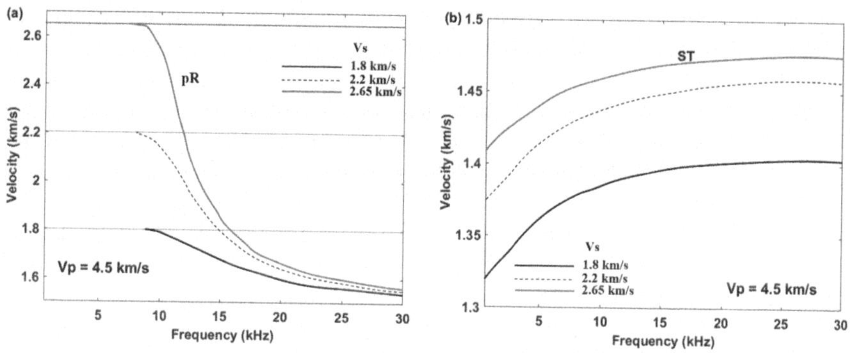

Fig. 2.12 Phase velocity dispersion curves of pR and ST modes for various formation velocities. Formations with Vp of 4500 m/s and various Vs. **a** Fundamental pR mode. **b** ST wave

Source Frequency

The frequency of a monopole source significantly affects waveforms in fast formations. The waveforms, at 3 m offset at different source frequencies, are shown in Fig. 2.11a (formation F1) and 2.11b (formation F2), where the amplitude of each waveform is normalized by its maximum value (listed on each plot). Different modes are marked in the plots. The arrival times of P- and S-waves are also marked. It is obvious that a high source frequency enhances the excitation of the P, S, and pR waves. Low frequency source enhances the ST wave.

Formation Velocities

Lowering the formation S-wave velocity with a fixed Vp results in a higher cutoff frequency for the pR waves and a lower velocity for the ST wave (Fig. 2.12). The

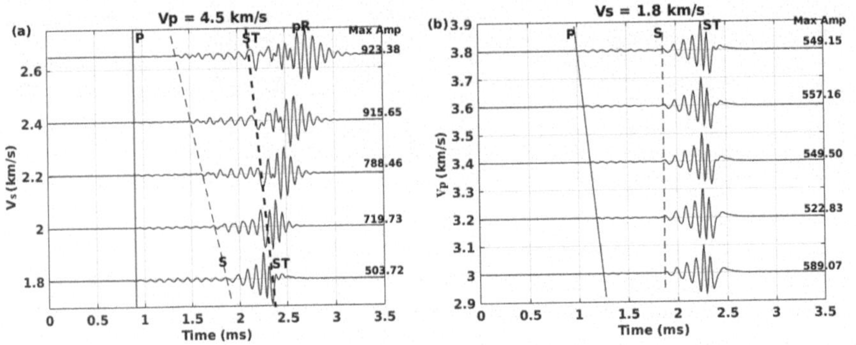

Fig. 2.13 Waveforms at 10 kHz with various formation velocities. **a** Formations with Vp of 4500 m/s and various Vs. **b** Formations with Vs of 1800 m/s and various Vp

relationship between the cutoff frequency and S-wave velocity is a linear function. The changing P-wave velocity changes the guided modes very little when the S-wave velocity is constant.

The waveforms (10 kHz source center frequency) with various formation velocities are shown in Fig. 2.13. Formations with low S-wave and large P-wave velocities generate obvious P- and S-waves. Only the S-wave velocity affects the pR waves.

2.2.2 *Dipole Source*

2.2.2.1 Modes

To measure S-wave velocities, a dipole source can be used to generate shear motion (White 1967; Kitsunezaki 1980; Zemanek et al. 1984). A dipole source generates a flexural motion of the borehole wall.

Figure 2.14 shows the phase velocity dispersion curves (formation F1) of the fundamental (marked as 0) and higher-order flexural modes (marked as 1, 2, and 3). All flexural modes exhibit normal dispersion where the phase velocity is equal to the formation S-wave velocity at the low frequency limit, and to fluid wave velocity at high frequency. The group velocity curve shown for the fundamental mode has an Airy phase, and this would be the last arrival (Fig. 2.15). A dipole source also generates a weak P-wave that rapidly attenuates (Fig. 2.15a).

For a formation with a lower velocity, F2 in Table 2.1, the dispersion curves are different from those of formation F1. Figure 2.16 shows the phase velocity dispersion curves. Although the cutoff frequency of the fundamental mode (marked 0) is similar to that in formation F1, the flexural wave velocity at high frequency is lower than the fluid wave velocity. The higher modes have larger cutoff frequencies than those in formation F1. The seismograms in Fig. 2.17 show two flexural modes. The Airy phase of the flexural wave is not as prominent as in formation F1. This is similar

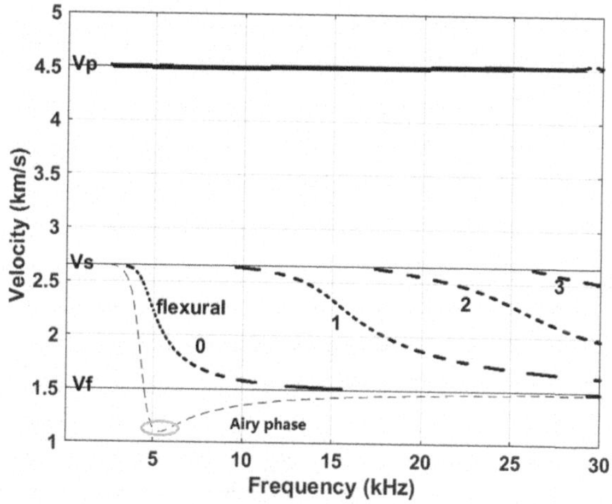

Fig. 2.14 Dispersion curves for four modes of flexural waves in fast formation F1. Thin curve is the group velocity of the fundamental flexural mode. Airy phase of the fundamental flexural is marked by an ellipse. All other curves show phase velocities

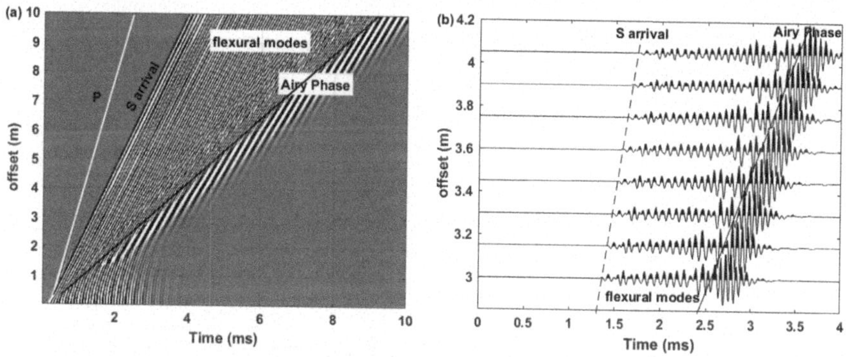

Fig. 2.15 Dipole waveforms in a fast formation (F1 in Table 2.1). **a** Waveforms with a dense receiver array (interval of 0.01 m). **b** Array waveforms

to the monopole case, where the Airy phase of the pR wave is strong in formation F1 but weak in formation F2. Another difference between formations F2 and F1 is that leaky P-waves are observable in formation F2. It is not surprising that a dipole source in a cylindrical borehole generates some P-waves.

Waves inside the borehole are illustrated by a snapshot and cross section of pressure (summation of normal stress) shown in Fig. 2.18a. The dipole polarization direction is along the x-axis in the figure. White star at x = 0 m, z = 0 m marks the center of the dipole source. The snapshot clearly shows the small amplitude P-wave in both formation and borehole. Two white lines mark the borehole wall. The small amplitude in the borehole illustrates the small amplitude leaky P wave acquired by

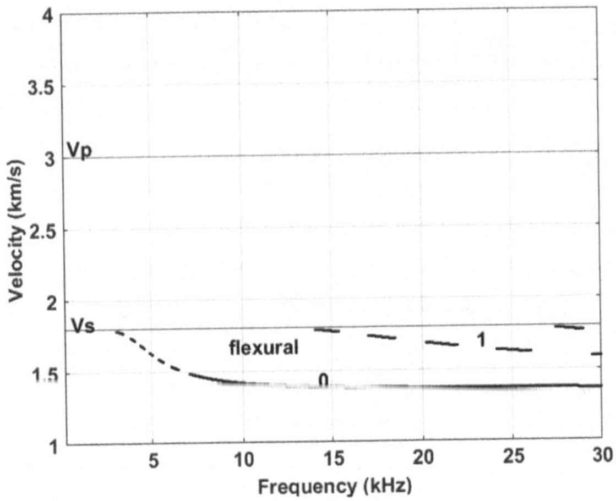

Fig. 2.16 Dispersion curves of the flexural wave in fast formation F2 in Table 2.1

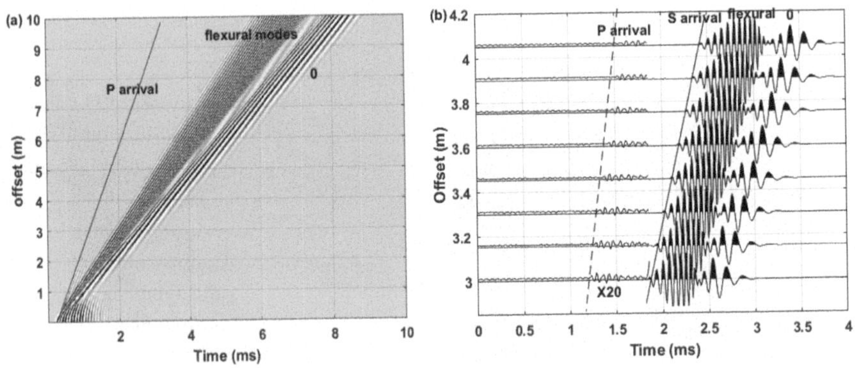

Fig. 2.17 Dipole waveform in formation F2 in Table 2.1. **a** Waveform section with a dense receiver array (interval of 0.01 m). **b** Array waveforms. P-wave is amplified 20 times. 0 in the figure is the fundamental flexural mode

the centralized array receiver. The amplitude distribution of dipole induced leaky P-waves at z = 3.546 m (Fig. 2.18b) is antisymmetric about borehole center. Behind the weak leaky P mode, the flexural modes with much larger amplitude arrive. The maximum amplitude is at the borehole wall (Fig. 2.18c). Amplitude rapidly decreases outside of the borehole. A low frequency source provides deeper penetration into the formation. In the borehole fluid, the maximum amplitude appears at 0.04 m away from the borehole center.

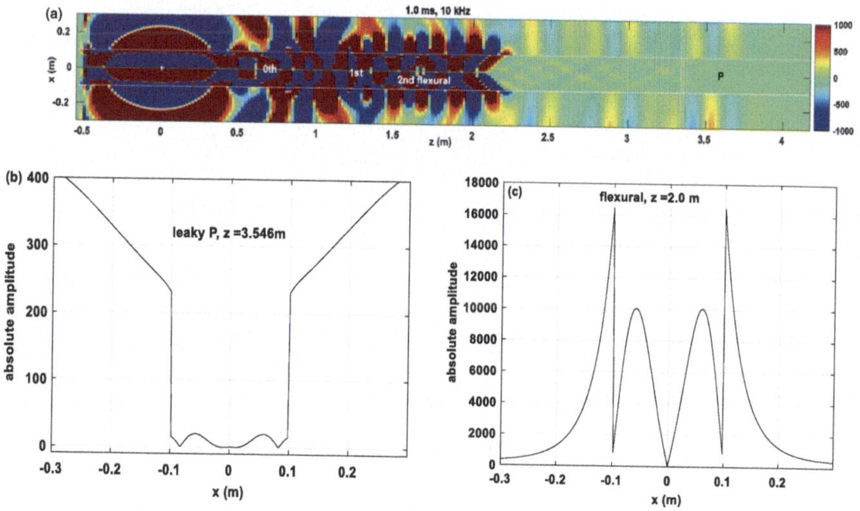

Fig. 2.18 **a** Dipole pressure (summation of normal stress) at 1 ms at 10 kHz in formation F1. **b**, **c** Absolute value of pressure from **a** as a function of radial position at two different z

2.2.2.2 Factors that Influence the Modes

Source Frequency

Dispersion curves, in Figs. 2.14 and 2.16, show that the higher modes have higher cutoff frequencies than the fundamental mode. This indicates that the source frequency is a factor that controls the mode excitation. The waveforms, at 3 m offset (source-receiver distance) at different source frequencies, are shown in Fig. 2.19a (formation F1) and 2.19b (formation F2).

Figure 2.19 indicates that there is only the fundamental flexural mode (marked as 0) when the source frequency is low (below 3 kHz). Higher modes appear when the source frequency increases. The amplitudes of the waveforms increase with the source frequency, and fundamental flexural mode is the maximum in each wave train. A higher source frequency excites a stronger P-wave.

Formation Velocities and Poisson's Ratios

Comparing the waveforms for formations F1 and F2 shown in Fig. 2.19a and 2.19b, decreasing the formation waves' velocity increases the amplitude of the P-wave. Figure 2.20 shows that lowering the S-wave velocity (with a fixed Vp) results in a lower cutoff frequency (frequency where phase velocity equals formation S-wave velocity) and decreases the dispersion of the fundamental flexural mode (Fig. 2.20a). Changing P-wave velocity slightly affects the flexural modes (Fig. 2.20b) when S-wave velocity is fixed. The lower P-wave velocity slightly increases the dispersion of

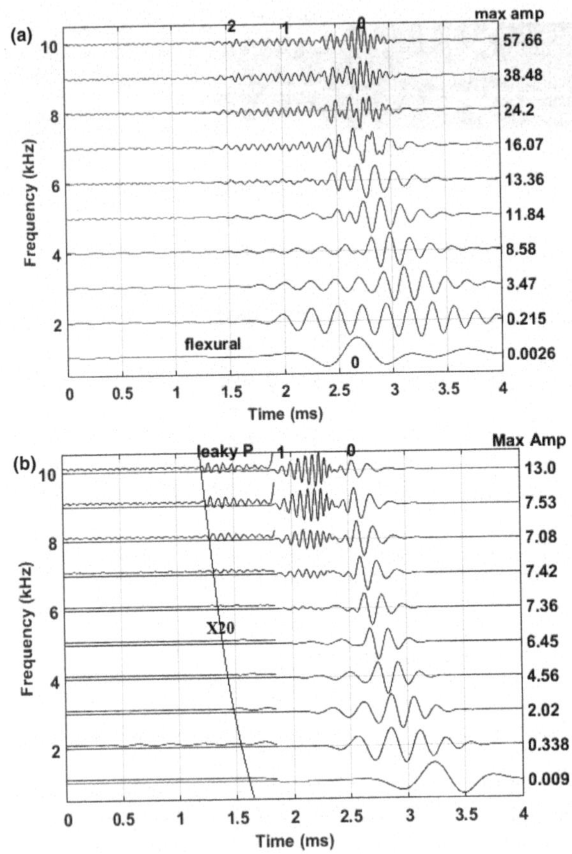

Fig. 2.19 Waveforms (3 m offset) for different source frequencies. Each trace is normalized by its maximum amplitude (listed on each plot). **a** Formation F1. **b** Formation F2. Flexural wave modes are labeled where 0 is the fundamental mode

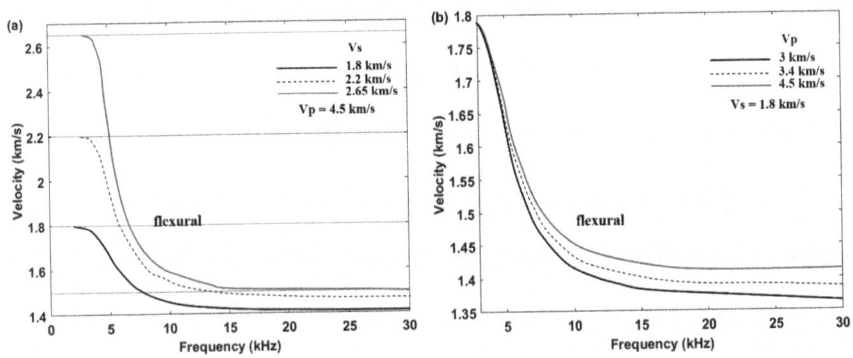

Fig. 2.20 Fundamental flexural mode phase velocity dispersion curves with various formation velocities. **a** Formations with Vp of 4500 m/s and various Vs. **b** Formations with Vs of 1800 m/s and various Vp

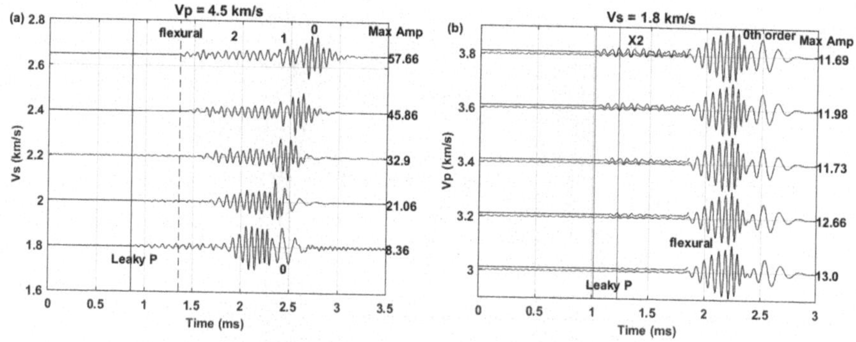

Fig. 2.21 Waveforms at 10 kHz with various formation velocities. **a** Formations with Vp of 4500 m/s and various Vs. **b** Formations with Vs of 1800 m/s and various Vp

the fundamental mode. This is similar to the influence of density, in which a formation with a low density generates a more dispersive fundamental flexural mode.

In summary, the S-wave velocity primarily controls the cutoff frequency, dispersion, and number of flexural modes.

Waveforms (a 10-kHz source frequency) with various formation velocities are shown in Fig. 2.21. A formation with low S-wave and high P-wave velocities (large Poisson's ratio) generates large leaky P-waves.

The borehole radius also affects the modes. Increasing the radius moves the flexural mode to a lower frequency. This is similar to the case of pR modes, discussed earlier.

2.2.3 *Field Data in Fast Formations*

In this section, we show some field data to illustrate the characteristics of monopole and dipole logs in a borehole surrounded by a fast formation. The iso-offset monopole and dipole waveforms are shown in Fig. 2.22. For the monopole measurement, the source frequency is 10 kHz, and offset from the source to the nearest receiver is 11 ft (3.35 m). For the dipole measurement, the source frequency is 4 kHz and the source-receiver offset is 10.25 ft (3.12 m).

The P, S, pR, and ST waves, in the monopole data, are identified and marked on the figure. The amplitude of the P-wave is smallest. The formation waves' velocity varies slightly at different depths, affecting the waveforms. The P and pR waves are affected most strongly. ST is least affected. A weak reflected ST wave can be seen at depth range from 3924 to 3928 m. This corresponds to a borehole wash out.

For the dipole waveforms, the arrival time of the flexural wave is almost the same as that of the S arrival in Fig. 2.22a. The highly dispersive fundamental mode is responsible for the long duration of the flexural wave. There are also some higher flexural modes.

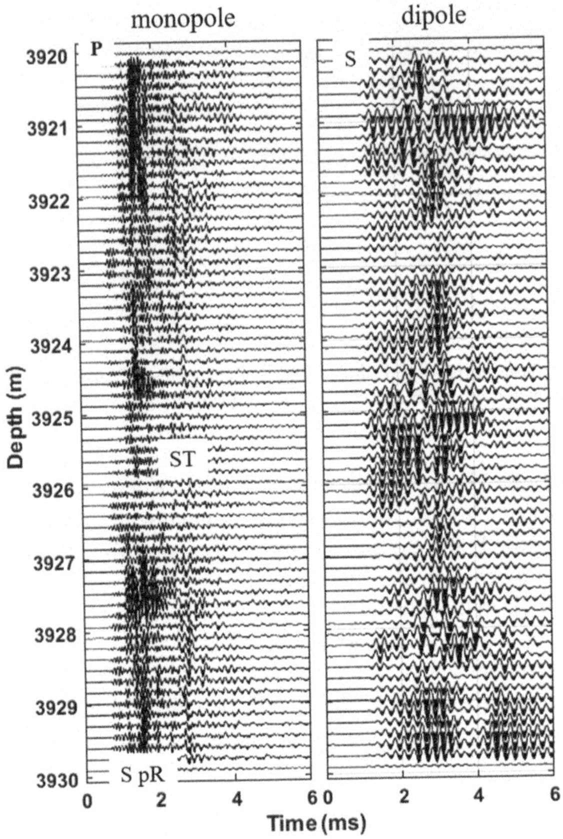

Fig. 2.22 Field examples of the monopole (left) and dipole (right) measurements in a borehole surrounded by fast formations. The source frequency of the monopole tool is 10 kHz. The waveforms for a fixed source-receiver offset (11 ft, 3.35 m) are shown as a function of depth. The source frequency of the dipole tool is 4 kHz. The waveforms for a fixed source-receiver offset (10.25 ft, 3.12 m) are shown as a function of depth

To observe the detail of the waveforms, the array waveforms at depth of 3922 m are shown in Fig. 2.23. From the monopole array waveforms, the arrival times of P, S, and ST modes can be picked. The P-wave has a small amplitude as expected. The ST wave has a large amplitude and low frequency. The dispersive pR waves, following the S arrival, are clearly observed. The dipole array waveforms are shown in Fig. 2.23b.

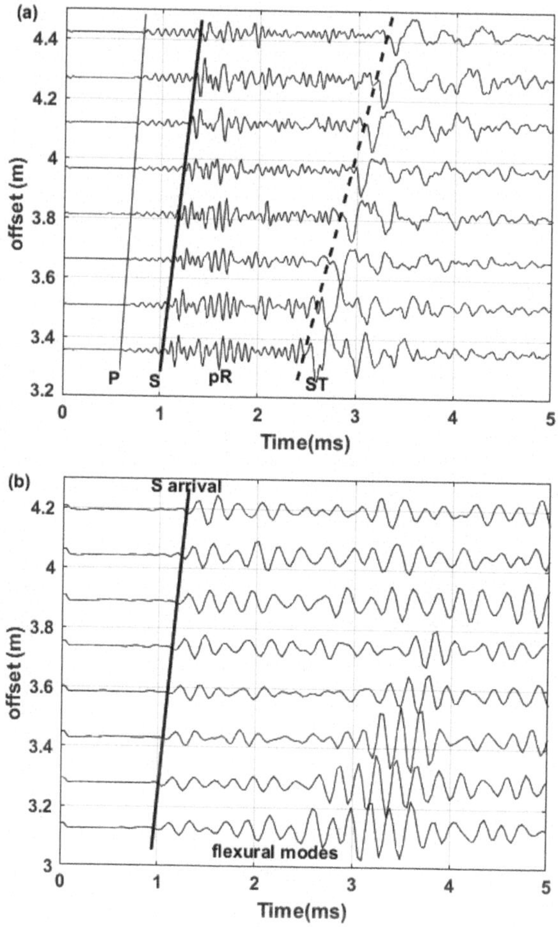

Fig. 2.23 Array waveforms at depth of 3922 m. **a** Monopole data. **b** Dipole data

2.2.4 Quadrupole Source

A quadrupole source, shown in Fig. 2.2d, generates the purest form of S-waves and gives the exact value of S-wave velocities at the cutoff frequency.

Figure 2.24 shows the dispersion curves and waveforms of quadrupole measurements for two fast formations. Velocity at the cutoff frequency is exactly that of the formation S-wave velocity. A clear difference between the quadrupole and dipole wavefields is that there is no P mode generated by a quadrupole source, and the amplitudes of the quadrupole waves are lower than those of the dipole waves. The dispersion curves in Fig. 2.24a show that a low source frequency (e.g. approximately 4 kHz) generates only the fundamental quadrupole (or "screw") mode propagating with the formation S-wave velocity at the low frequency limit.

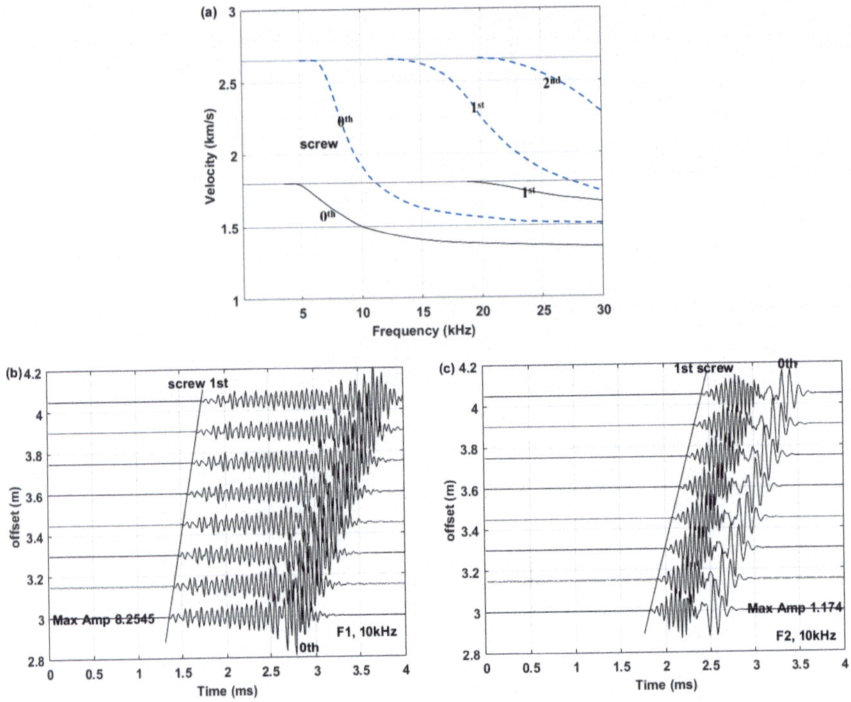

Fig. 2.24 Quadrupole wavefield in fast formations. **a** Dispersion curves. Black lines are for formation F2 and blue lines for formation F1. **b** Waveforms in formation F1. **c** Waveforms in formation F2

2.3 Wavefields in Slow Formations

2.3.1 Monopole Source

2.3.1.1 Modes

Slow formations are widely encountered in shallow, offshore environments, in shales, and in semi-consolidated silts and sands. A number of studies have investigated waves in slow formations (Cheng and Toksöz 1981; Chang and Everhart 1983; Tichelaar and van Luik 1995). Monopole sources in slow formations, when Vs < Vf, generate only the leaky P and Stoneley waves.

Figure 2.25 shows the dispersion curves for two slow formations (S1 and S2 in Table 2.1). Figures 2.26 and 2.27 show the waveforms. There are only two waves: leaky P and ST, and both are dispersive. There are no pR waves. The phase velocity of the leaky P mode goes to the borehole fluid wave velocity at the high-frequency limit and to formation Vp at the low frequency cutoff. The formation P-wave velocity can be determined from the leaky P mode when the tool frequency is low—approximately

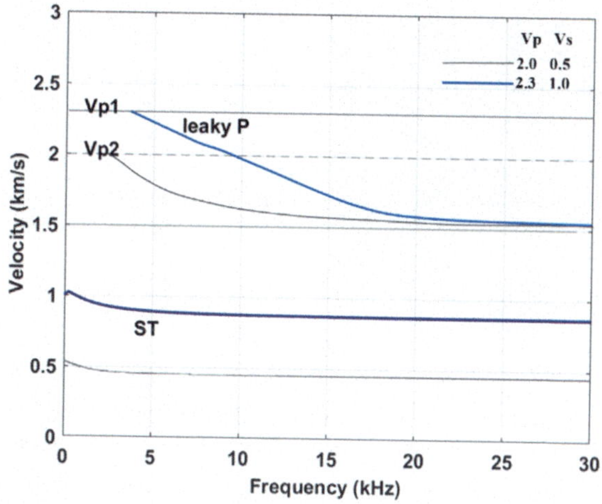

Fig. 2.25 Phase velocity dispersion curves in a fluid-filled borehole surrounded by two different slow formations. Blue and black curves are for formations S1 and S2. Formation velocities are listed on the plot

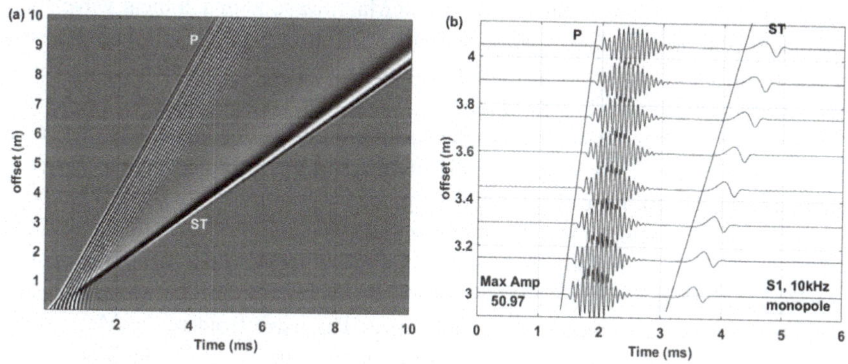

Fig. 2.26 Monopole waveforms in a slow formation (S1 in Table 2.1, Vp = 2300 m/s, and Vs = 1000 m/s) with a 10-kHz source frequency. **a** Dense display. **b** Selected array waveforms

2 kHz (Hornby and Pasternark 2000). The influence of the source frequency on waveforms is discussed in detail later in this chapter.

The ST wave on the seismograms are small because of the high source frequency (10 kHz). The ST wave velocity ranges from the formation S-wave velocity at low frequency and decreases slightly at high frequency (Fig. 2.25). The zero frequency ST wave velocity is given by Biot (1952) and White (1965) as,

$$V_T = \frac{V_f}{\sqrt{1 + \frac{\rho_f V_f^2}{\rho V_s^2}}} \tag{2.29}$$

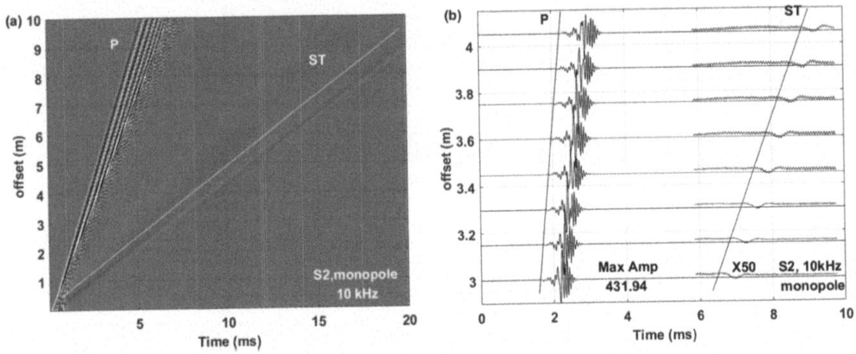

Fig. 2.27 Monopole waveforms in a slow formation (S2 in Table 2.1, Vp = 2000 m/s, and Vs = 500 m/s) with a 10-kHz source frequency. **a** Dense display. **b** Selected array waveforms. ST mode is amplified 50 times

When S-wave velocity is very low (Vs < 0.7 Vf), the amplitude of the ST wave becomes very small, making the S-wave velocity determination difficult (Cheng and Toksöz 1982).

In a borehole in a slow formation, a strong Mach wave with a conical wavefront, propagates with an apparent S-wave velocity (Cheng et al. 1992). Pressure distributions of different modes in formation S1 are given by the wavefield snapshot in Figs. 2.28a (10 kHz) and 2.29 (2 kHz).

Figure 2.28a shows the snapshot of the wavefield illustrating the clearly separated leaky P and ST waves. The amplitude distributions of the different modes along the borehole radial direction are also shown in Fig. 2.28. The amplitudes of leaky P modes decrease with the distance from the borehole wall. The rapidly decreasing amplitude of the leaky P limits the depth of investigation in slow formations. Although most of the energy of the ST is trapped in the borehole, the leakage of the ST wave propagates as the moving source to radiate the Mach wave. The wave front of the Mach wave is not obvious in Fig. 2.28a due to the high frequency. In Fig. 2.29, the Mach wave becomes apparent at lower frequency (2 kHz).

2.3.1.2 Factors that Influence the Modes

Source Frequency

As we mentioned in Sect. 2.3.1.1, frequency is critical for both leaky P and ST modes generation. A low frequency source generates large ST and small leaky P-waves.

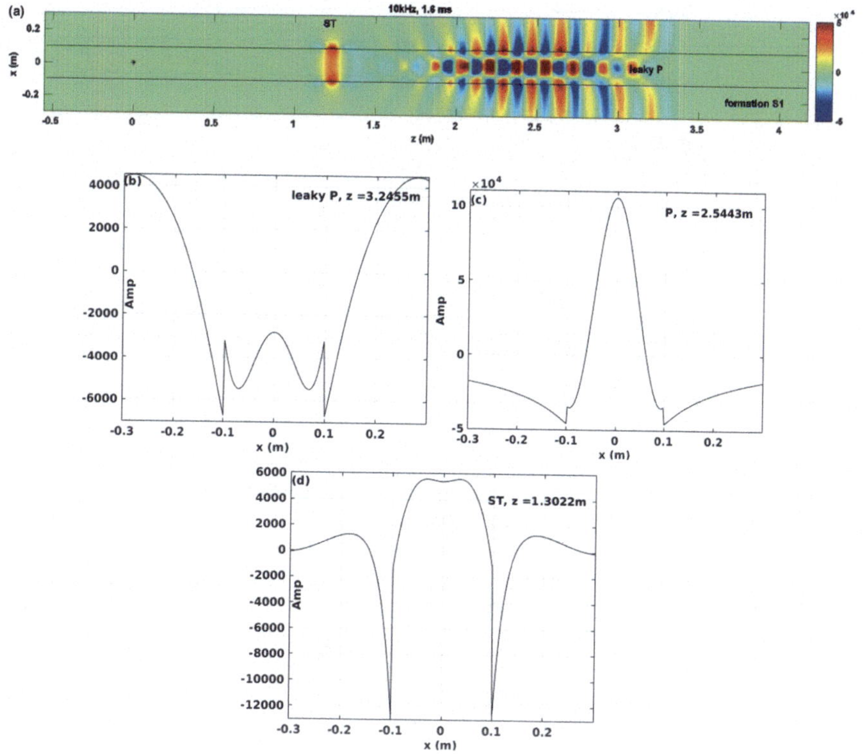

Fig. 2.28 **a** Monopole wavefield snapshot at 1.6 ms (10 kHz source) in formation S1. **b–d** Amplitudes as a function of radial distance at various source-receiver offsets. Note the different scales

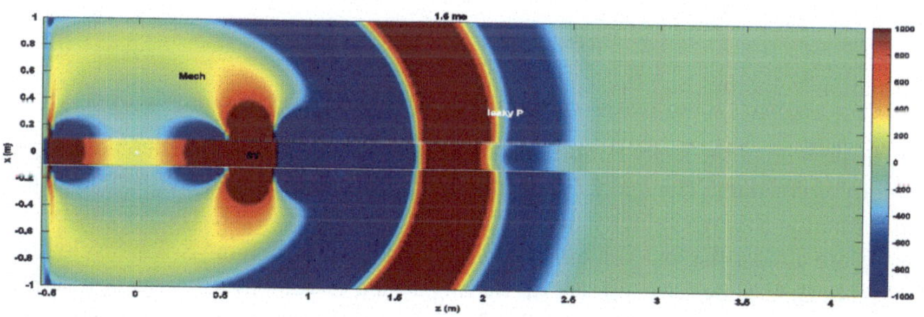

Fig. 2.29 Monopole pressure wavefield snapshot at 1.6 ms (2 kHz source) in the formation S1 illustrating the Mach wave

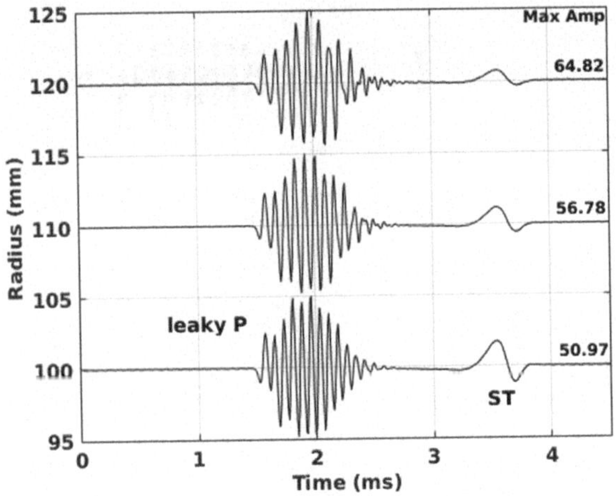

Fig. 2.30 Monopole waveforms (3 m offset, 10 kHz) in formation S1 with different borehole radii: 10, 11, and 12 cm

Borehole Radius

Here we show the effects of borehole radius by calculating synthetics. An example is shown in Fig. 2.30. Note the maximum amplitudes listed above each trace. Similar to the effect of borehole radius on the dispersion curves of the pR modes in the fast formations (Fig. 2.10), a large borehole radius moves the dispersion curves of the leaky P modes to a low frequency range. It increases the amplitudes of leaky P modes (compare leaky P waveform at the radius of 100 and 120 mm in Fig. 2.30). The ST amplitude increases slightly when radius increases.

Formation Velocities and Poisson's Ratios

The dispersion of the leaky P modes is affected by P-wave velocity when the S-wave velocity is constant. The leaky P dispersion curves move to lower frequencies when the formation S-wave velocity decreases, as shown in Fig. 2.31. The ST dispersion curves are almost flat and velocities decrease with the decreasing S-wave velocity.

Figure 2.32 shows the effects of varying P and S velocities on the waveforms at source frequency of 2 kHz. Leaky P modes in a formation of large Poisson's ratio include more modes and larger amplitude than when Poisson's ratio is smaller. The amplitude and dispersion of leaky P-waves may be used for S-wave velocity determination in cases when S-wave velocities are very low (Toksöz et al. 1984b).

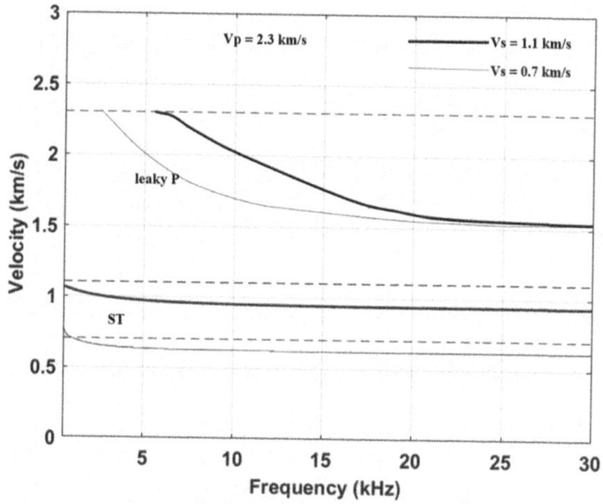

Fig. 2.31 Dispersion curves of leaky P and ST waves with various S-wave velocities when Vp is fixed. The dashed lines are the formation S-wave velocities

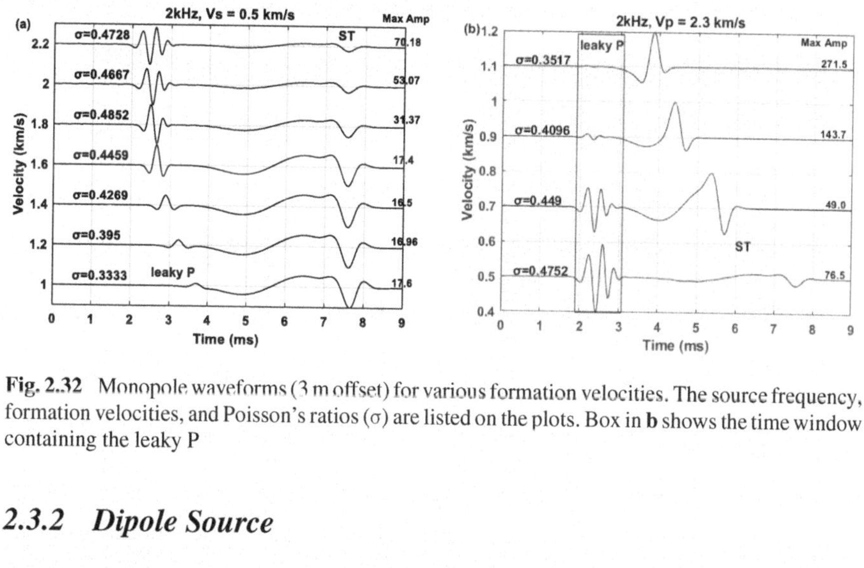

Fig. 2.32 Monopole waveforms (3 m offset) for various formation velocities. The source frequency, formation velocities, and Poisson's ratios (σ) are listed on the plots. Box in **b** shows the time window containing the leaky P

2.3.2 Dipole Source

2.3.2.1 Modes

The dispersion curves and waveforms for dipole sources in two slow formations are shown in Figs. 2.33, 2.34 and 2.35. The S-wave velocity in the slow formation is difficult to determine from monopole measurements. A dipole source that generates flexural waves works much better than a monopole source for S-wave velocity determination.

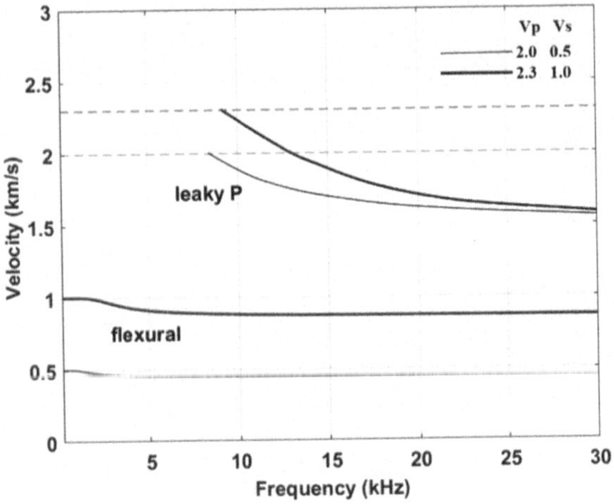

Fig. 2.33 Phase velocity dispersion curves for a dipole source in a fluid-filled borehole in slow formations. Thick and thin curves are for formations S1 and S2 in Table 2.1. The velocities are listed on the plot. Cutoff frequencies of leaky P-waves are higher than the monopole case in Fig. 2.25

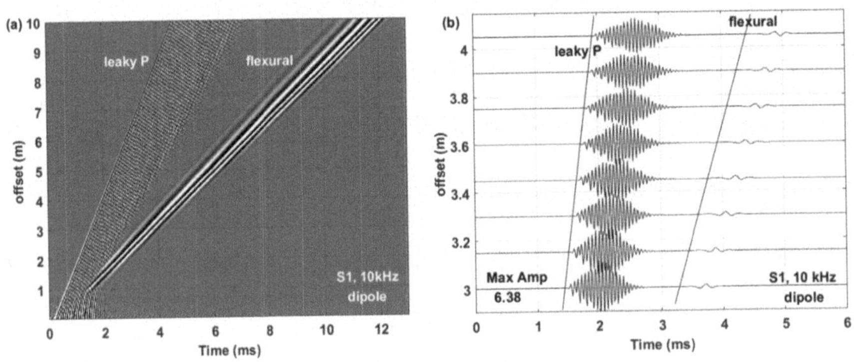

Fig. 2.34 Dipole waveforms in a slow formation (S1 in Table 2.1) with a 10-kHz source frequency. **a** Dense display. **b** Selected array waveforms

Leaky P-waves are efficiently generated by a high-frequency monopole and dipole source. Velocity of the leaky P wave reaches the formation Vp at the cutoff frequency and decreases to the fluid wave velocity at high frequency. The formation flexural mode has the S-wave velocity at a cutoff frequency (approximately 2 kHz). It is possible to obtain Vp from the leaky P and Vs from flexural waves at their cutoff frequencies. However, the flexural waves have small amplitudes in very slow formations.

A wavefield snapshot is shown in Fig. 2.36 for formation S1. In the 3DFD simulation, the dipole source is polarized along the x direction, and the wavefronts of

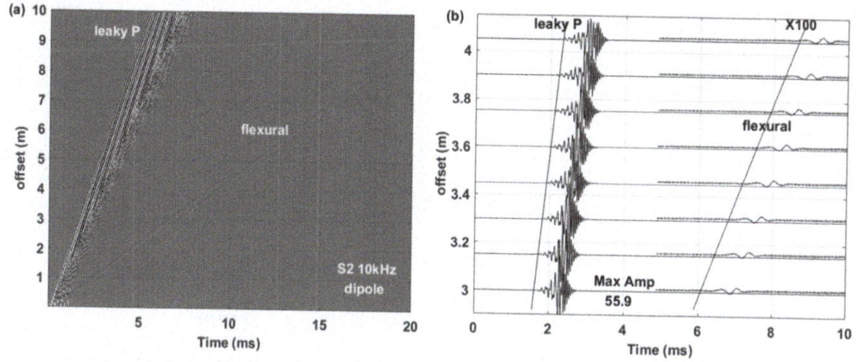

Fig. 2.35 Dipole waveforms in a slow formation (S2 in Table 2.1) with a 10-kHz source frequency. **a** Dense display. **b** Selected array waveforms

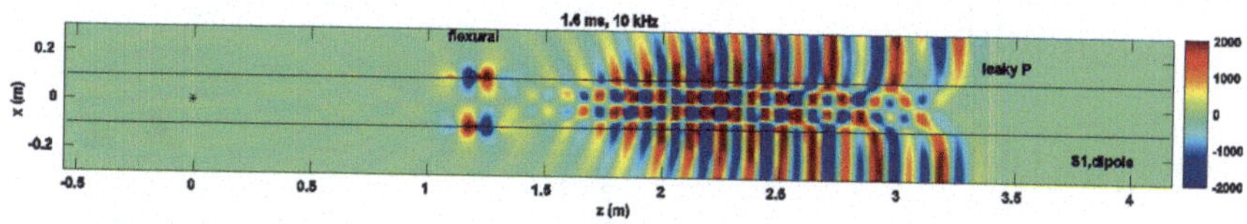

Fig. 2.36 Dipole wavefield snapshot at 1.8 ms (10 kHz) in the formation S1

the leaky P-waves from 1.5 to 3.3 m along the z direction (borehole axial direction) exhibit dipole characteristics of antisymmetry about the x-axis.

2.3.2.2 Factors that Influence the Modes

Source Frequency

Waveforms in Fig. 2.37 illustrate a high source frequency, which generates strong leaky P and weak flexural modes. The leaky P-waves are suppressed if the source frequency was below the cutoff frequency. A low frequency source generates large flexural and weak leaky P-waves. One can utilize a low frequency dipole to measure the S-wave velocity in slow formations. In the very slow formation, such as formation S2, it is possible to utilize the weak leaky modes at low frequency (Fig. 2.37b) to determine the formation P-wave velocity.

Borehole Radius

The effect of the borehole radius on the dispersion curves of the leaky P modes is the same as that in the monopole case, i.e., a large borehole radius moves the dispersion

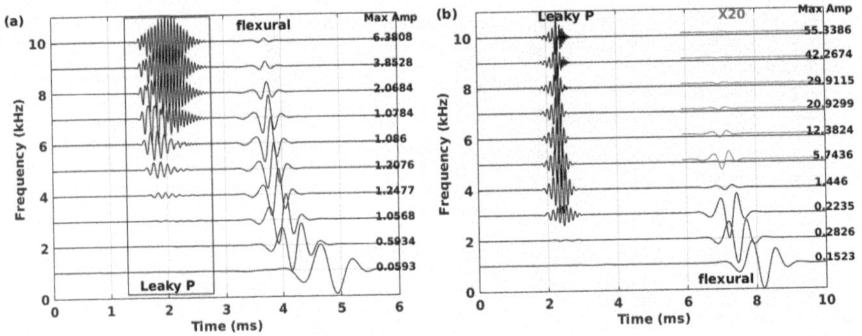

Fig. 2.37 Dipole waveforms (3 m offset) at various source frequencies. **a** Formation S1, **b** Formation S2

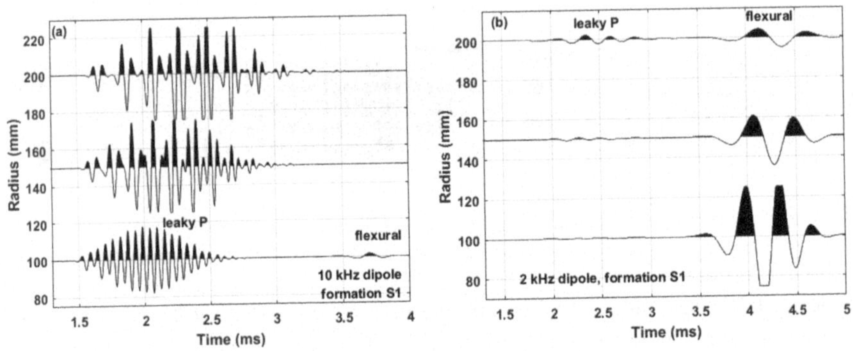

Fig. 2.38 Dipole waveforms at a 3-m source-receiver separation for various borehole radii in formation S1. **a** 10 kHz. **b** 2 kHz

curves to a lower frequency and increases the amplitude. It decreases the relative amplitude of the flexural wave (Fig. 2.38a).

Formation Velocities and Poisson's Ratio

The influence of the formation velocities on the dispersion curves is similar to those in the monopole case except for the influence of the formation S-wave velocity on the flexural cutoff frequency. See Fig. 2.39 for example waveforms. The cutoff frequency of the flexural wave decreases with decreasing S-wave velocity. This indicates that a low source frequency is needed to measure S-wave velocity in a formation with very low S-wave velocity.

In formations with large P-wave and small S-wave velocities, corresponding to a large Poisson's ratios, there are large leaky P and small flexural waves. However, when the source frequency is low, the amplitude of the flexural wave is not strongly affected, but the leaky P is suppressed. Therefore, it is possible to use a low frequency

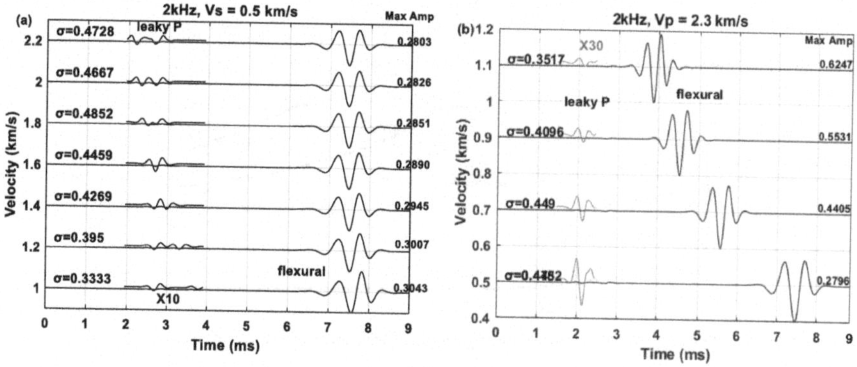

Fig. 2.39 Dipole waveforms (3 m offset) with various formation velocities. Source frequencies, formation velocities, Poisson's ratios (σ) are listed on the plots. Amplitudes of leaky P-waves are amplified 10 and 30 times in **a** and **b**, respectively

dipole to measure both formation P- and S-wave velocities from leaky P and flexural waves with a dispersion correction (Tichelaar and van Luik 1995). The dispersion correction methods will be discussed in Chap. 3.

2.3.3 *Quadrupole Source*

2.3.3.1 Modes

Quadrupole sources generate two modes, leaky P and screw (or formation quadrupole) waves in both formations S1 and S2. The leaky P exhibits strong dispersion (see Fig. 2.40) and has a higher cutoff frequency and a larger amplitude (Figs. 2.41 and 2.42) than the leaky P modes in the monopole and dipole cases. The screw wave has a weaker dispersion than the ST and flexural modes in slow formations. The pressure snapshot (Fig. 2.43) illustrates the amplitude distribution of the leaky P and screw modes. The amplitude distribution along the borehole radial direction of the leaky P (not shown) is different from that in the monopole and dipole cases.

2.3.3.2 Factors that Influence the Modes

Source Frequency

Similar to the monopole and dipole cases, a high-frequency quadrupole source excites strong leaky P and weak screw modes (Fig. 2.44a), although the maximum amplitudes of the waveforms at different frequencies are smaller than those in the monopole and

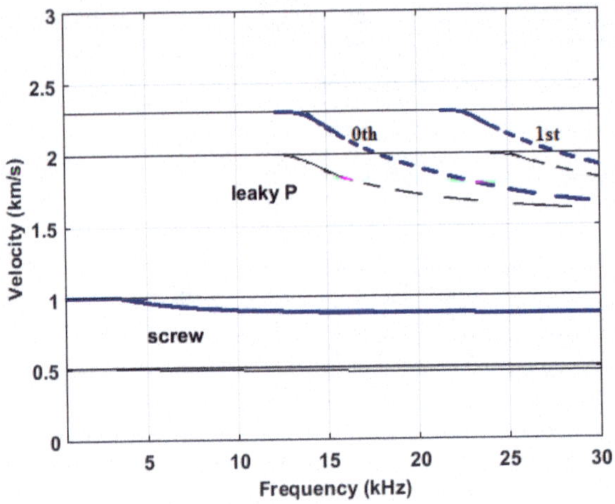

Fig. 2.40 Mode phase velocity dispersion curves in a fluid-filled borehole surrounded by two different slow formations. Blue and black curves are for formations S1 and S2 in Table 2.1. Borehole radius is 0.1 m

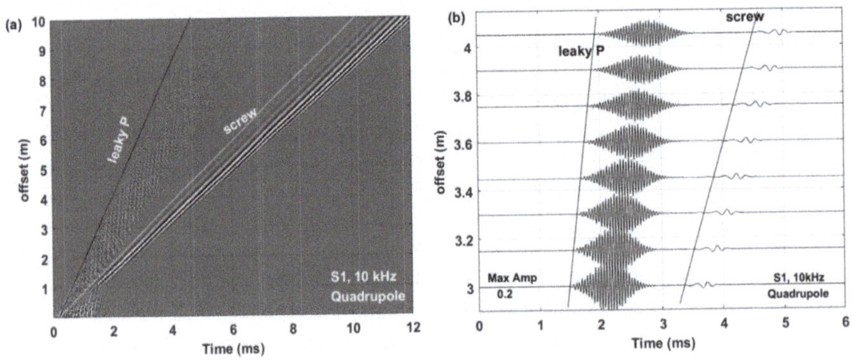

Fig. 2.41 Quadrupole waveforms in a slow formation (S1 in Table 2.1) with a 10-kHz source frequency. **a** Dense display. **b** Selected array waveforms

dipole cases. The screw mode can still be discerned at 10 kHz in formation S1. However, the influence of the source frequency on the screw wave in formation S2 (Fig. 2.44b) is more severe than that in formation S1. Conversely, the leaky P in formation S1, decreases more rapidly with the increasing source frequency than that in formation S2. These characteristics are similar to the dipole case in Fig. 2.37.

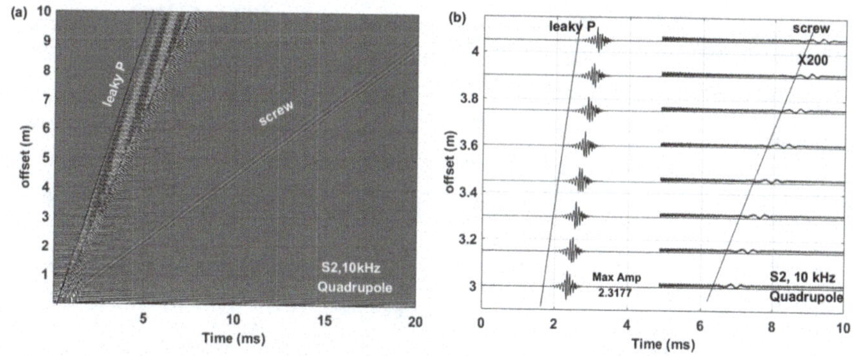

Fig. 2.42 Quadrupole waveforms in a slow formation (S2 in Table 2.1) with a 10-kHz source frequency. **a** Dense display. **b** Selected array waveforms

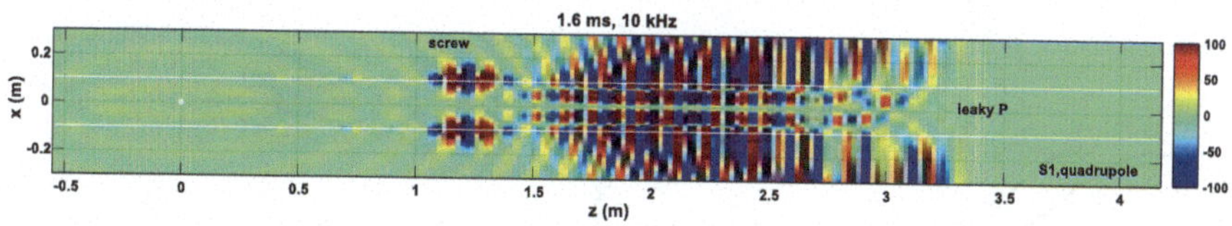

Fig. 2.43 Quadrupole wavefield snapshot at 1.6 ms (10 kHz) in the formation S1

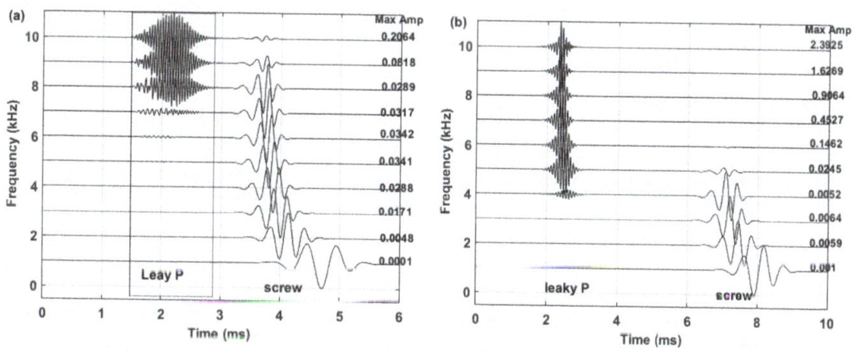

Fig. 2.44 Quadrupole waveforms (3 m offset) at various source frequencies. **a** Formation S1. **b** Formation S2

Formation Velocities and Poisson's Ratio

The influence of the formation velocities on the dispersion curves are similar to those in the dipole case. The only difference is that the cutoff frequencies of leaky P and formation screw modes in the quadrupole case are higher than those of the leaky P and flexural modes in the dipole case. The effects of formation waves' velocity on the waveforms are shown in Fig. 2.45. Similar to the monopole and dipole cases,

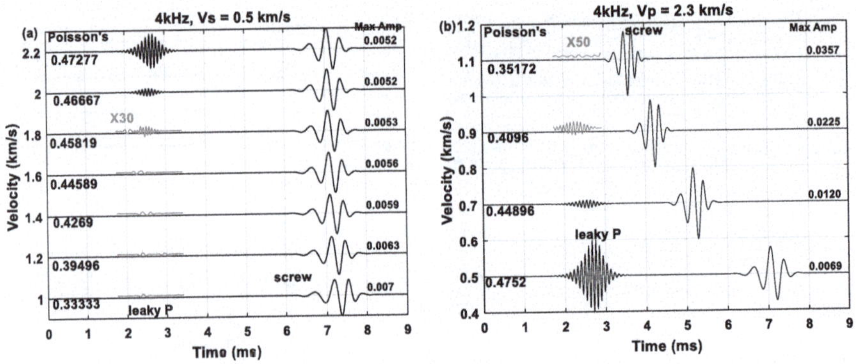

Fig. 2.45 Quadrupole waveforms (3 m source-receiver distance) with various formation velocities. Source frequencies and velocities are listed on the plots

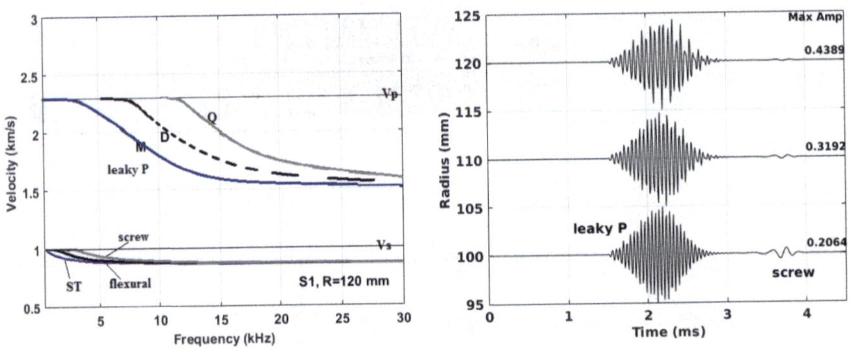

Fig. 2.46 Left: dispersion curves of the multipole modes in an enlarged borehole (radius of 120 mm) surrounded by the formation S1. Letters M, D, and Q are monopole, dipole, and quadrupole. Right: the waveforms (3 m offset) at 10 kHz in formation S1 with various borehole radii

the formation with a large Poisson's ratio generates good leaky P and a small screw mode. The two modes are discernable even in a formation with very large Poisson's ratio (0.4752 for a formation of Vp = 2300 m/s, Vs = 500 m/s) when the source frequency is low (Fig. 2.45d). Therefore, it is possible to utilize a low frequency quadrupole tool to measure both P- and S-wave velocities in a very slow formation.

Borehole Radius

The dispersion curves for monopole, dipole, and quadrupole sources in an enlarged borehole (radius of 120 mm) surrounded by formation S1 are shown in Fig. 2.46. Comparing Fig. 2.46 with Fig. 2.40 shows that a large borehole radius moves the dispersion curves of the leaky P modes towards the lower frequency. This is similar to the monopole and dipole cases. The dispersion curve of the screw mode is differently affected. Its relative amplitude becomes smaller when borehole radius increases and makes it unidentifiable at a high source frequency (Fig. 2.46 right). The quadrupole source has the higher cutoff frequency for both the leaky P and screw waves. A low frequency quadrupole tool would be the best choice for velocity measurements in a slow formation. The second choice is the low frequency dipole tool as described by Tichelaar and van Luik (1995).

2.3.4 Field Data in Slow Formations

Similar to Sect. 2.2.3, the field examples of monopole and dipole data in a slow formation are shown in Fig. 2.47. The tool configurations and source frequencies are the same as those in Sect. 2.2.4. In the monopole data, the dispersive leaky P-wave is strong while the ST wave is nearly invisible because the source frequency of the monopole tool is high (10 kHz). The P-wave velocity obtained from leaky P is approximately 2500 m/s. For the dipole data, both leaky P and flexural modes can be observed. These two modes are strongly dispersive. The leaky P-wave is mostly in the high frequency range. The lower frequency flexural wave has larger amplitude. The S-wave velocity obtained from flexural wave is approximately 1100 m/s. The waveforms from 1273 to 1277 m depth are different from other depths. The ST and flexural waves are strong. The reason is that the zone is saturated with gas and its Poisson's ratio is smaller than at other depths.

The monopole and dipole array waveforms at 1285 m are shown in Fig. 2.48a and b, respectively. Leaky P in the monopole array waveforms is dispersive and strong. The ST is small and has low frequency. There are multiple modes in the leaky P part. The flexural wave is clear and is dispersed in the dipole data.

Cross-dipole measurements (Fig. 1.6) were conducted in this section for formation anisotropy determination (Fig. 2.49). The data acquisition depth is from 1262 to 1318 m (Track 2). The waveforms are displayed at every 4.5 m. Track 1 shows the orientation of the X-dipole (solid) and borehole diameter (dashed). Track 3 shows the waveforms at inline receivers (XX and YY). There are two waves: leaky P at 1400 μs as the first arrival and formation flexural wave. Track 4 shows the slowness (reciprocal of velocity) measured from the XX waveforms. The leaky P-wave slowness (black curve) is between 100 and 120 μs/ft. The slowness of the S-wave is approximately 280 μs/ft (blue curve). The S-wave velocity is lower than the borehole fluid wave velocity (slowness of 185 μs/ft). The crossline waveforms (XY and YX) are shown in Track 5. Track 6 shows the magnitude of the anisotropy obtained from the cross-dipole measurements. Track 7 is the inverted azimuth of the fast S-wave with respect to the orientation of the X-dipole (blue curve) and to the north (red curve). Track 8 shows the principal fast (black curve) and slow S (red curve) waves after removing the leaky P-wave. Results show that a strong and stable anisotropy appears at depths from 1280 to 1300 m with an average magnitude of 8%. The average azimuth of the fast S-wave is 85° East of North. The anisotropy is also strong at depths from 1262

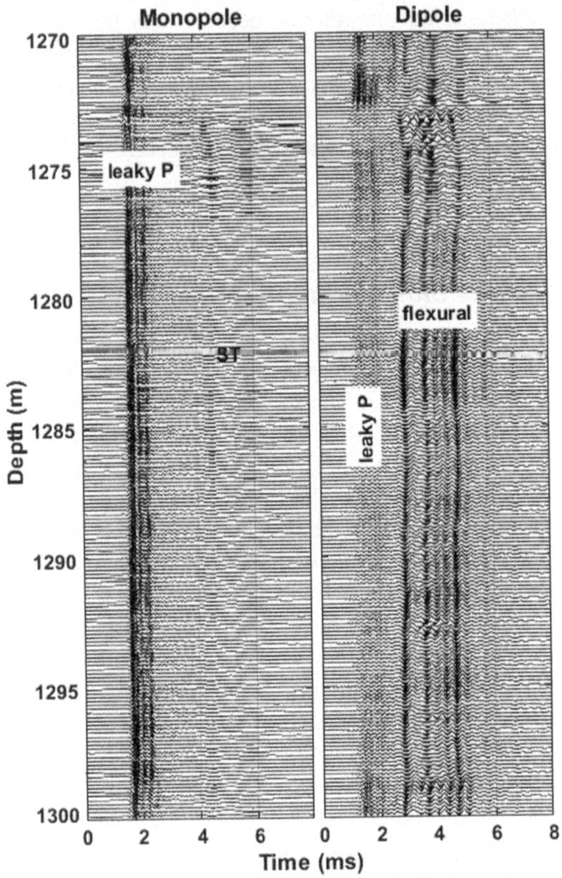

Fig. 2.47 Monopole (left) and dipole (right) measurement in a borehole surrounded by a slow formation. The source frequency of the monopole and dipole tools are 10 and 4 kHz, respectively. The waveforms at the first receiver (11 ft offset, 3.35 m) at different depths are shown for the monopole case. The waveforms at the first receiver (10.25 ft offset, 3.12 m) at different depths are shown for the dipole case

to 1273 m. The average anisotropy magnitude is 10% and the average azimuth of the fast S-wave is 130° East of North.

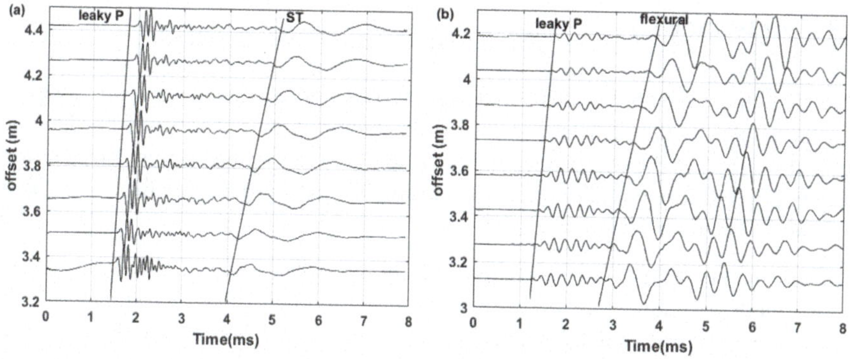

Fig. 2.48 Array waveforms at depth of 1285 m. **a** Monopole data. **b** Dipole data

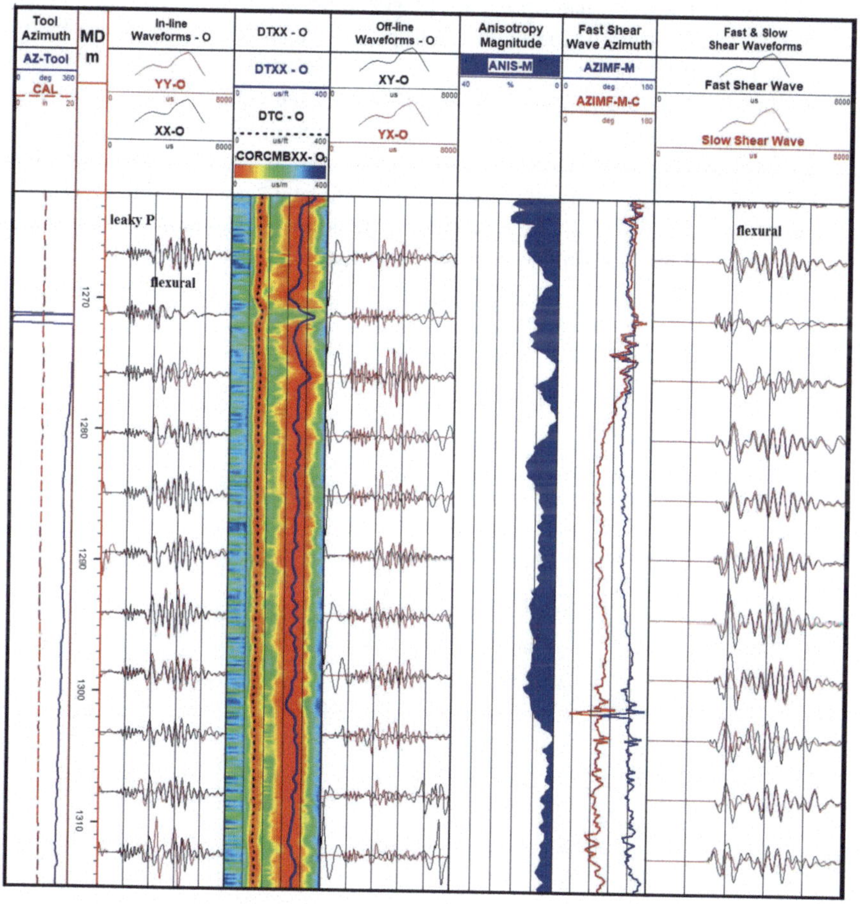

Fig. 2.49 Sample waveforms, velocity analysis, and anisotropy inversion from cross-dipole measurement in a well at depth from 1262 to 1318 m. Well is the same as the one whose waveforms are shown in Fig. 2.47

2.4 Summary

This chapter covered the fundamental aspects of seismic/acoustic wave propagation in a borehole. Synthetic waveforms and dispersion curves were used to demonstrate the effects of the formation velocities, borehole radius, source type, and source frequency on different modes. Field data examples in different formations were also presented. The main conclusions are,

(1) In a borehole in a fast formation, different sources generate different waves. A monopole source generates P, S, pseudo Rayleigh (pR), and Stoneley (ST) waves. The pR and ST waves are affected by the borehole radius, source frequency, and formation velocities. A dipole source generates leaky P and flexural waves. The leaky P-waves are very small and are only generated at high frequencies. A quadrupole source generates a formation quadrupole (or "screw") wave. Both flexural and quadrupole waves reach the S-wave velocity at their cutoff frequencies.

(2) In a borehole in a slow formation, where $Vs < Vf$, there are two types of waves: leaky P-waves and S-related waves (ST in monopole, flexural in dipole, and screw in quadrupole). They are affected by the source frequency and formation Poisson's ratio. The leaky P-waves at low frequency can be used to determine the P-wave velocity. The S-related modes, including ST, flexural, and screw, are only affected by the formation S-wave velocity.

Chapter 3
Data Processing Methods for Borehole Acoustics

Borehole acoustic data need to be processed to determine the formation waves' velocity. Here we present some of the most commonly used methods to extract the different modes and to determine the formation velocities. In the early history of acoustic logging, where there were a few receivers in the array, the arrival times were used for velocity determination. With digital recording and expanded arrays, elaborate methods were developed. In this Chapter, several of the most commonly used methods are discussed. In Sect. 3.1, velocity determination methods in the time domain are discussed. In Sect. 3.2, the methods for extracting the dispersion features of different modes in the frequency domain are covered. These methods are based on parametric spectrum estimation and include extended Prony's method (EPM) and Forward/Backward Averaging Matrix-Pencil (FBAMP) method. The nonparametric spectrum estimation methods, including Weighted Spectrum Semblance (WSS), Filtered Frequency Semblance (FFS), and Amplitude and Phase EStimation (APES), are presented in Sect. 3.3. Synthetic and field data are used to illustrate the features of different data processing methods.

3.1 Time Domain Velocity Determination

3.1.1 Traditional Methods

Information extracted from borehole waveforms includes the arrival times, P- and S-wave velocities and attenuations. In acoustic logging, typically a receiver array with fixed interval between receivers is used. Velocity along the borehole axis was initially obtained by determining the time delays at different receivers. Figure 3.1 shows array waveforms in formation F1 (see Table 2.1). The arrivals of different modes are marked with circles. Table 3.1 lists the picked arrival times. Velocities can be obtained from a time-distance plot. The velocities, for P, S, and ST waves are

H. Wang et al., *Borehole Acoustic Logging—Theory and Methods*, Petroleum Engineering, https://doi.org/10.1007/978-3-030-51423-5_3

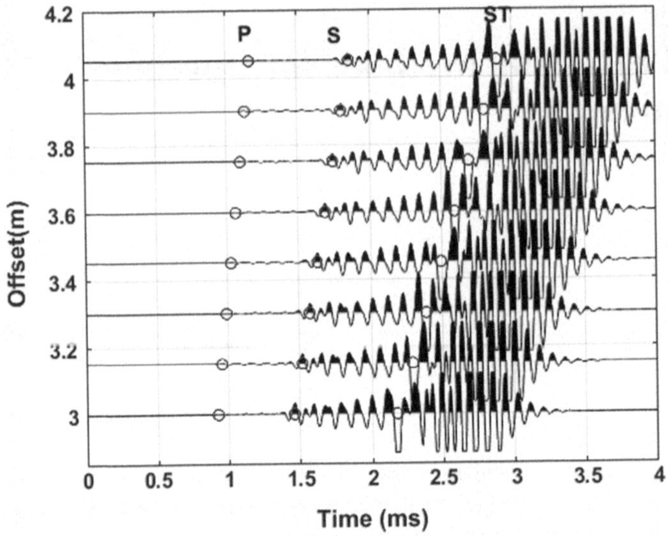

Fig. 3.1 Monopole array waveforms in formation F1 (see Table 2.1). Circles are picked arrival time for different modes

Table 3.1 Parameters for the LWD model T_P, T_S, and T_{ST} are picked arrival times for P, S, and ST waves

Offset (m)	T_P (ms)	T_S (ms)	T_{ST} (ms)
3	0.9277	1.46	2.178
3.15	0.957	1.519	2.29
3.3	0.9912	1.572	2.388
3.45	1.025	1.631	2.495
3.60	1.06	1.689	2.593
3.75	1.094	1.743	2.695
3.90	1.128	1.802	2.803
4.05	1.162	1.86	2.896

4495 m/s, 2630 m/s, and 1468 m/s, respectively. P and S velocities are close to the true velocities of the model (Vp = 4500 m/s, and Vs = 2650 m/s).

Currently, a commonly used method for array processing to improve velocity estimation is the velocity-time semblance or the slowness-time coherence (STC), which are based on cross-correlation (Kimball and Marzetta 1984). Let us assume there are N receivers in the array located at distance of z_1, z_2 … z_N from the source along the borehole axis. There are p (p < N) modes in the waveform. The waveform at the receiver 1 (at distance of z_1 from the source) can be expressed by,

$$R_1(t-\tau) = \sum_{j=1}^{p} S_1^j\left(t-\tau_j\right), \tag{3.1}$$

where τ_j is the arrival time of mode j at receiver 1. S^j_1 corresponds to different modes at receiver 1, such as, P, S, pR, and ST. The waveform at receiver m ($1 \leq m \leq$ N) is as follows:

$$R_m(t - \tau - \Delta t) = \sum_{j=1}^{p} S^j_m\left(t - \tau_j - \frac{z_m - z_1}{V_j}\right), \tag{3.2}$$

where Δt is the time delay and it is different for different modes. If the interval between any two adjacent receivers is constant, equal to d, then $z_m - z_I = (m - 1)d$. The time delays can be obtained from the cross-correlation of the waveforms at two or more receivers:

$$\rho(V_j, \tau_j) = \frac{\int_{\tau}^{\tau+T_w}\left[\sum_{m=1}^{N} R_m(t - \tau - \Delta t)\right]^2 dt}{N\sum_{m=1}^{N}\int_{\tau}^{\tau+T_w}[R_m(t - \tau - \Delta t)]^2 dt}, \tag{3.3}$$

where T_w is the time window of the anticipated duration of the mode. A portion of, or the entire, waveform can be spanned by the windows with different slopes (V_j) in a specified velocity range and different arrival time τ_j, to obtain a series of coherences $\rho(V_j, \tau_j)$ ranging from 0 to 1. Figure 3.2a, b illustrate the window spanning and corresponding coherences on the array waveforms.

The projection of the corresponding coherences on the 2D time-velocity plane is shown as a "correlogram". The contour plot of the correlogram is shown in Fig. 3.2c. The arrival time and velocity of the modes can be obtained by finding the peak semblance values. In practice, the process is converted to find the maximum value of coherence in the projection of the 2D contour plot on the velocity axis (as shown in Fig. 3.2d). The arrival time τ_j can be determined by searching the corresponding V_j and $\rho(V_j, \tau_j)$ in the contour plot. In practice, the reciprocal of the velocity, slowness, is used to cover a large range of formation velocities.

The semblance method presented in Eq. 3.3 is very robust and can work with multiple wave types and with weak arrivals, giving reliable results if the recorded waves are not dispersed.

Figure 3.3 shows monopole and dipole field data acquired in a fast formation. The velocity and arrival of P, S, and ST waves can be obtained from the velocity-time semblance plot, in Fig. 3.3c, for monopole waveforms. The velocities of P, S, and ST waves are 7160, 3865, and 1395 m/s. For the dipole data, in Fig. 3.3d, there are two high coherence areas because the dispersive flexural wave has a large velocity range. The first one, close to 3865 m/s, is the low frequency component of the flexural wave. The velocity is close to the S-wave velocity, marked with a dashed line. The second one, a lower velocity and large arrival time peak, is the high-frequency component of the flexural wave. The velocity ranges from 3865 m/s to 2285 m/s. This agrees with the dispersion characteristic of the flexural waves in a fast formation.

Figure 3.4 shows the monopole and dipole waveforms acquired in a slow formation. The velocity-time semblance results are also shown. The velocities, obtained

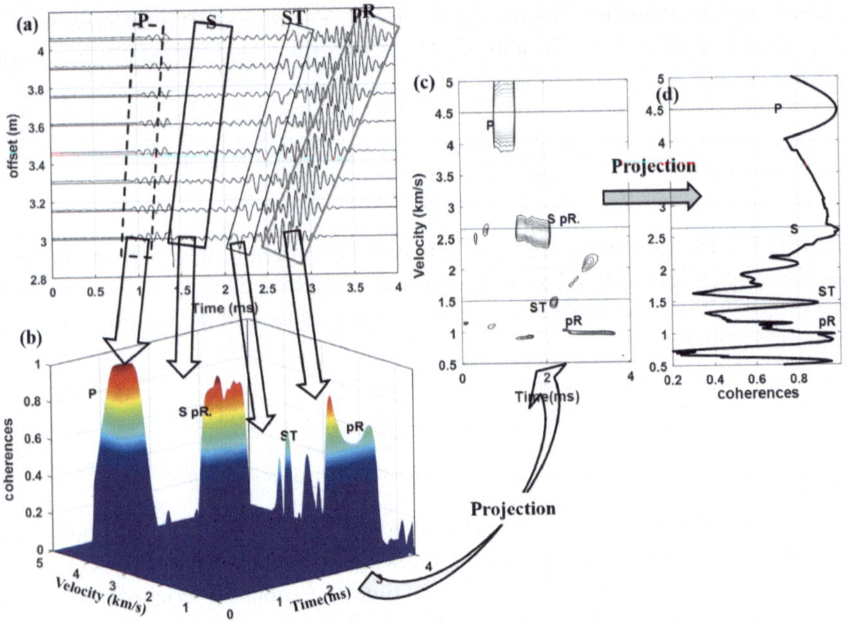

Fig. 3.2 The application of the velocity-time coherence method for the simulated monopole array waveforms in fast formation F1 in Table 2.1. **a** Synthetic monopole array waveforms. P, S, ST, and pR modes are marked with rectangles of different slopes. **b** Calculated coherences of different modes in **a**. **c** Projection of **b** in time-velocity plane. **d** Velocity determination according to the maximal coherence

from the monopole data in Fig. 3.4c, are 2500 m/s and 940 m/s for P and ST velocities, respectively. Because there is no refracted S-wave in the slow formation, the S-wave velocity cannot be obtained in Fig. 3.4c. The S-wave velocity, obtained from the flexural wave in the dipole log data shown in Fig. 3.4d, is 1115 m/s. The P-wave velocity obtained from monopole data is marked by a dashed line in Fig. 3.4d.

It is clear that if the waveforms are highly dispersive, such as the leaky P-wave in Fig. 3.4d, the peak value of the coherence is located at the velocity of the Airy phase (see definition in Sect. 2.2.1) of the dispersion curve. In this case, the semblance velocities should be corrected for the dispersion effect. This leads to the dispersive velocity-time semblance method (Kimball 1998) where dispersion curves are calculated using inputs, such as borehole radius, formation velocities, fluid wave velocity, and densities of fluid and formation. In field applications, it may be difficult to calculate precise dispersion curves because of the unknown or uncontrollable parameters such as the variation of the borehole fluid properties, tool eccentering, and borehole breakouts.

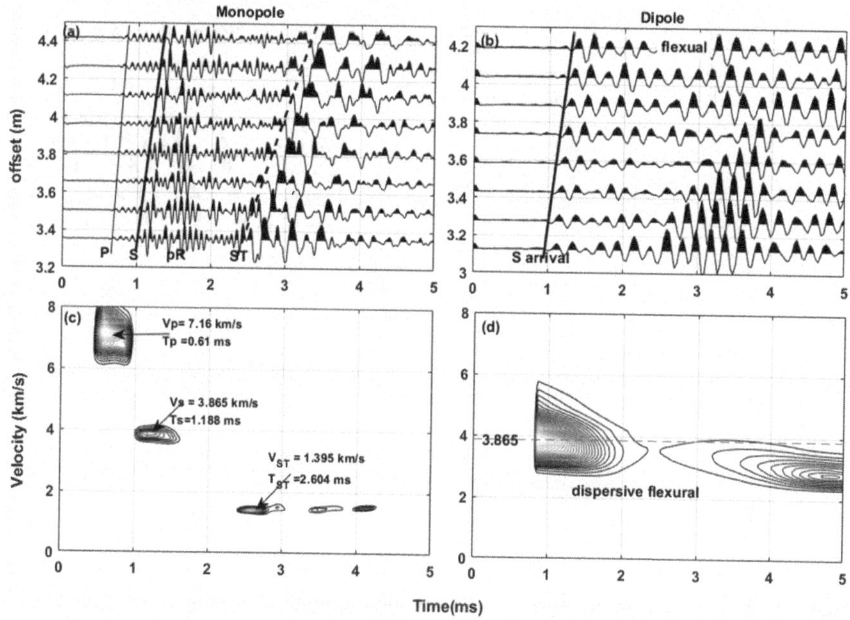

Fig. 3.3 Field data examples in a fast formation. **a** Monopole data. **b** Dipole data. **c** Velocity-time semblance for monopole data. **d** Velocity-time semblance for dipole data. The calculated P and S velocities, as shown in **c**, are 7160 m/s and 3865 m/s. The S velocity is marked with a dashed line in **d**

3.1.2 Time Semblance with Multiple Scale Wavelets

Because the velocity of each mode varies with frequency, a multiple scale semblance method that combines the multiple scale wavelet transform and the time semblance methods can be used to determine the formation velocities (Li et al. 2016). The dual tree complex wavelet transform (DTCWT) was introduced to improve the conventional wavelet transform (Kingsbury 2001).

The DTCWT (ψ) is composed of two separate real wavelets as shown below,

$$\psi(\mathrm{t}) = \psi_r(\mathrm{t}) + i\psi_i(\mathrm{t}). \tag{3.4}$$

Here, ψ_r and ψ_i are real and imaginary parts of DTCWT, respectively. They form a Hilbert transform pair to ensure the linear phase of the wavelet filter.

Figure 3.5 shows a four-level decomposition process from DTCWT analysis, which has two separate trees of real filters. Trees a and b produce the real and imaginary parts of complex coefficients, respectively. $S(t)$ refers to the original signal. $g_0(t)$ and $g_1(t)$ are the low-pass/high-pass filter pair for tree a. $h_0(t)$ and $h_1(t)$ refer to the low-pass/high-pass filter pair for tree b. The downward pointing arrow represents the down-sample process. x_{0a} and x_{1a} are the scaling and wavelet coefficients at

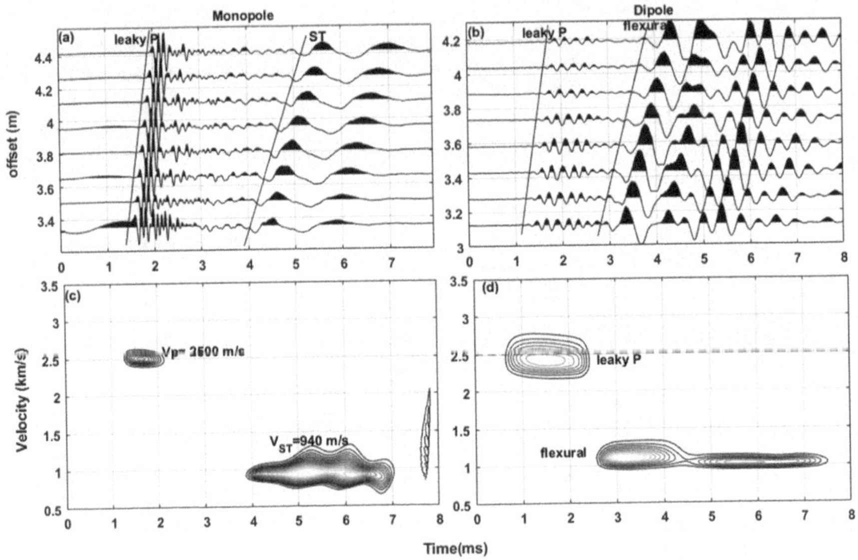

Fig. 3.4 Field data examples in a slow formation. **a** Monopole data. **b** Dipole data. **c** Velocity-time semblance for monopole data. **d** Velocity-time semblance for dipole data. The calculated P and S velocities are 2500 m/s, from **c**, and 1115 m/s from **d**. The P-wave velocity is marked with a dashed line in **d**

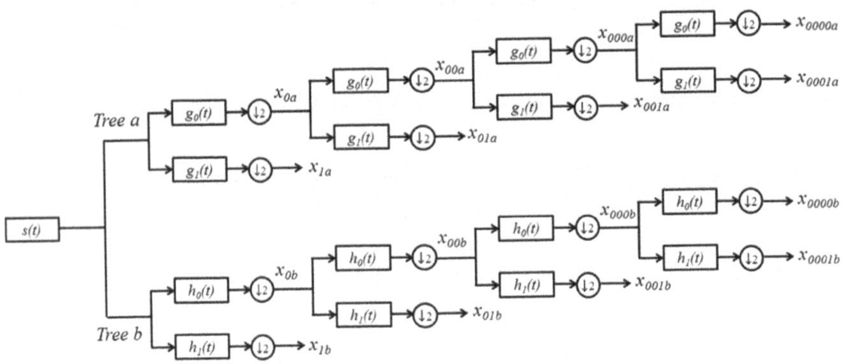

Fig. 3.5 Schematic diagram of DTCWT decomposition. DTCWT employs two different real discrete wavelet transforms, the first (tree a) gives the real part of the complex coefficients while the second (tree b) gives the imaginary part. $s(t)$ refers to the original signal. $g_0(t)$ and $g_1(t)$ refer, respectively, to the low-pass/high-pass filter pair for tree a. $h_0(t)$ and $h_1(t)$ refer, respectively, to low-pass/high-pass filter pair for tree b. Downward pointing arrow represents the down-sample process and associated number represents the down-sample factor

level 1 for tree *a*. x_{0b} and x_{1b} are the scaling and wavelet coefficients at level 1 for tree *b*. The two sets of low-pass filters are used to construct the DTCWT that satisfies a half-sample delay condition to provide the perfect reconstruction of the original signal. The data can be recovered from either the real or imaginary parts of the original signal. For simplicity, the scaling component at level *l* is named as the wavelet component at level *l + 1*. The original signal *s(t)* can be expressed as,

$$\mathrm{s(t)} = \sum_{l=1}^{L+1} x_l(t) \tag{3.5}$$

where x_l refers to the wavelet component at level *l*. *t* indicates the time sample. *L* stands for the decomposition layer. More details can be found in Kingsbury (2001) and Selesnick et al. (2005).

The specific steps for multiscale time semblance method are listed below:

1. Obtain the wavelet components $WC_l^m(t)$ at each level by applying the DTCWT to array waveforms. *l* is index of level, *m* is the receiver number and *t* is time.
2. Extract the common source gathers for each level as the array waveforms at level *l*.
3. Calculate the time-velocity semblance correlograms of the array waveforms at each level.

Waveforms from the 10 kHz dipole measurement in slow formation S1 (Fig. 2.34b) are used to investigate features of the multiscale time semblance method. The wavelet components at selected levels, after 4-level decomposition by DTCWT, are shown in Fig. 3.6. Theoretically, the input waveform is equal to the sum of wavelet components from level 1 to level 5. The wavelet components at levels 1 and 2 are similar and mainly include the high-frequency leaky P modes. The flexural wave is small at those two levels. Level 4 has only a flexural wave. Although leaky P and flexural waves simultaneously exist in the wavelet component at level 3, they are in different frequency bands, high frequency for leaky P and low frequency for flexural wave.

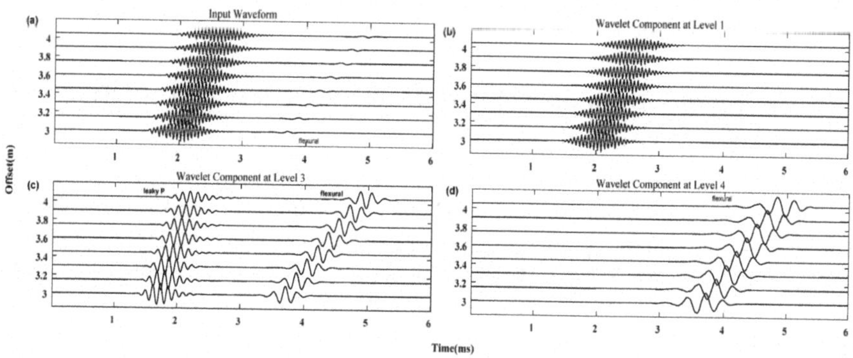

Fig. 3.6 Input waveforms **a** and wavelet components at selected levels

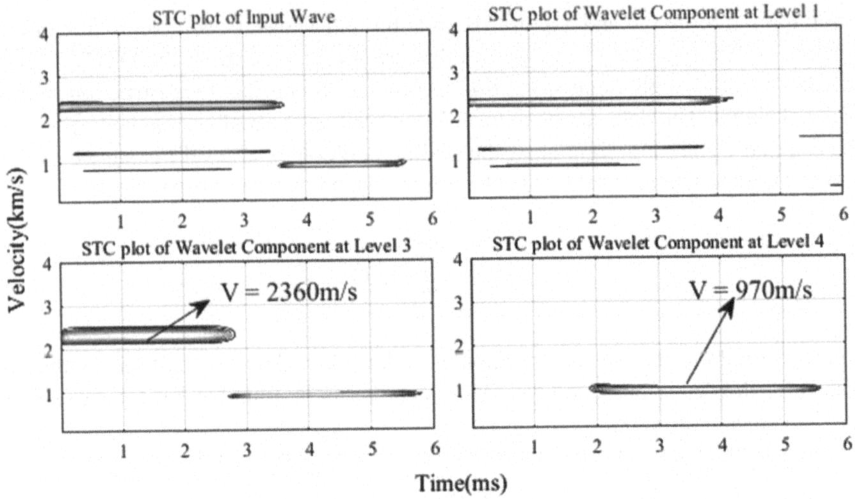

Fig. 3.7 STC plots of input waveforms and wavelet components of selected levels

Figure 3.7 shows the velocity-time semblance plots for the waveforms in Fig. 3.6. The time semblance plot at level 1 mainly contains the leaky P-wave. The plot at level 3 gives a good estimation (~2360 m/s) of the P-wave velocity and those at level 4 gives the S-wave velocity (~970 m/s).

This is an ideal example showing that multiscale semblance analysis enables the determination of formation P- and S-wave velocities. In general, the velocities estimated from multiscale time semblance method are very good.

In addition, the DTCWT may reduce (attenuate) random noise and enhance the coherence of the analyzed signal, which leads to improved time semblance correlograms. Because both leaky P and flexural waves are dispersive, the formation velocities should be obtained from dispersion analysis. However, multiscale time semblance method can be viewed as an alternative when the dispersion analysis is not available.

3.2 Frequency Domain Velocity Determination: Parametric Spectrum Estimation

Time semblance methods in Sect. 3.1 have limitations for processing dispersive data. Several methods, such as Maximum-likelihood Method (MLM) (Capon 1969), extended Prony Method (EPM) (Lang et al. 1987), Holomorphic Processing (Ellefsen et al. 1993), Weighted Spectrum Semblance (WSS) (Nolte et al. 1997), Forward/Backward Averaging Matrix Pencil (FBMP) (Ekstrom 1995), Filtered Frequency Semblance (FFS) (Rao and Toksöz 2005), have been developed to obtain the dispersion curves. In the frequency domain, Eq. 3.2 can be written as,

$$\tilde{R}_m(\omega) = \sum_{j=1}^{p} \tilde{R}_m^j(\omega) = \sum_{j=1}^{p} S_m^j(\omega)\exp[i\omega(\tau_j + \frac{(m-1)d}{V_j(\omega)})]. \tag{3.6}$$

If attenuation $\alpha_j(\omega)$ is included, the above equation can be expressed as follows,

$$\tilde{R}_m(\omega) = \sum_{j=1}^{p} A_m^j S_1^j(\omega) e^{i\varphi_j} z_j^{(m-1)}, \tag{3.7}$$

where $A_m^j = e^{-\alpha_j(\omega)(m-1)d}$ is the attenuation which is a function of frequency and distance. φ_j is the initial phase and $z_j = \exp[i\omega d/V_j(\omega)]$ is the phase shift of a receiver relative to the adjacent receiver.

The velocity is included only in the phase of the signal $z_j^{(m-1)}$. Therefore, the methods in the frequency domain are pursuing the phase shifts between different receivers. Basically, they use cross-correlation to determine the phase shift. For example, the Fourier coefficients are computed by cross-correlating the signal with the sine and cosine functions. The methods can be separated into parametric and nonparametric methods. In this section, the methods based on the parametric estimation, including EPM and FBMP, are presented.

Using Eq. 3.7, the spectrum of the waveform at receiver m can be expressed as,

$$\tilde{R}_m(\omega) = \sum_{j=1}^{p} b_j z_j^{(m-1)}, \tag{3.8a}$$

$$b_j = A_m^j |\tilde{R}_1^j(\omega)|. \tag{3.8b}$$

By comparing the constructed spectrum $\tilde{R}_m(\omega)$ and raw spectrum $R_m(\omega)$, the estimation of the b_j, z_j, and p can be obtained. The velocity and attenuation at a given frequency can be determined from the estimated z_j and b_j, respectively. However, the direct method to minimize the errors between the constructed data and raw data involves a nonlinear least-square equation. One approach is to convert the nonlinear problem to a linear problem using the methods based on the linear superposition theorem of the complex numbers. Polynomial and Matrix-pencil methods are two commonly used methods. The polynomial method, proposed by Prony (1795), has been extended to reduce the sensitivity to noise (Lang et al. 1987; Liu et al. 2008), and named as extended Prony method. The Matrix-pencil method was improved by Ekstrom (1995) as the Forward/Backward Averaging Matrix Pencil method (FBMP). It has higher computation efficiency and better statistic on z_j than those in the polynomial method. Below, these two methods are presented in detail.

3.2.1 Extended Prony's Method (EPM)

This method is a spectrum analysis method based on the mathematics model (Prony 1795) to describe the array data by employing the linear superposition on the complex coefficient of sine functions.

From Eqs. 3.8a and 3.8b the spectrum at receiver $m - l$ $(p \leq m \leq N, 0 \leq l \leq p)$ can be constructed by,

$$\tilde{R}_{m-l}(\omega) = \sum_{j-1}^{p} b_j z_j^{(m-l-1)}. \tag{3.9}$$

By multiplying both sides of Eq. 3.9 by coefficient a_l, and summing for l values from 0 to p gives,

$$\sum_{l=0}^{p} a_l \tilde{R}_{m-l}(\omega) = \sum_{l=0}^{p} a_l z_j^{(p-l)} \sum_{j=1}^{p} b_j z_j^{(m-p-1)}. \tag{3.10}$$

Prony (1795) recognized that the exponentials z_j ($j = 1, \ldots, p$) are the solutions of the following polynomial ($a_0 = 1$),

$$\psi(z) = \prod_{l=1}^{p} (z - z_j) = \sum_{l=0}^{p} a_l z_j^{p-l} = 0. \tag{3.11}$$

When Eq. 3.10 is equal to zero, the constructed term $\tilde{R}_m$ at receiver m $(p \leq m \leq N)$ can be expressed as,

$$\tilde{R}_m = -\sum_{l=1}^{p} a_l \tilde{R}_{m-l}, \tag{3.12}$$

Assuming the error between the constructed spectrum $\tilde{R}_m$ and raw spectrum $R_m(\omega)$ is e_m,

$$R_m = \tilde{R}_m + e_m, \tag{3.13}$$

Thus the raw spectrum R_m at receiver m can be expressed as,

$$R_m = -\sum_{l=1}^{p} a_l R_{m-l} + \sum_{l=0}^{p} a_l e_{m-l}; \; (p \leq m \leq N), \tag{3.14}$$

Now the problem is converted to find the coefficients a_l to minimize $\sum_{m=p}^{N} |e_m|^2$, which is still a nonlinear problem.

The extended Prony method (EMP) (Marple 1987; Kay 1988; Zhang 1996) converts the nonlinear process to a linear problem by simplifying $\sum_{l=0}^{p} a_l e_{m-l}$ as ε_m. Then Eq. 3.14 can be expressed as,

$$\sum_{l=0}^{p} a_l \mathrm{R}_{m-l} = \varepsilon_m;\ (p \leq m \leq N). \tag{3.15}$$

By minimizing the $\sum_{m=p}^{N} |\varepsilon_m|^2$, the coefficients a_l can be obtained. Defining an objective function (***Ja***),

$$\mathrm{J}(a_l) = \min_{a_l} \sum_{m=p}^{N} |\varepsilon_m|^2 = \min_{a_l} \sum_{m=p}^{N} \left| \sum_{l=0}^{p} a_l \mathrm{R}_{m-l} \right|^2, \tag{3.16}$$

the procedure can be implemented by,

$$\sum_{l=0}^{p} a_l r(n,l) = 0, n = 1, 2, \ldots, p, \tag{3.17}$$

where $\mathrm{r}(n,l) = \sum_{m=p}^{N} R_{m-l} R_{m-n}\ (0 \leq l \leq p)$.

Equation 3.17 can be expressed by the following matrix expression,

$$\begin{bmatrix} r(1,0) & r(1,1) & \cdots & r(1,p) \\ r(2,0) & r(2,1) & \cdots & r(2,p) \\ \vdots & \vdots & \ddots & \vdots \\ r(p,0) & r(p,1) & \cdots & r(p,p) \end{bmatrix} \begin{bmatrix} 1 \\ a_1 \\ \vdots \\ a_p \end{bmatrix} = \begin{bmatrix} 0 \\ 0 \\ \vdots \\ 0 \end{bmatrix}, \tag{3.18}$$

After the a_l are determined from Eq. 3.18, the poles z_j in Eq. 3.11 can be used for the velocity determination. Then the attenuation can also be determined from b_j in Eqs. 3.8a and 3.8b using the following least-squares method,

$$\boldsymbol{b} = \left(\mathbf{Z}^H \mathbf{Z}\right)^{-1} \mathbf{Z}^H \boldsymbol{R}, \tag{3.19}$$

where $[\ ]^{\mathrm{H}}$ denotes the conjugate transpose operator. Matrix **Z**, and vectors **b** and **R** are,

$$\boldsymbol{Z} = \begin{bmatrix} 1 & 1 & \cdots & 1 \\ z_1 & z_2 & \cdots & z_p \\ \vdots & \vdots & \ddots & \vdots \\ z_1^{N-1} & z_2^{N-1} & \cdots & z_p^{N-1} \end{bmatrix}, \tag{3.20a}$$

$$\boldsymbol{b} = \left[b_1 \; b_2 \cdots b_p \right]^{\mathrm{T}}, \tag{3.20b}$$

$$\boldsymbol{R} = \left[R_1(\omega) \; R_2(\omega) \cdots R_N(\omega) \right]^{\mathrm{T}}, \tag{3.20c}$$

where $[\;]^{\mathrm{T}}$ is the transpose operator.

The application the EPM to the array data consists of 6 steps.

(1) For a given frequency ω, a matrix of $r(n, l)$ in Eq. 3.18 is built. The mode number p (an integer) is usually set as half of the number of receivers or a larger number.
(2) Using the SVD (singular-value decomposition)-TLS (Total least squares) algorithm, the real mode number p is estimated from the rank of the matrix built in step 1, and the coefficients $a_1, a_2, \ldots, a_p$ are obtained.
(3) The complex $z_1, z_2, \ldots, z_p$ are determined from the polynomial Eq. 3.11.
(4) Parameters $b_1, b_2, \ldots, b_p$ are determined from Eq. 3.19 after the matrix **Z** and vector **R** are built.
(5) The amplitude A_m^j, initial phase φ_j, attenuation coefficient α_j, and velocity V_j ($V_j = k_z^j/\omega$) are given below,

$$\left.\begin{aligned} A_m^j &= \left|b_j\right| \\ \varphi_j &= \arctan\left[\mathrm{Im}\left(b_j\right)/\mathrm{Re}\left(b_j\right)\right]/\left(2\pi d\right) \\ \alpha_j &= \ln\left|z_j\right|/d \\ k_z^j &= \arctan\left[\mathrm{Im}\left(z_j\right)/\mathrm{Re}\left(z_j\right)\right]/2\pi d \end{aligned}\right\}, \quad j = 1, \ldots, p \tag{3.21}$$

(6) Repeat steps from 1 to 5 until all frequencies are covered. Then the values of velocity, attenuation, and amplitude as a function of frequency are obtained.

3.2.2 *Forward/Backward Averaging Matrix Pencil (FBAMP)*

FBAMP is another method to estimate the parameters b_j, z_j, and p in Eqs. 3.8a and 3.8b. In this method, a Hankel Matrix (Eq. 3.22) is used.

$$Y = \begin{bmatrix} R_1 & R_2 & \cdots & R_{p+1} \\ R_2 & R_3 & \cdots & R_{p+2} \\ \vdots & \vdots & \ddots & \vdots \\ R_{N-p} & R_{N-p+1} & \cdots & R_N \end{bmatrix}_{(N-p)\times(p+1)}, \tag{3.22}$$

Here the number p is usually set as half number of the receivers N or more. The two new matrices $\mathbf{Y_1}$ and $\mathbf{Y_2}$ can thus be obtained using the first p and last p columns of the matrix $\mathbf{Y}$ in Eq. 3.22.

Those two matrices are,

$$Y_1 = Z_1 B Z_2, \tag{3.23a}$$

$$Y_2 = Z_1 B Z_0 Z_2. \tag{3.23b}$$

Here Z_0, B, Z_1, and Z_2 are

$$Z_0 = diag\left[z_1 \; z_2 \cdots z_p \right], \tag{3.24a}$$

$$Z_1 = \begin{bmatrix} 1 & 1 & \cdots & 1 \\ z_1 & z_2 & \cdots & z_p \\ \vdots & \vdots & \ddots & \vdots \\ z_1^{N-p-1} & z_2^{N-p-1} & \cdots & z_p^{N-p-1} \end{bmatrix}, \tag{3.24b}$$

$$Z_2 = \begin{bmatrix} 1 & z_1 & \cdots & z_1^{p-1} \\ 1 & z_2 & \cdots & z_2^{p-1} \\ \vdots & \vdots & \ddots & \vdots \\ 1 & z_p & \cdots & z_p^{p-1} \end{bmatrix}, \tag{3.24c}$$

$$B = diag\left[b_1, b_2, \cdots, b_p \right] \tag{3.24d}$$

The complex exponentials z_j are the generalized eigenvalues of the matrix-pencil $Y_2 - zY_1$,

$$([Y_1] - z[Y_2])e = 0 \tag{3.25}$$

The poles z_j can be obtained from Eq. 3.25. Other parameters such as attenuation, amplitude, and initial phase can be obtained from step 5 in the EPM (in Sect. 3.2.1). In this method, S_i and S_{max} are the ith and maximum singular values of matrix $\boldsymbol{Y}$ and the ratio S_i/S_{max} is used as a measure of noise. If the ratio is less than a certain value, the singular values are considered to be noise contributions. Therefore, this method is less affected by noise than that in the EMP. Another advantage of this method is its faster computational speed than EMP.

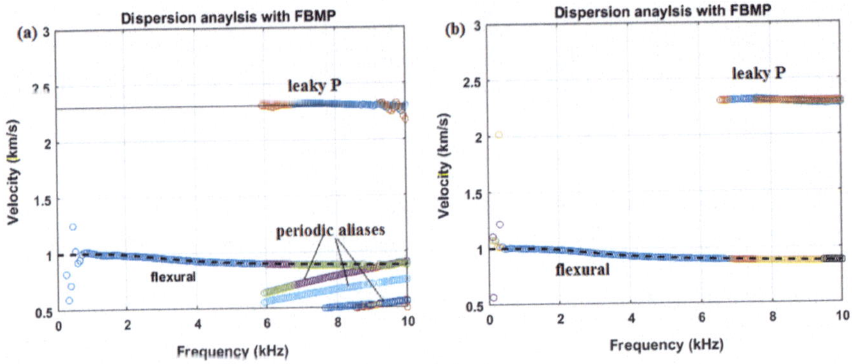

Fig. 3.8 Dispersion analyses with FBAMP. The array waveform is the same as Fig. 3.6a. Black dashed-lines are analytical dispersion curves. **a** Result from the array waveforms with offset from 3 to 4.05 m and interval of 0.15 m. **b** Result from the dense array waveforms with offset from 3 to 10 m and interval of 0.01 m

Figure 3.8 shows the dispersion (circles) extracted from the monopole waveforms, shown in Fig. 3.6a, using FBAMP. The analytical dispersion curves (black curves) are plotted. The extracted dispersion curves and analytical curves agree well except where there is some aliasing. The aliasing can be suppressed by a dense receiver array (Fig. 3.8b).

3.3 Frequency Domain Velocity Determination: Nonparametric Spectrum Estimation

3.3.1 Frequency-Wavenumber (F-K)

In Eq. 3.7, the phase delay of a mode is a linear function of slowness ($s_j(\omega) = 1/V_j(\omega)$). For a given frequency, the cross-plot between the phase difference between two adjacent receivers and the receiver locations should be a line and the slope of the line is the velocity. The Frequency-wavenumber (F-K) transform can be used to determine the dispersion. As expressed in Eq. 2.16, the term $P(\omega, k_z)$ includes information about both the frequency and velocity that are used for synthetic waveform calculation (Toksöz et al. 1984b; Schmitt and Bouchon 1985). For field data analysis, a frequency-wavenumber transform can be applied to array waveforms to produce the spectra. The time domain data can be converted into the frequency domain using the forward Fourier transform and the distance can be converted into the wavenumber domain by an inverse Fourier transform. This method is commonly used in near surface seismology. The power spectrum in the frequency-wavenumber domain of the array waveforms in Fig. 3.6 is shown in Fig. 3.9. The slope of the high amplitude portions of the plot gives the velocity. The leaky P modes are distributed between the Vp

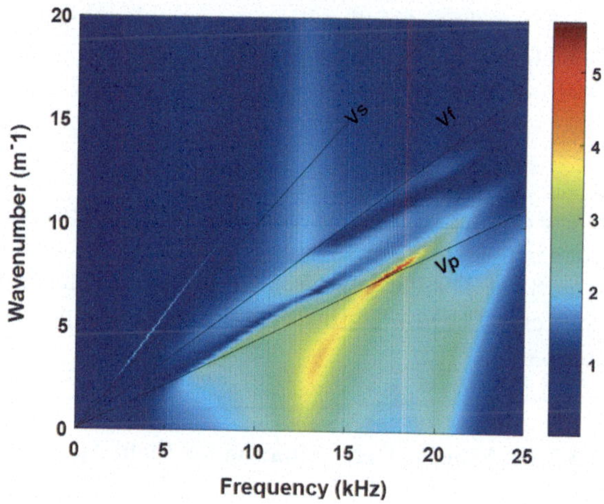

Fig. 3.9 The power spectrum in the frequency-wavenumber domain of the array waveforms in Fig. 3.6

and Vs lines. The weak flexural wave has a velocity lower than the S-wave velocity (marked with solid line Vs). A transform approach (such as the High-Resolution Radon Transform, see Sect. 7.3.1.3) can be used to convert the power spectrum from the frequency-wavenumber domain to the frequency-phase velocity domain.

This method is straight forward and easy to implement. The only shortcoming is the lower resolution when only a limited number of receivers is used.

3.3.2 *Weighted Spectrum Semblance (WSS)*

WSS is a frequency domain analog of the STC (Nolte et al. 1997). There are two steps, spectral semblance and applying weights for different coherences. Equation 3.26 defines the spectral semblance function which is the analog of the Eq. 3.3 of the semblance in the time domain.

$$\rho(\mathrm{V}, \omega) = \frac{\left|\sum_{m=1}^{N} R_m^*(\omega) z^{m-1}\right|}{\sqrt{N \sum_{m=1}^{N} R_m^*(\omega) R_m(\omega)}}, \tag{3.26}$$

where "*" denotes complex conjugate.

The spectral weighting $W(\omega_\xi, \omega_\zeta)$ is added to the result obtained from Eq. 3.26 to enhance the data resolution and to reduce noise,

$$\mathrm{F}\left(V_j, \omega\right) = \sum_{\xi=l-\zeta}^{l+\zeta} W(\omega_\xi, \omega_\zeta)\rho(V_j, \omega), \tag{3.27}$$

The weighting function $W(\omega_\xi, \omega_\zeta)$ employs a Gaussian function to give maximum weight to the estimated frequency,

$$W(\omega_\xi, \omega_\zeta) = \exp(-\frac{[\omega_\zeta - \omega_\xi]^2}{2\sigma^2}), \tag{3.28}$$

where $\sigma = N_w \Delta\omega$, and N_w are points in the spectrum. $\Delta\omega$ is the frequency interval. For example, if the eight adjacent points are selected for the weighting processing, $\sigma = 8\Delta\omega$. For the given velocity range, the maximum V can be obtained from coherence $F(V, \omega)$. The dispersion curve can be produced by finding the $V(\omega)$ at all frequencies.

3.3.2.1 Velocity Determination for Field Applications

In the field application, the speed of the velocity determination from the coherence plot is critical. The dispersion curves are usually converted into an objective function of velocity. There are two main methods to obtain the objective function from the coherence plot, the probability density function (Huang and Yin 2005) and fitting method (Tang et al. 2009).

Probability density function method

As discussed in Chap. 2, the guided wave dispersion curve at low frequency is relatively flat and approaches to the formation waves' velocity. Huang and Yin (2005) considered that there is more population ("point") at the flat part than that at the sharp part in a narrow frequency range. Therefore, a histogram can be generated by projecting the discrete dispersion curves to the velocity axis.

In addition, the histogram with a color map plot in the depth and slowness domain can be used as a quality control map as it reflects the true slowness population in the waveform.

Fitting method

Tang et al. (2009) considered that the spurious histogram peaks caused by data noise and/or mode interference may mislead the edge detection. They used a hyperbolic tangent function to fit the coherence because it can simulate the guided wave dispersion curves at both low and high frequencies. There are four variables in the arctangent function,

$$s(\omega) = s_0 + \delta s \cdot \mathrm{atan}(x\omega - y), \tag{3.29}$$

where s is the slowness. Parameters x and y are used to determine the variation of the fitting function with frequency. The parameter x denotes the gradient of the curve with frequency, while the parameter y determines the frequency where the curve has large gradient. These two parameters control the shift and stretching of the fitting curve. s_0 and δs are the selected slowness and perturbation defining the slowness range,

$$s(\omega) = \begin{cases} s_0 - \delta s, \omega \to 0 \\ s_0 + \delta s, \omega \to \infty \end{cases}, \tag{3.30}$$

The fitting error is minimized through minimizing the mean-square deviation between the theoretical curve and dispersion data.

After a fit to the dispersion curve is obtained, a slowness probability density function is defined for quality control,

$$\mathrm{SPD(s)} = \frac{1}{\omega_M - \omega_1} \int_{\omega_1}^{\omega_M} \exp[-(s - s_d(\omega))/\sigma_s] d\omega \tag{3.31}$$

where M is the number of frequency points from ω_1 to ω_M in the dispersion data S_d. σ_s is the standard deviation of the data.

3.3.2.2 Limitations of WSS and Possible Solutions

There are p modes in a waveform. According to Eqs. 3.8a and 3.8b the spectrum of waveform can be predicted by $\tilde{R}_m(\omega) = \sum_{j=1}^{p} b_j z_j^{(m-1)}$ and $b_j = A_m^j |\tilde{R}_1^j(\omega)|$. In Eq. 3.26, if the determined velocity V in $\rho(V, \omega)$ is equal to the phase velocity of mode j ($j = 1, \ldots, p$), the phase term $z_j^{(m-1)}$ of spectral $R_m^j(\omega)$ will be canceled in $R_m^*(\omega) z^{m-1}$. If the modes' number p is 1, $R_m^j(\omega) = R_m(\omega)$ and then the maximum coherence of $\rho(V, \omega)$ is 1. If the modes' number p is greater than 1, $\tilde{R}_m^j(\omega) < \tilde{R}_m(\omega)$ and then the obtained coherence $\rho(V, \omega)$ may not reach a peak value. Therefore, WSS is not suited for the case where there is more than one mode in the waveform at the same frequency. Because only the numerator of Eq. 3.26 affects the resolution, the vector form of the numerator is as follows:

$$\left| \sum_{m=1}^{N} \tilde{R}_m^*(\omega) z^{m-1} \right| = \left| R(\omega) z^H \right|, \tag{3.32}$$

where $\boldsymbol{R}(\omega) = [R_1(\omega), R_2(\omega), \ldots, R_N(\omega)]$ and $\boldsymbol{z} = [1, z, \ldots, z^{N-1}]$. The right part of Eq. 3.32 projects the waveform spectrum $R(\omega)$ to the base function $z(\omega, s)$, and the absolute amplitude of the corresponding coefficients is the output. This operator is the periodogram in spectral analysis based on the Fourier Transform. The treatment of the finite sequence as the windowed truncation of the sampled infinite sequence definitely introduces the leakage of the power spectrum in the main beam and thus results in the low resolution and periodic signal in the dispersion plot.

Similar to Fig. 3.8, the extracted dispersion characteristic with WSS is also shown in Fig. 3.10. The synthetic dipole waveforms (10 kHz source) in formation S1 are used (see Fig. 3.6 upper-left). The analytical dispersion curves, dashed curves, are

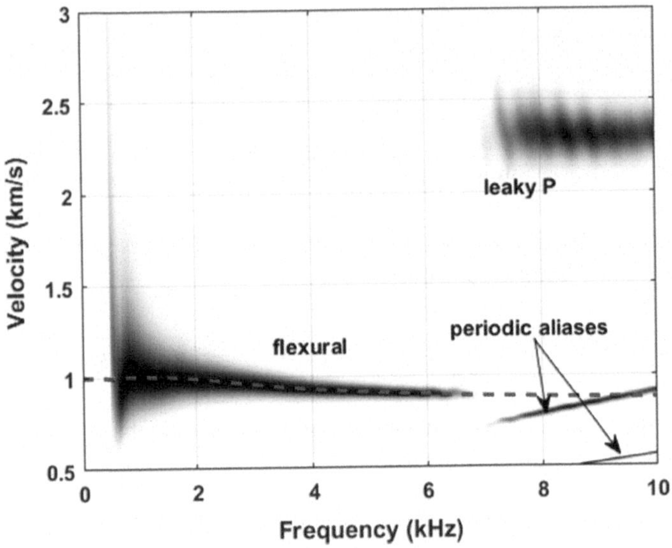

Fig. 3.10 Dispersion analyses with WSS. The array waveform is shown in Fig. 3.6. Dashed curves are analytical dispersion curves

plotted. A difference between Figs. 3.8 and 3.10 is that there is smaller frequency range for the flexural mode in Fig. 3.10. The reason is that the WSS cannot determine more than one mode at the same frequency. However, the EBMP can find more than one mode from 6 to 15 kHz.

Another issue is the poor resolution of WSS at low frequencies (see the dispersion characteristics below 5 kHz for the flexural wave in Fig. 3.10). As seen from Eq. 3.26, WSS determines the small phase shift between two signals of similar wavelets by the cross-correlation method. However, cross-correlation is a very insensitive method when the phase-shift is much less than the signal period, which increases the uncertainty of the measured dispersion at low frequencies. The uncertainty becomes larger when noise exists in the data.

To overcome the resolution limitation, Zheng and Hu (2017a) proposed a nonlinear signal comparison (NLSC) method to obtain the phase shift. In their method, the linear comparison in the numerator of Eq. 3.26 is converted into an exponential term which exponentially increases the small phase difference and thus improves the resolution at low frequencies.

3.3.3 *Amplitude and Phase Estimation (APES)*

The parametric estimation methods use known or estimated knowledge of number of wave modes. It is difficult to know the exact value of modes number p when the SNR (signal noise ratio) is low. Therefore, it is necessary to find a method which is not based on parameter estimation. However, as discussed in Sect. 3.3.2, the weighted spectral semblance (WSS) introduced by Nolte et al. (1997) is a nonparametric method. It can only identify the mode with the maximum coherence at a given frequency when multiple modes exist. The APES (Amplitude and Phase EStimation) method uses an adaptive filter bank (Stoica et al. 1999) so it does not require prior knowledge of the exact number of wave modes. Li et al. (2015) used this method for characterizing the dispersion of array data. This method minimizes the interference from other modes and noise.

The APES method is similar to the EPM and FBAMP methods. APES is also based on the linear superposition theorem of the complex coefficients of the signal.

In this method, an $M \times p$ matrix $\mathbf{X}_m$ is built based on the complex spectra from receivers m to $m + M - 1$ $(1 < m + M - 1 < N)$,

$$\mathrm{X_m} = \left[\mathrm{R_m}, \mathrm{Rm} + 1, \ldots, \mathrm{R_{m+M-1}}\right]^{\mathrm{T}}. \tag{3.33}$$

Using Eqs. 3.8a, 3.8b and 3.14, the matrix can be expressed as follows,

$$\mathrm{X_m} = \mathbf{ZS_m} + \mathbf{E_m}, \tag{3.34}$$

where $\mathbf{Z} = \left[\mathbf{Z}_1 \cdots \mathbf{Z}_p\right] = \begin{bmatrix} z_1^0 & \cdots & z_p^0 \\ \vdots & \ddots & \vdots \\ z_1^{(M-1)} & \cdots & z_p^{(M-1)} \end{bmatrix}$, $S_m = \begin{bmatrix} b_1 z_1^{(m-1)} \\ \vdots \\ b_p z_p^{(m-1)} \end{bmatrix}$, and the noise vector $\mathbf{E_m} = [e_m, e_{m+1}, \ldots, e_{m+M-1}]^{\mathrm{T}}$.

This matrix can be further disassembled into the following form,

$$\mathrm{X}_m = b_1 z_1^{(m-1)} \mathrm{Z}_1 + \tilde{\mathrm{Z}}_m \tilde{\mathrm{S}}_m + E_m = b_1 z_1^{(m-1)} \mathrm{Z}_1 + \tilde{\mathrm{X}}_m, \tag{3.35}$$

where $\tilde{Z}_m = \begin{pmatrix} z_2^0 & \cdots & z_p^0 \\ \vdots & \ddots & \vdots \\ z_2^{(M-1)} & \cdots & z_p^{(M-1)} \end{pmatrix}$ and $\tilde{S}_m = \begin{bmatrix} b_2 z_2^{(m-1)} \\ \vdots \\ b_p z_p^{(m-1)} \end{bmatrix}$.

This disassembly separates the original matrix into two parts, one $(b_1 z_1^{(m-1)} \mathrm{Z}_1)$ carrying the velocity and amplitude of mode 1, and the other $(\tilde{\mathrm{X}}_m)$ containing interference from all other components and noise.

An Mth order Finite Impulse Response (FIR) filter $F = [F_0, F_1 \ldots F_{M-1}]^{\mathrm{T}}$ is applied to the matrix $\boldsymbol{X}_m$,

$$\mathrm{F}^H \mathrm{X}_m = \mathrm{F}^H [b_1 z_1^{(m-1)} \mathrm{Z}_1] + \mathrm{F}^H \tilde{\mathrm{X}}_m, \tag{3.36}$$

Generally, to filter out interference and pass the mode j without distortion, the filter meets the condition $F^H Z_j = 1$. Equation 3.36 can be expressed as follows,

$$\mathrm{F}^H \mathrm{X}_m = b_j z_j^{(m-1)} + \mathrm{F}^H \tilde{\mathrm{X}}_m, \tag{3.37}$$

The design of the filter is converted to minimize the following objective function,

$$\mathrm{J}(\mathrm{F}, \mathrm{b}_j) = \mathrm{E}\left[\left|\mathrm{F}^H \mathrm{X}_m - b_j z_j^{(m-1)}\right|^2\right], \tag{3.38a}$$

With the condition that $F^H Z_j = 1$. Then the expectation **E** in Eq. 3.38a can be replaced with the average of the sequence $\mathrm{X_M}$, $\mathrm{X_{M+1}}$ … $\mathrm{X_N}$ with the length of L ($L = N - M + 1$)

$$\mathrm{J}(\mathrm{F}, \mathrm{b}_j) = \frac{1}{L}\sum_{m=M}^{N}\left|\mathrm{F}^H \mathrm{X}_m - b_j z_j^{(m-1)}\right|^2, \text{ subject to } \mathrm{F^H Z_j} = 1, \tag{3.38b}$$

By minimizing the above objective function, the noise and other components are suppressed and the amplitude and phase of mode j is estimated. Because the $\mathrm{F^H X_m} = \mathrm{X_m^T F^*}$, the expansion of $\mathrm{J(F, b_j)}$ is as follows:

$$\mathrm{J}(\mathrm{F}, \mathrm{b}_j) == \left|b_j - \mathrm{F}^H \mathrm{g}(\mathrm{V}_j)\right|^2 + \mathrm{F}^H \hat{X} F^* - F^H \mathrm{g}(\mathrm{V}_j)\mathrm{g}^H(\mathrm{V}_j)F^*, \tag{3.39}$$

Here $\hat{X} = \frac{1}{L}\sum_{m=M}^{N} X_m X_m^H$, and $\mathrm{g}(\mathrm{V}_j) = \frac{1}{L}\sum_{m=M}^{N} X_m z_j^{i(m-1)}$, and a covariance matrix estimation for noise is defined as $\hat{Q}(\mathrm{V}_j) = \hat{X} - \mathrm{g}(\mathrm{V}_j)\mathrm{g}^H(\mathrm{V}_j)$.

The minimum of the objective function can be obtained by letting $b_j = \mathrm{F}^H \mathrm{g}(\mathrm{V}_j)$ and the Eqs. 3.38a and 3.38b is converted to,

$$\min_F \mathrm{F}^H \hat{Q} \mathrm{F}^*, \text{ subject to } \mathrm{F^H} \mathbf{Z}_\mathrm{j} = 1. \tag{3.40}$$

The filter is obtained,

$$\mathrm{F} = \frac{\hat{Q}^{-1}\mathbf{Z}_j}{\mathbf{Z}_j^H \hat{Q}^{-1}\mathbf{Z}_j}, \tag{3.41}$$

and the amplitude of the mode j is estimated,

$$\mathrm{b}_j = \frac{\mathbf{Z}_j^H \hat{Q}^{-1}\mathrm{g}(\mathrm{V}_j)}{\mathbf{Z}_j^H \hat{Q}^{-1}\mathbf{Z}_j}. \tag{3.42}$$

Scanning the velocity of different modes, $|b_j(V)|$ will be nearly zero and only reach the peak when V is equal to V_j. Then the number of modes can be determined by determining the number of peaks and the velocity and amplitude can also be determined from the location and amplitude of the peaks.

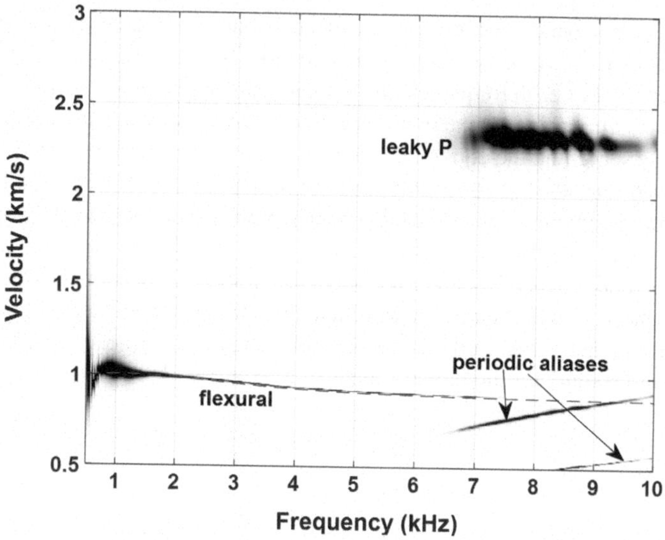

Fig. 3.11 Dispersion analyses with APES. The array waveforms are shown in Fig. 3.6

The implementation of the APES generally includes four steps.

First, the covariance matrix estimates of the signal ($\hat{X} = \frac{1}{L}\sum_{m=M}^{N} X_m X_m^H$) and noise ($\hat{Q}[V_j] = \hat{X} - g[V_j]g^H[V_j]$) are calculated at a given frequency based on the spectra $R_m(\omega)$, where m is the receiver index from 1 to the total receiver number N. The M is an integer and usually set as the half of N.

Second, the elements of the filter are determined from Eq. 3.41 for a given velocity range. The amplitude b_j of the mode j is obtained through Eq. 3.42.

Third, a 2-dimensional profile of amplitude $|b(\omega, V)|$ as a function of frequency and velocity is obtained by scanning all frequencies through repeating steps 1 and 2.

Finally, the number of modes is determined by counting the number of peaks and the velocity and amplitude for each mode can also be determined from the location and amplitude of the peak. Figure 3.11 shows the extracted dispersion result with APES for the simulated waveforms shown in Fig. 3.6.

3.3.4 *Filtered Frequency Semblance (FFS)*

Because the methods in Sect. 3.2.1 can only apply for non-dispersive data and the WWS method in the Sect. 3.3.2 has limitation for multiple modes at the same frequency, Rao and Toksöz (2005) proposed a direct method utilizing the time semblance on band-pass filtered data to obtain the velocity of multiple modes at the same frequency. In this method, the raw data are bandpass filtered through multiple narrow frequency bands and the output of each band is treated as non-dispersed data. Non-dispersive velocity analysis methods like STC can be applied to estimate phase velocities. For each frequency, the velocity and arrival time matrix is obtained. Therefore, a 3-dimensional matrix including frequency, velocity, and arrival time is produced. By searching the coherence value at each frequency at a given time range, the dispersion relationship can be obtained. Another benefit of the method is that it can accommodate both dispersive and non-dispersive modes. Figure 3.12 shows the workflow for the method.

In this method, the bandwidth of the bandpass filter is proportional to the selected frequency where there is a large bandwidth at high frequency and a small bandwidth at the low frequency. This limits the resolution at the high frequency and this results in the "mode-kissing" for the higher modes (e.g. Wang et al. 2015a), where different

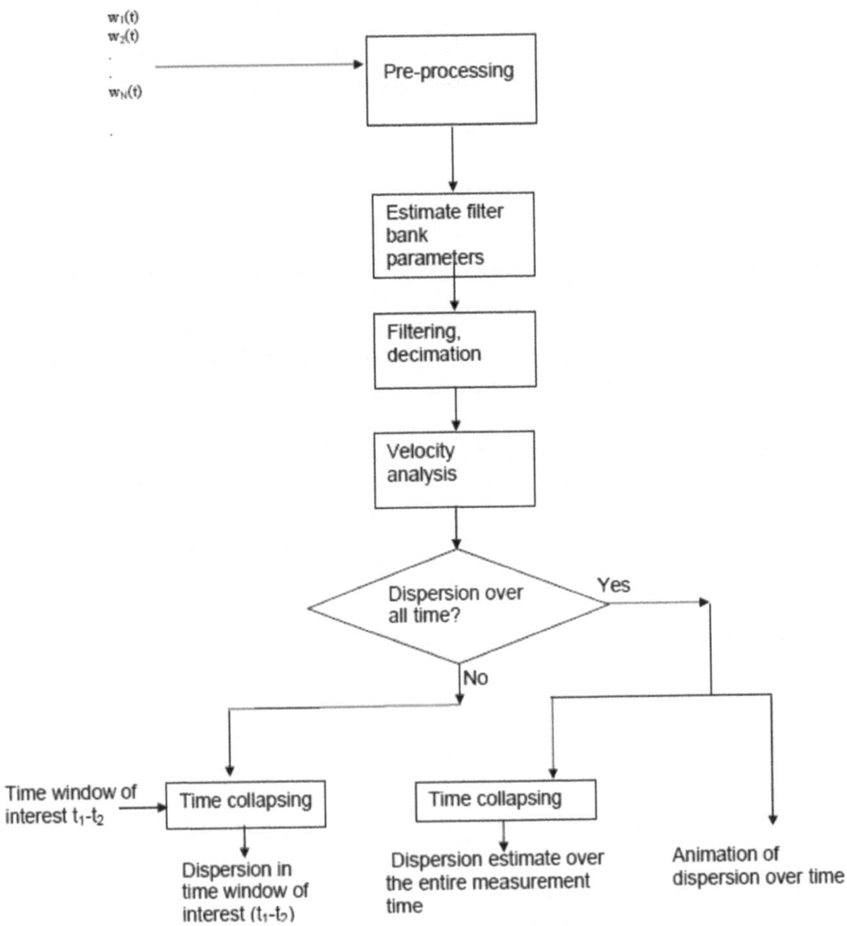

Fig. 3.12 Flow Chart of FFS processing (Rao and Toksöz 2005)

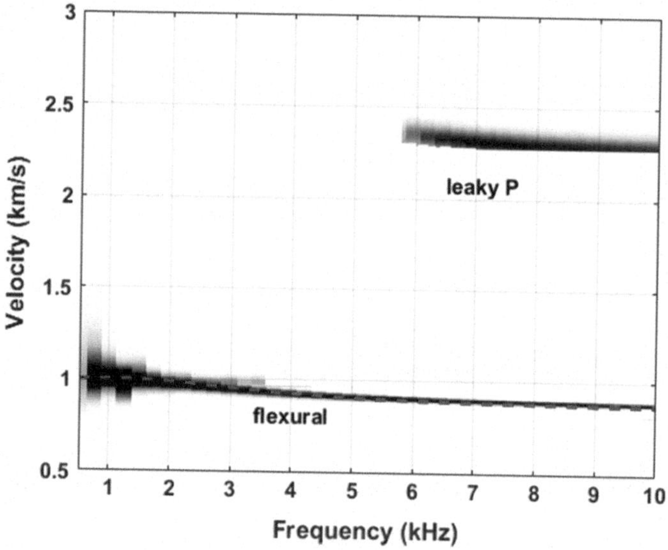

Fig. 3.13 Dispersion analyses with FFS. The array waveforms are shown in Fig. 3.6. Dashed-lines are analytical dispersion curves

modes are mixed with each other. However, the dispersion at low frequency is more important than that at the high frequency because the velocity at the low-frequency limit is usually the formation velocity. A potential method to improve the resolution at high frequencies is to determine an adaptive bandwidth for each frequency based on the frequency-time analysis, similar to the method used in Sect. 3.1.

Figure 3.13 shows the extracted dispersion characteristic with FFS for the 10 kHz dipole measurement in formation S1 (waveforms shown in Fig. 3.6). The analytical dispersion curves (red dashed-lines) are plotted. The good fit between the extracted and analytical dispersion distinguishes the FFS from other methods.

FFS is applied to field data shown in Fig. 3.4b and the dispersion result is shown in Fig. 3.14 (contour plot). The blue dashed curve is the analytical dispersion curve for a flexural wave in a borehole model having Vp = 2500 m/s, Vs = 1150 m/s, and borehole radius of 120 mm. It is clear that the dispersion characteristics of both leaky P and flexural waves are extracted well. The formation S-wave velocity is 1150 m/s, which is slightly larger than the velocity obtained by velocity-time semblance, as shown in Fig. 3.5b.

3.4 Summary

In this chapter, data processing methods for obtaining mode dispersion from array data have been presented. The phase velocity can be obtained by different methods.

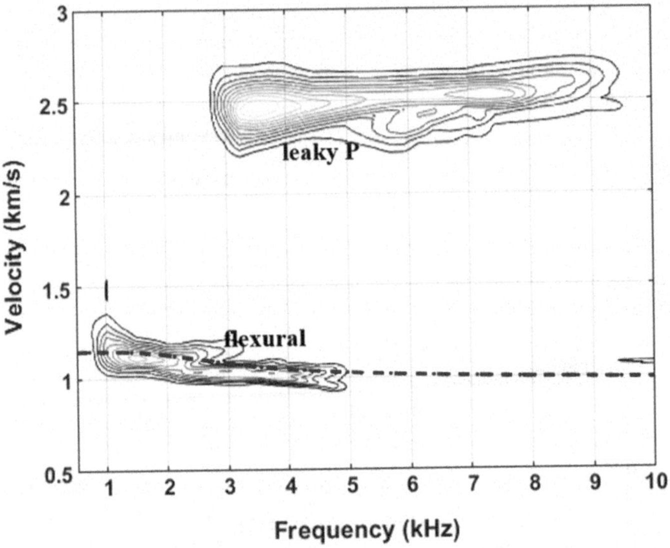

Fig. 3.14 Dispersion result obtained by applying FFS to field data shown in Fig. 3.4b. Blue dashed curve is an analytical solution for a borehole model: Vp = 2500 m/s, Vs = 1150 m/s, and borehole radius is 120 mm

For non-dispersive data, the most commonly used method is velocity-time semblance. This method avoids picking arrival times and can accurately determine the phase velocities for different modes. If the data are strongly dispersive, an improved velocity-time semblance method based on different wavelet scales may helpful. The dispersion analysis methods are effective for dispersive array data. They can be divided into two different categories: those based on parametric spectrum estimation and those based on non-parametric methods. The principles of those methods have been presented. The performances of the dispersion analysis methods have been compared using synthetic array waveforms where dispersion characteristics can be calculated analytically. The velocities of field data sets have been calculated by both velocity-time semblance and by dispersion analysis methods.

Chapter 4
Wave Propagation in a Cased Borehole and Cement Bond Evaluation

In many boreholes, casing is used and cemented to seal the borehole wall and to stabilize the borehole. Figure 4.1 shows a schematic diagram of a cased hole. The annulus between the casing and formation may be filled with fluid or cement. Table 4.1 lists the geometrical and elastic parameters of a well-cemented cased hole model. The casing properties used in this chapter are for an industry standard $9^{5/8}$-in. (244.5 mm) casing. Different from the open hole case discussed in Chap. 2, the casing, fluid annulus and/or cement introduce a barrier between the logging tool and formation. Multiple layers complicate the wavefields.

4.1 Waves in Open and Cased Boreholes

Figure 4.2a shows synthetic waveforms in an open hole in slow formation S1 (Table 4.1). The radius of the open borehole is 101.6 mm. A 13 kHz Tsang and Rader type source time function (see Sect. 2.1.3.3) is used. Parameter α of the source time function is 0.5, and the bandpass filter is from 5 to 18 kHz. The array waveforms consist of large amplitude leaky P-waves (marked as P in Fig. 4.2a) and a small amplitude ST wave. Figure 4.2b shows synthetic waveforms in the well-cemented cased hole. The borehole radius is the same as the radius of the open hole in Fig. 4.2a. The outer radii of fluid and casing are 47 and 57.15 mm, respectively. A small radius accentuates the waveform differences between the open and cased holes. Synthetic waveforms in the cased and cemented borehole, shown in Fig. 4.2b, are completely different from those of the uncased borehole (Fig. 4.2a). Leaky P-waves are much smaller in the cased borehole because of the decreased fluid column radius that causes the cutoff frequency of the leaky P-wave to reach higher values (see Figs. 2.30 and 2.31). In the cased borehole, the waveforms are dominated by the ST wave. Figure 4.2c shows the profiles of synthetized waveforms with a 4.75-m (15 ft) source-receiver spacing in the open and cased sections. The data above 100 ft simulate the logs in an open hole. The data below 100 ft are for a cased hole. The space

H. Wang et al., *Borehole Acoustic Logging—Theory and Methods*,
Petroleum Engineering, https://doi.org/10.1007/978-3-030-51423-5_4

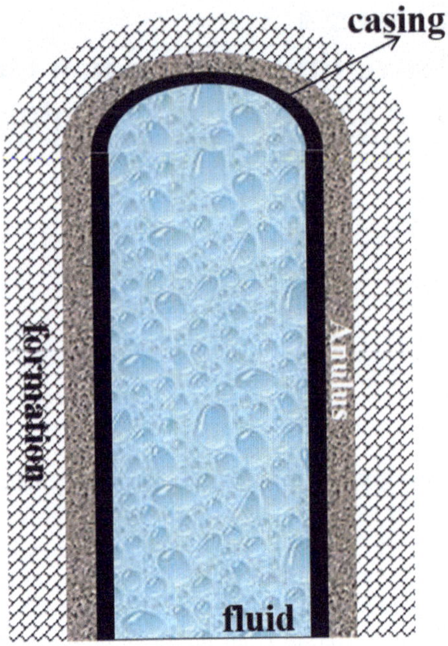

Fig. 4.1 Schematic diagram of a cased hole

Table 4.1 Elastic parameters for the model. Formation F1 is a fast formation and S1 is a slow formation

Medium	Vp (m/s)	Vs (m/s)	Density (kg/m^3)	Outer radius (mm)
Fluid	1500	0	1000	108
Steel	5500	3170	8300	122
Cement	3000	1730	1800	170
F1	4500	2650	2400	∞
S1	2300	1000	2000	∞

between traces is 1 ft. The differences between cased and uncased sections are very significant. Figure 4.3 shows a field data example of logs acquired by a monopole tool with the source-receiver separation of 4.57 m (Fig. 3.6 in Tubman 1984). The borehole is cased above 100 ft. The formation is "slow" with a P-wave velocity of 2290 m/s. The waveforms, between 150 and 200 ft in the uncased section, show the P, S, pR, and ST waves. Between depths of 100 and 150 ft, the S-wave velocity is very low. The pR and ST waves disappear and there is only a strong P-wave. This is a typical feature of monopole waveforms in slow formations. The waveforms are similar to the synthetic waveforms shown in Fig. 2.26. In the cased section, above 100 ft, there are small P and strong Stoneley (ST) waves, similar to the synthetics shown in Fig. 4.2a, c. The P-wave amplitude is smaller than in other sections.

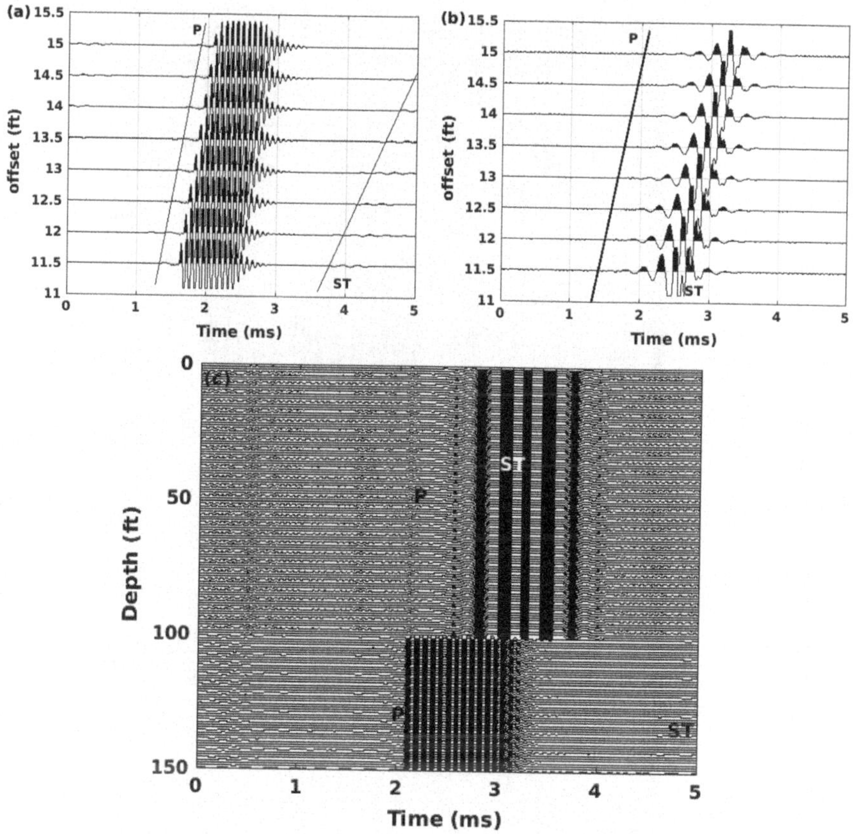

Fig. 4.2 Comparison between synthetic waveforms in an open hole (**a**) and a cased and cemented borehole (**b**). **c** Monopole logging profile created using the last trace: above 100 ft from (**a**) and below 100 ft from (**b**). A 13 kHz Tsang and Rader type source (see Sect. 2.1.3.3 , $\alpha = 0.5$ with a band-pass filter from 5 to 18 kHz) is used as the source time function. Elastic parameters are listed in Table 4.1. Formation is S1. The radius of the open hole is 101.6 mm. The outer radii of the fluid and casing are 47 and 57.15 mm, respectively. Attenuation is added for each of the media, where the quality factor for fluid is 20 and for the formation P- and S- waves, the quality factor is 60. For the steel, the quality factors for both P- and S-waves are 1000

Figure 4.4 shows another example of field data from a borehole in a semiconsolidated formation before and after the borehole was cased (Fig. 3.7 in Tubman 1984). The same monopole tool was used for both cased and uncased logs. The P-wave velocities are low with observable variations. In the uncased example, leaky P-waves are dominant and the ST wave is not recorded when its arrival time is greater than 5 ms (see the velocity dispersion curve in Fig. 2.25). In the cased hole, the ST waves are prominent. The P-wave amplitudes are smaller than those recorded in the uncased hole.

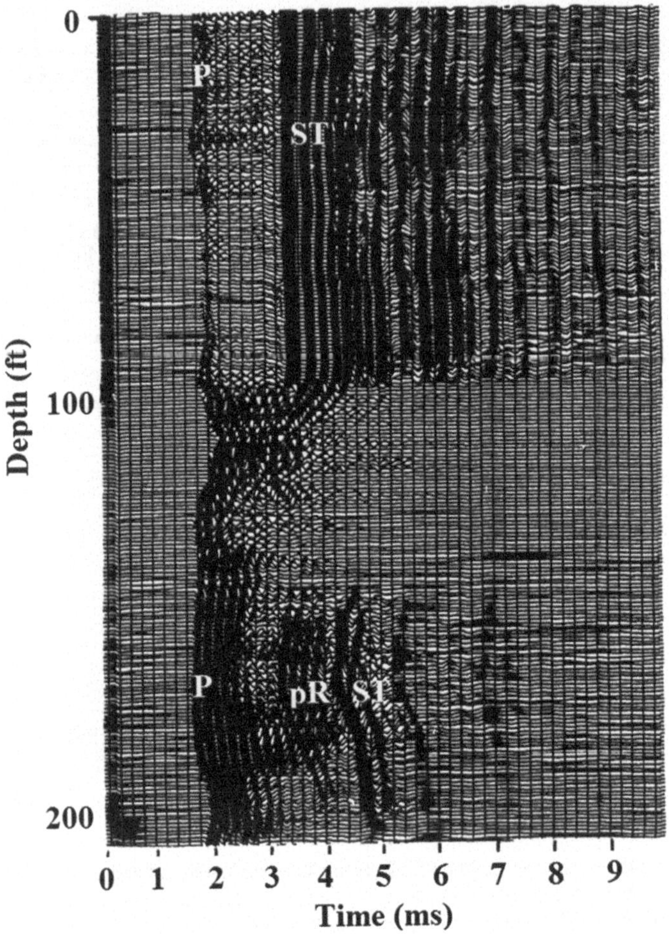

Fig. 4.3 Data collected in a borehole by a monopole tool with a 4.57 m separation between source and receiver. There is one trace every foot. The borehole is cased above 100 ft. Figure is modified from Fig. 3.6 in Tubman (1984)

The above synthetics and field examples illustrate that the casing and cement significantly alter the borehole wavefields. In the following sections, the effects of different factors on the wavefields in cased holes are discussed.

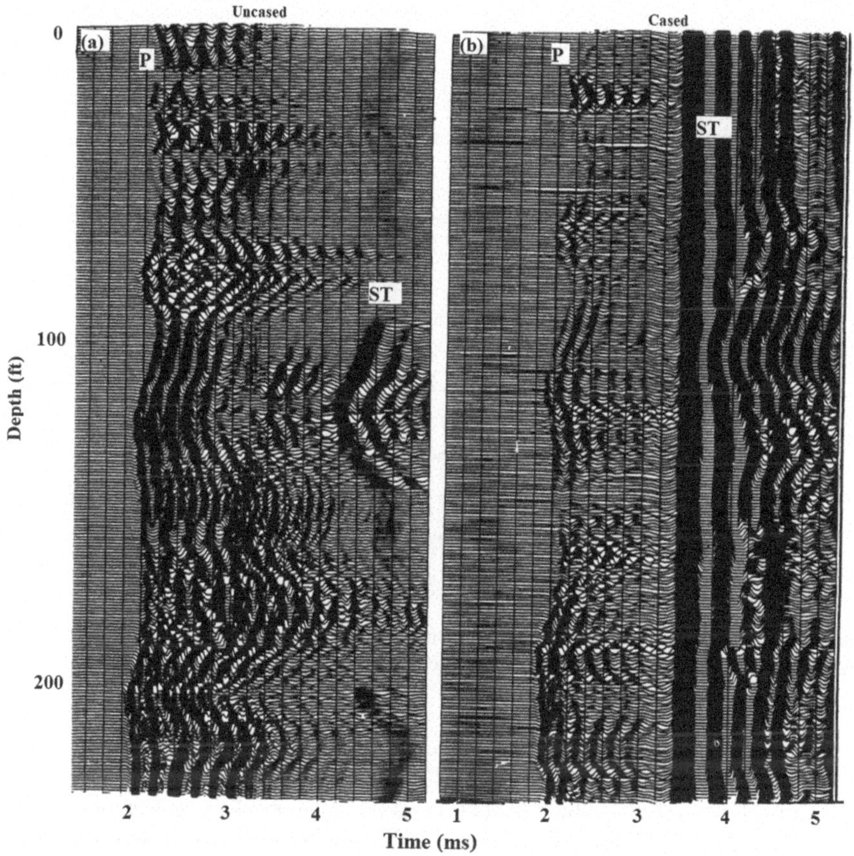

Fig. 4.4 Monopole measurements in a borehole before casing (**a**) and after casing and cementing (**b**). The source-receiver separation is 4.57 m. There is one trace every foot. Figure is modified from Fig. 3.7 in Tubman (1984). The formation is slow. There are no pseudo-Rayleigh waves

4.2 Wave Modes in a Free Pipe

4.2.1 Sonic Frequencies

An uncemented casing behaves like a free pipe in a fluid. Waves generated in the pipe are trapped and propagate as guided modes. The number of the modes is infinite. The modes can be grouped into three categories: L (longitudinal), T (torsional), and F (flexural) (Cawley et al. 2002; Edwards and Gan 2007). Figure 4.5 shows the motion of the three mode types. L modes axisymmetrically expand and contract the pipe. T modes rotate the pipe and F modes distort the pipe.

Parameters *m* and *n* are used to designate the modes in radial and azimuthal pipe directions, respectively. The three categories can be denoted as L(n, m), T(n, m),

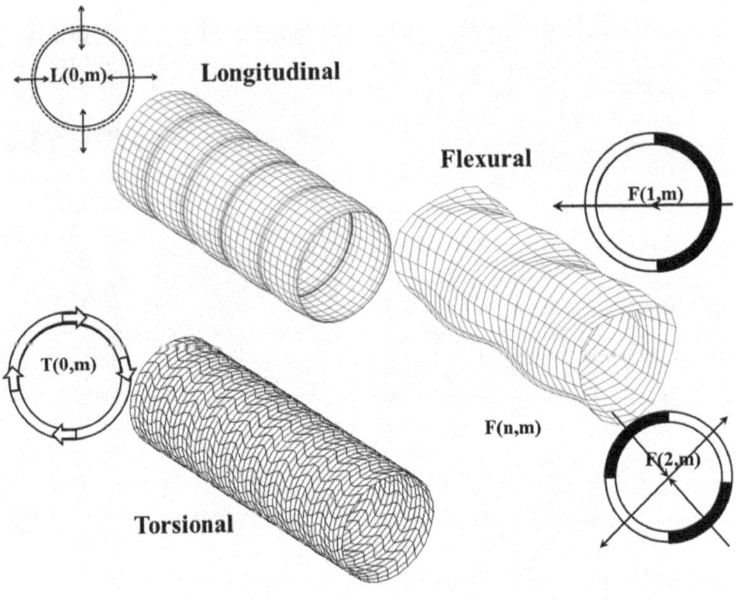

Fig. 4.5 Modes in a free casing in a fluid. Parameters m and n are used to designate the modes in the radial and azimuthal directions of the pipe, respectively

and F(n, m) (n, m = 0, 1, 2, 3…). Parameter n is 0 in the T and L modes and those modes are designated by L(0, m) and T(0, m) (m = 1, 2, 3 …). For the F modes, n ranges from 1 to infinity, where the F(1, m) and F(2, m) modes correspond to dipole (flexural) and quadrupole (screw) modes in the pipe, respectively. For convenience, the letters 'F' and 'Q' are used to denote dipole and quadrupole modes, respectively.

Figure 4.6 shows the dispersion curves of the L and F modes (Wang and Fehler 2018a, b; Zhang et al. 2016) when the pipe is in water. The calculation of the dispersion curves is similar to the open hole case in Sect. 2.1.3, where the dispersion curves come from the solutions of taking the determinant of $[m_{ij}]$ as zero. The matrix $[m_{ij}]$ is determined from the boundary condition. Appendix A2 shows the matrix used for acoustic logging while drilling which also works for free casing conditions. The pipe properties are listed in Table 4.1. The two ST waves, inside and outside the casing, are marked as ST1 and ST2 [slow ST in Plona et al. 1992] waves). In the L category, there is L(0, 1) (marked as L1 in Fig. 4.6a), the fundamental mode and a series of higher modes such as L2, L3, and L4. The dipole (F) and quadrupole (Q) modes also have fundamental and numerous higher modes as shown in Fig. 4.6b, c.

With a source inside the pipe, most of the energy becomes trapped inside the pipe body, which is illustrated by the pressure snapshots in Fig. 4.7 (calculated by a 3D finite difference method). The source position is marked with a white star and the inner and outer boundaries of the casing are marked with white solid lines. The figure demonstrates the modes propagating in and around the pipe.

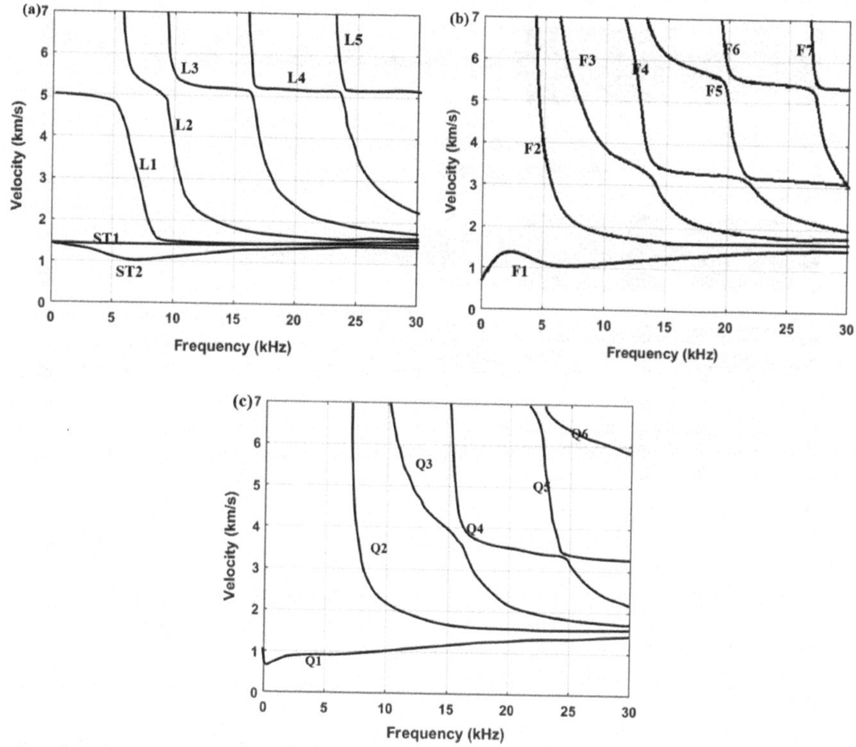

Fig. 4.6 Dispersion curves of the L, ST, F, and Q modes when the pipe is in water. F designates the flexural mode and Q is the quadrupole (i.e., screw) mode

Casing modes (L1, L2, L3, and others in Fig. 4.6a) at offsets of 1.5–4 m are visible when a 10 kHz monopole source is used (Fig. 4.7a). These modes leak energy into the fluid, both inside and outside of the casing. Two ST waves follow the L modes. A dense receiver array, with a 0.01 m interval, is used to record the waveforms from the source to the top of the model along the z-axis. The waveforms are shown in Fig. 4.8a. The three visible modes marked with lines are the L casing modes, ST1, and ST2. Although the ST2 mode is mainly propagating outside of the casing, the mode can be detected in the inner fluid column at a low frequency.

Figure 4.7b, c show snapshots when the source center frequencies are 2 kHz and 10 kHz, respectively, for a dipole source. There are two casing flexural modes propagating on the pipe: the flexural casing mode F1 with a large amplitude and the much faster but smaller F2. The modes become complicated when the source frequency is high (10 kHz) as shown in Fig. 4.7c. Higher-order casing modes arrive before mode F1. In Fig. 4.8c, F1 and higher-order modes such as F2, F3, F4, and F5, are identified. F1 is the slowest casing mode and is seen in the latest part of the wave train. All the F modes have cutoff frequencies except the F1 mode. The source

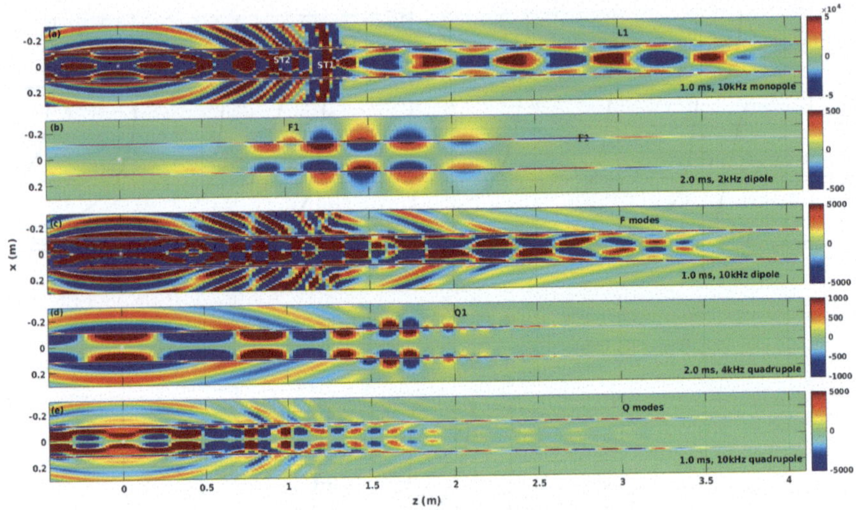

Fig. 4.7 Pressure snapshots (at different times) of the monopole, dipole and quadrupole source in a casing immersed in fluid. The dipole source polarization is along the x direction. Source type and frequency are listed on each plot. Source is located at the white star

frequency controls the mode excitation. Higher frequency sources excite higher-order modes (Wang and Fehler 2018b).

The wavefields of the quadrupole source are similar to those of the dipole case. There is only the fundamental casing mode Q1 (F(2, 1)) at a low source frequency (Fig. 4.7d). With a high-frequency source, higher-order modes arrive ahead of the Q1 mode (Fig. 4.7e). All Q modes have cutoff frequencies (low frequency limits). The cutoff frequency of the fundamental mode Q1 is very low (below 1 kHz). There are few higher modes that arrive before the slowest Q1 (Fig. 4.8d), which is when the source frequency is at 10 kHz.

Pipe modes are affected by the size and properties of the pipe (stiffness) and the source frequency. Figures 4.9 and 4.10 show the dispersion curves of different modes for different pipe sizes. Only low-order modes are displayed in the figures.

From Fig. 4.9, it is obvious that a larger ratio between the outer and inner radii (OR/IR) moves the modes to a higher frequency when the pipe thickness is constant. The pipe thickness significantly affects the ST2, F1, and Q1 modes (Fig. 4.10). The excitation of ST2 becomes more difficult when the pipe thickness increases. The velocity of mode F1 increases with pipe thickness, and the velocity is sensitive to the outer radius at low frequencies (below 3 kHz) and sensitive to the inner radius at high frequencies (above 3 kHz). A thicker pipe moves the cutoff frequency of Q1 to a higher frequency. The pipe thickness is not the controlling factor for the L modes, or higher-order F and Q modes. Instead, the radii variation controls the modes where either an increasing inner or increasing outer radius moves these modes to a lower frequency range, except for the Q modes.

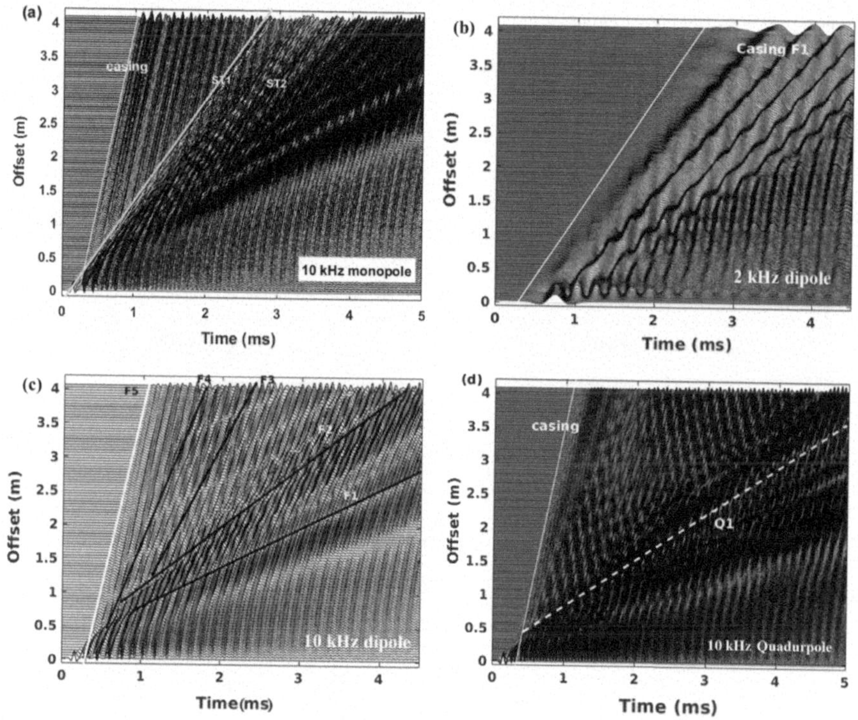

Fig. 4.8 Dense array waveforms for mulitpole measurements in a pipe immersed in fluid. The nearest offset from the source position and the interval between receivers is 0.01 m. The source frequency and source type are listed in the figures

The computations for pipes in fluid are important for the design of logging tools. For monopole logging, the L modes that propagate along the tool body can obstruct the P-wave velocity measurement, while slots on the tool body can suppress the L modes (Tang and Cheng 2004, p. 224).

In the dipole measurement, the existence of the F1 mode, with no cutoff frequency, overlaps with the formation flexural mode (Sects. 2.2.2 and 2.3.2) and affects determination of the formation S-wave velocity. In the wireline case, this issue can be mitigated using an isolator, including an attenuator, and by decoupling the receiver from the rigid tool body to suppress the F1 mode. Suppressing the F1 mode allows one to measure the S-wave velocity from the formation flexural wave (Cowles et al. 1994).

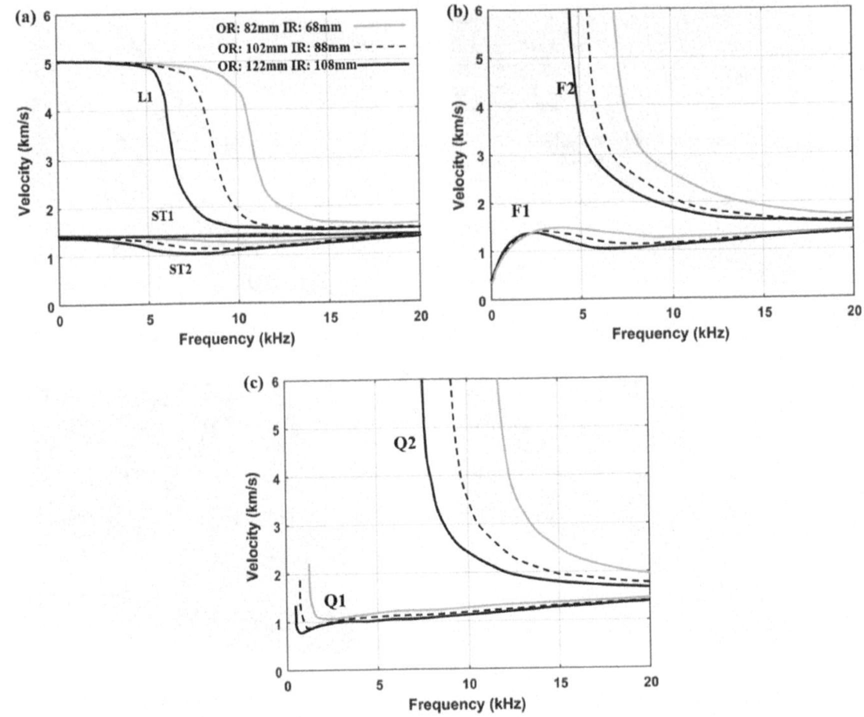

Fig. 4.9 The effect of the pipe radii on the dispersion curves of different modes when the pipe thickness is constant. **a** L and ST modes; **b** F modes (F(1, m)); and **c** Q modes (F(2, m))

4.2.2 Ultrasonic Frequencies

Ultrasonic frequencies are often used for cement bond evaluation. When the source frequency moves to the ultrasonic range (more than 80 kHz), the effect of the curvature of the pipe can be ignored. The pipe may be treated as a plate. The axially propagating modes may be treated as Lamb waves (Lamb 1917; Brekhovskikh 1960; Cawley et al. 2002; Edwards and Gan 2007). There are two families of Lamb waves, known as symmetric (*S*) and antisymmetric (*A*) modes. The displacements of S modes are symmetric with $x = 0$ (center of the plate) and the particle motions are normal to the boundary (Fig. 4.11a). The displacements of A modes are antisymmetric about $x = 0$ (Fig. 4.11b).

The dispersion equations for Lamb waves in a 2d thick plate immersed in a fluid are given by Eqs. 4.1a and 4.1b (Rose 1999). Equations 4.1a and 4.1b are the period equations for the S and A modes, respectively.

$$\left(k_z^2 + s^2\right)^2 coth(pd) - 4k_z^2 ps \cdot coth(sd) - i\frac{\rho_f}{\rho}\frac{pk_s^2}{\sqrt{k_f^2 - k_z^2}} = 0, \tag{4.1a}$$

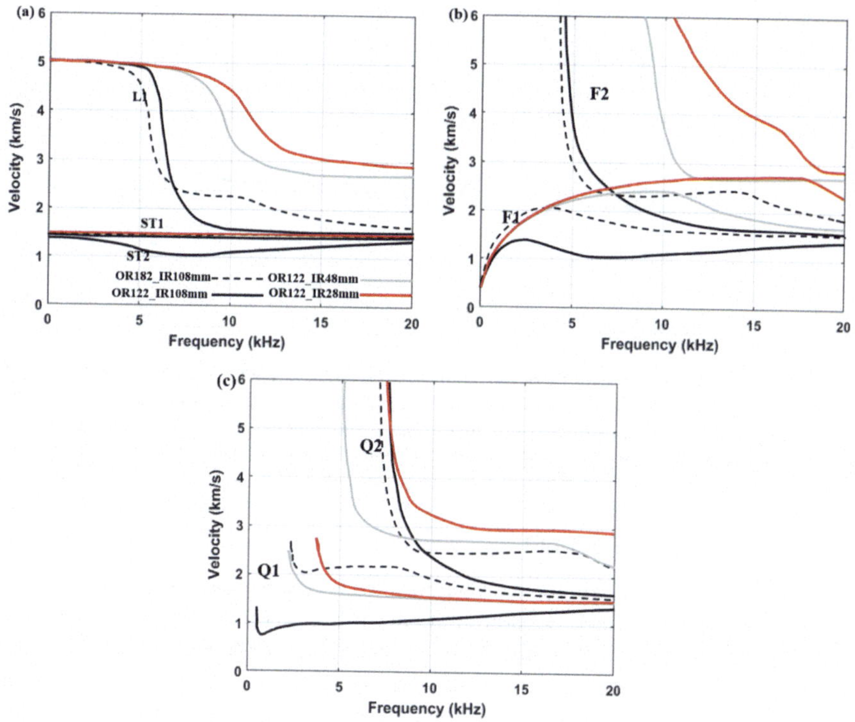

Fig. 4.10 The effect of pipe thickness on the dispersion curves of different modes when one pipe radius is constant. **a** L and ST modes. **b** F modes (F(1, m)). **c** Q modes (F(2, m))

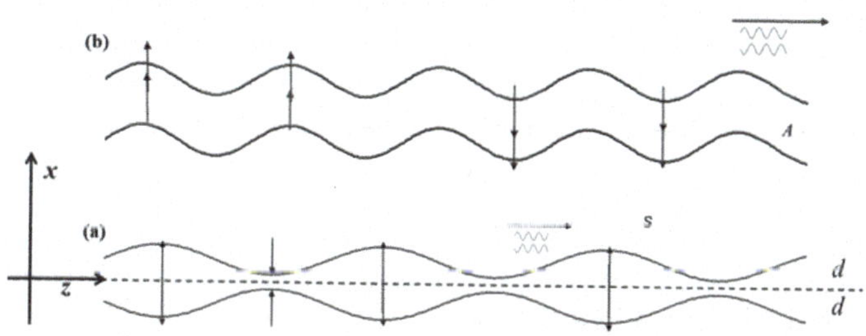

Fig. 4.11 Displacements (horizontal) and particle motion (vertical) in a thin plate. **a** Symmetric Lamb waves. **b** Antisymmetric Lamb waves. The thickness of the plate is 2d and the origin of x is the center of the plate

$$\left(k_z^2 + s^2\right)^2 tanh(pd) - 4k_z^2 ps \cdot tanh(sd) - i\frac{\rho_f}{\rho}\frac{pk_s^2}{\sqrt{k_f^2 - k_z^2}} = 0, \quad (4.1b)$$

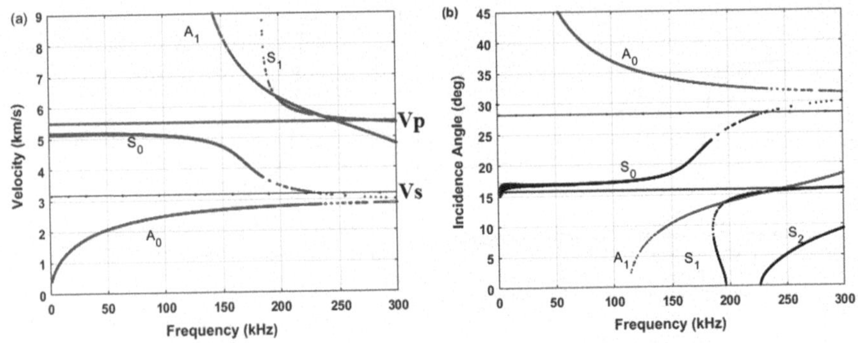

Fig. 4.12 **a** Dispersion curves of modes in a pipe (14 mm thick) immersed in fluid. **b** Relationship between the incident angles and frequencies of different modes (Critical angles for P-wave (15.83°) and S-wave (28.24°) are marked with gray lines). Black and gray curves are dispersion curves for the S and A modes, respectively. Vp and Vs are velocities of the pipe material

where k_z is the wavenumber of Lamb waves in the z direction; $p = \sqrt{k_z^2 - k_p^2}$ and $s = \sqrt{k_z^2 - k_s^2}$. k_p and k_s are the wavenumbers of P- and S-waves, respectively.

Figure 4.12a shows the dispersion curves derived from Eq. 4.1a and 4.1b when the plate thickness is 14 mm. The dispersion curves move to a higher frequency range when pipe thickness decreases. All the modes, except S_0 and A_0, have a cutoff frequency (low frequency limit). Another method to illustrate ultrasonic modes in a pipe is to transform the frequency-velocity dispersion plot in Fig. 4.12a to a frequency-incidence angle plot, which is shown in Fig. 4.12b. The incidence angle in Fig. 4.12b is calculated using asin (Vf/V), where asin is the arcsin function and V is the phase velocity of different modes. The first and second critical incidence angles calculated from Snell's law are shown with solid horizontal lines. The first critical angle is slightly above 15°. A1, S1, S2, and A2 can be excited when incidence angle is less than the first critical incident angle, but A0 can only be excited when the incidence angle is greater than the second critical angle.

4.3 Wavefields in a Cased Hole

Wavefields in a pipe immersed in fluid are discussed in Sect. 4.2. In this section, we describe the theoretical and practical aspects of acoustic logging in a cased hole. The annulus between the casing and formation can be fully or partially filled by fluid, or by cement, as shown in Fig. 4.13. The topics covered in this section include acoustic logging in cemented well (annulus fully filled with cement), free casing (annulus completely filled with fluid), and (partially) poorly cemented wells.

The elastic parameters for different materials used for modeling are given in Table 4.1. The surrounding formation is F1. A cylindrical fluid channel is present, in

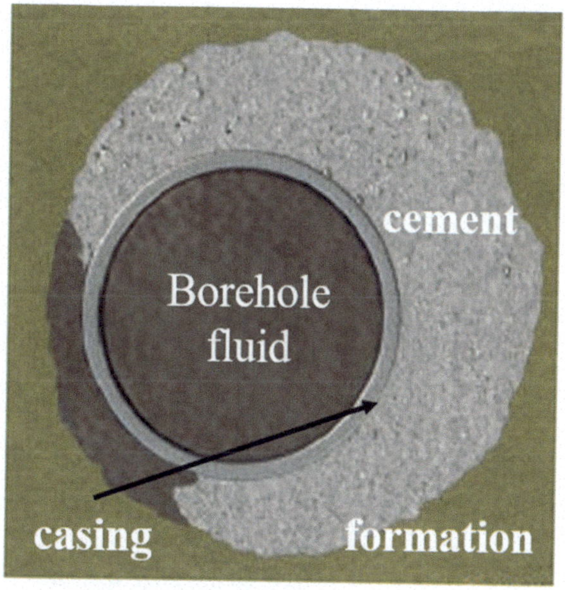

Fig. 4.13 An example where the cement is partially replaced by fluid (Modified from Cameron 2013)

the annulus between the casing and formation and is described by four parameters. Two of the parameters describe the location of the channel: radial position of the channel $\boldsymbol{r}$ and the azimuthal position $\boldsymbol{\varphi}$. The dimensions of the channel are described by the channel's thickness $\boldsymbol{d}$ and its angular extent $\boldsymbol{\theta}$ (Fig. 4.14). The four parameters are varied to investigate their influences on the wavefields.

4.3.1 Monopole Source

Figure 4.15 shows pressure snapshots at 0.8 ms for a series of cement bonding conditions: (a) well bonded ($\boldsymbol{r} = 122$ mm, $\boldsymbol{d} = 0$ mm, $\boldsymbol{\theta} = 0°$, and $\boldsymbol{\varphi} = 0°$), (b) free casing ($\boldsymbol{r} = 122$ mm, $\boldsymbol{d} = 48$ mm, $\boldsymbol{\theta} = 360°$, and $\boldsymbol{\varphi} = 0°$), (c) partially cemented well ($\boldsymbol{r} = 122$ mm, $\boldsymbol{d} = 16$ mm, $\boldsymbol{\theta} = 360°$, and $\boldsymbol{\varphi} = 0°$) and (d) partially cemented well with fluid interface at the cement-formation interface ($\boldsymbol{r} = 154$ mm, $\boldsymbol{d} = 16$ mm, $\boldsymbol{\theta} = 360°$, and $\boldsymbol{\varphi} = 0°$). The two innermost lines at approximately x = 0 are the inner boundaries of the casing, the two outermost lines are boundaries between the borehole wall and the formation, and the other two lines are the outer boundary of the casing. The source position is marked by a white star. Casing, P, S, pR, and ST modes are marked in the figure.

In a well-cemented cased hole (Fig. 4.15a), casing L modes are the fastest modes. These modes are very small and do not affect the formation P- and S- waves recorded in the borehole fluid. Figure 4.16 shows the waveforms and velocity semblance results in the time and frequency domains. Casing waves are too small to be visible. P, S, pR, and ST waves are seen in the array waveforms and marked in Fig. 4.16a. The

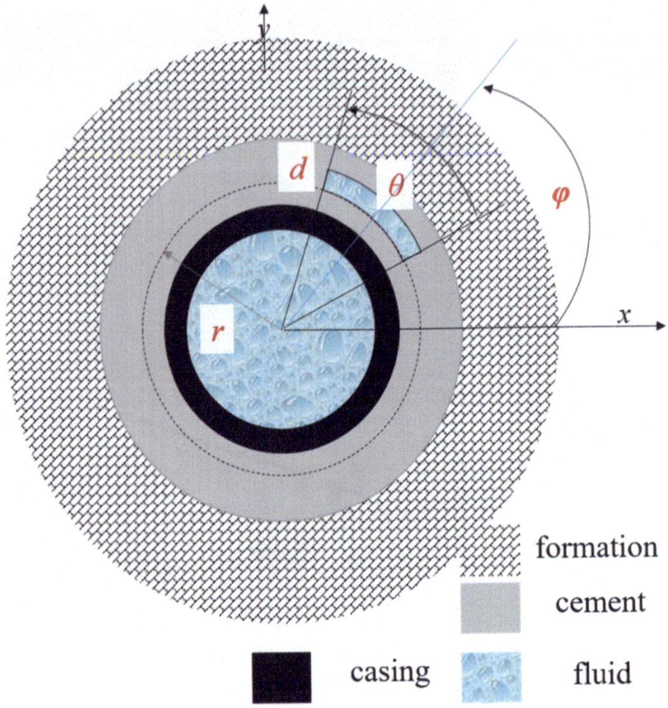

Fig. 4.14 Horizontal section perpendicular to the borehole showing positions of fluid, casing, cement and formation (Fig. 1 in Wang et al. 2018). The 4 parameters that are used to describe a fluid channel in the cement are illustrated

velocity-time semblance in Fig. 4.16b gives the velocities of the modes. The modal dispersion curves are plotted as dashed curves on the dispersion analysis contour plots (from the FFS described in Chap. 3) in Fig. 4.16c.

The wavefield changes when the cement is completely replaced by fluid (Fig. 4.15b). Casing modes leak into the fluid from both sides of the pipe (see section z = 2.5 to 3 m). The first arrival in the borehole is a strong leaky casing mode. The P-wave is difficult to see. The S-wave is also impacted by the casing modes. There are two ST modes: one is inside the casing, and the other is in the fluid annulus outside the casing. The wave modes in the time sequence (Fig. 4.17a) are the casing (different order L modes), poorly defined S, dispersive pR waves, ST1, ST2 and the Airy phase of the pR wave. The ST inside the casing has a higher velocity and the ST in the fluid between the casing and formation has a lower velocity, as shown in Fig. 4.17b, c.

For the case where cement is partially replaced by fluid to form a fluid annulus, the wavefield snapshot is shown in Fig. 4.15c. The wavefields are similar to the "no cement" case in Fig. 4.15b. One difference between these two models is the amplitudes of different modes. Another is that the ST2 wave in Fig. 4.15c is slower than the wave in Fig. 4.15b. Figure 4.18 shows the waveforms and velocity determination.

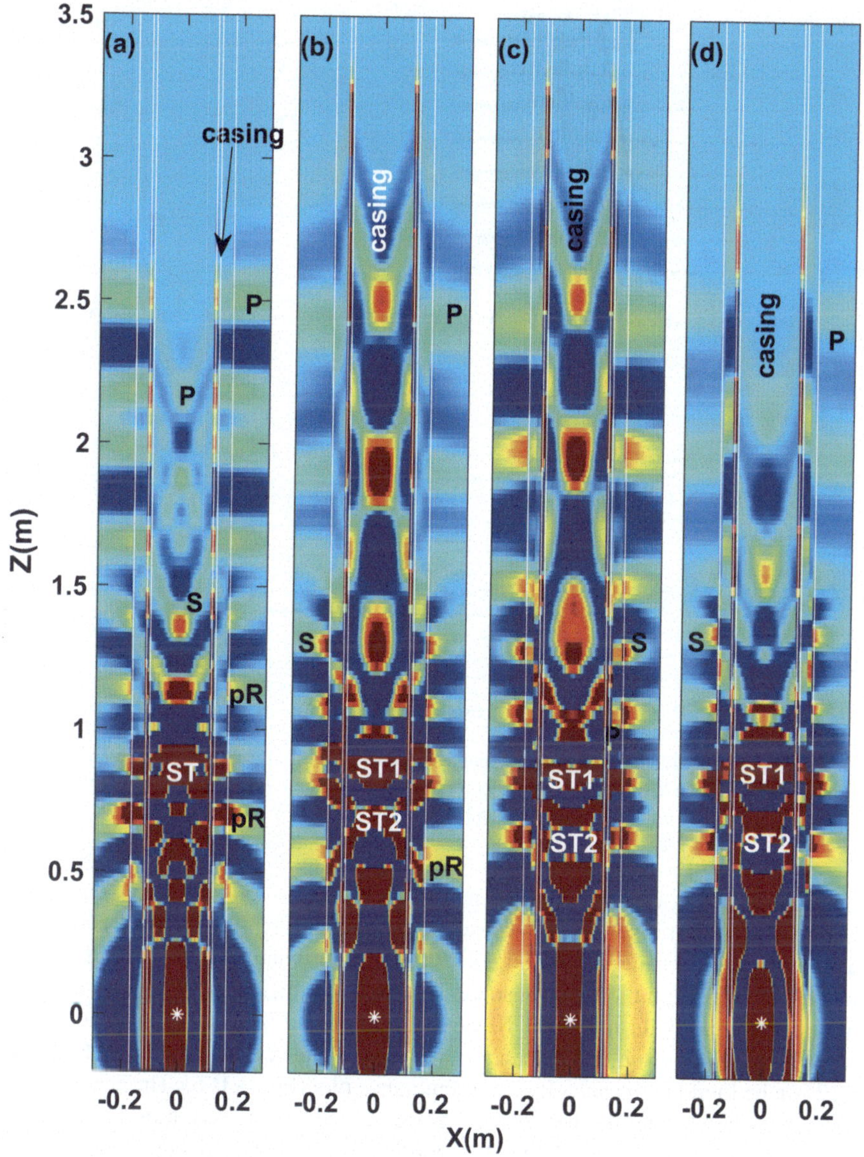

Fig. 4.15 Vertical section through a 60-cm diameter model (including formation) with a 10-kHz monopole source centered at z = 0 m, showing acoustic waves at t = 0.8 ms for (a) good cement ($\boldsymbol{r}$ = 122 mm, $\boldsymbol{d}$ = 0, $\boldsymbol{\theta}$ = 0°, and $\boldsymbol{\varphi}$ = 0°), (b) no cement ($\boldsymbol{r}$ = 122 mm, $\boldsymbol{d}$ = 48 mm, $\boldsymbol{\theta}$ = 360°, and $\boldsymbol{\varphi}$ = 0°), (c) fluid annulus at the casing-cement interface ($\boldsymbol{r}$ = 122 mm, $\boldsymbol{d}$ = 16 mm, $\boldsymbol{\theta}$ = 360°, and $\boldsymbol{\varphi}$ = 0°), and (d) fluid annulus at the cement-formation interface ($\boldsymbol{r}$ = 154 mm, $\boldsymbol{d}$ = 16 mm, $\boldsymbol{\theta}$ = 360°, and $\boldsymbol{\varphi}$ = 0°). The two innermost lines at approximately x = 0 are the inner boundaries of the casing, the two outermost lines are the boundaries between the borehole wall and formation, and the other two lines are the outer boundary of the casing. The source position is marked by a white star

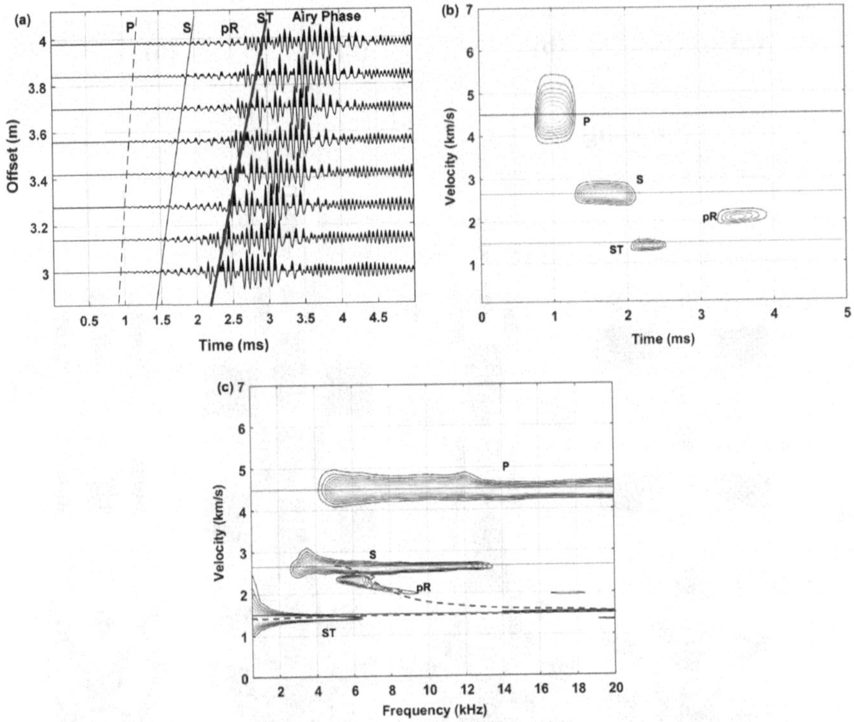

Fig. 4.16 **a** Array waveforms in a well-cemented cased hole with a centralized receiver and a point monopole source (10 kHz) located in the borehole fluid. **b** Velocity-time semblance contour plot. **c** Frequency analysis contour plot. Fig. b and c are from Fig. 5 in Wang and Fehler (2018a)

The first arrival is the casing wave, and it has the same arrival time as that in the free casing case in Fig. 4.17a. Because the fluid layer next to the casing is thin, the amplitude of the casing wave in Fig. 4.18a is smaller than that in Fig. 4.17a. Casing modes interfere with the P-wave, so the P-wave is not distinguishable. It is not easy to discern the arrival time of the S-wave. The ST1 and ST2 waves are not easy to identify due to the strong interference of pR waves. The velocities of S, pR, ST1, and ST2 can be found in the velocity-time semblance plot (Fig. 4.18b). The casing-wave velocity is larger than the formation P-wave velocity. The separation between the two ST waves becomes larger than those in the free casing model due to the slower ST2. Different modes can be identified by dispersion analysis in Fig. 4.18c.

If the fluid annulus exists at the cement-formation interface, the casing waves become slower. In this case, the first arrivals could be mistaken for formation waves, as shown in Fig. 4.15d. The P-wave in the formation is weaker than that in the well-bonded case. Figure 4.19 shows the waveforms and velocity analysis. The casing wave arrival from Fig. 4.18a is shown with a red line. The visible first arrival is between the casing wave and the formation P-wave. The dispersive modes consist of different L modes propagating in the casing and cement, including L1, L2, and L3

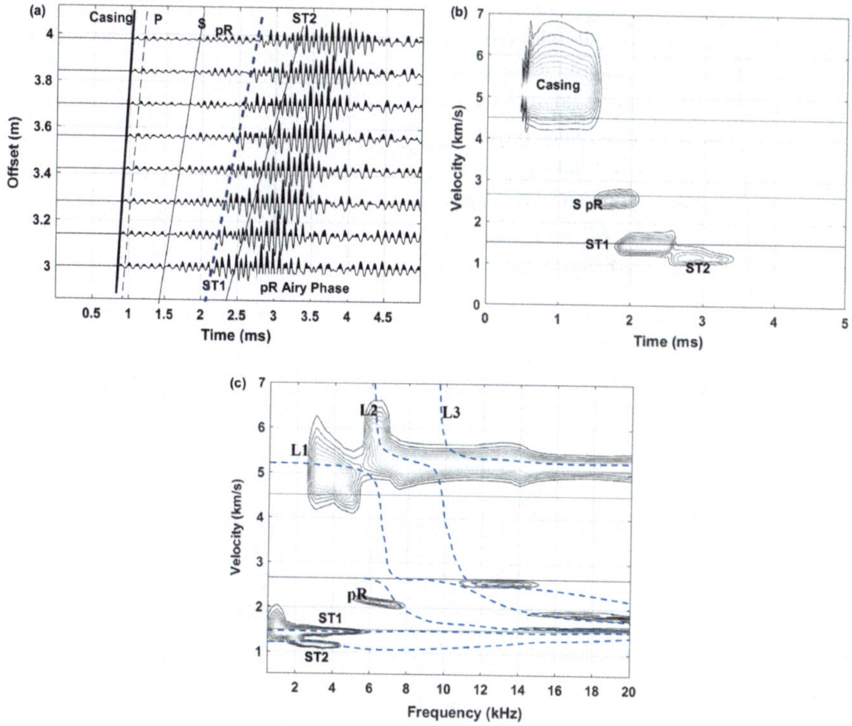

Fig. 4.17 Is the same as Fig. 4.16 except for the case where cement is completely replaced by fluid. Fig. b and c are from Fig. 5 in Wang and Fehler (2018a)

in Fig. 4.19c. The P-wave velocity cannot be determined. The S-wave velocity can be determined from the semblance and from the velocity of pR wave at the cutoff frequency, as shown in Fig. 4.19b. The appearance of ST2 can be used to distinguish the model from the well-bonded cement case in Fig. 4.15a (Wang and Fehler 2018a).

The ST2 wave is sensitive to the thickness of the fluid in the annulus. There is no significant difference between the ST2 waves in Fig. 4.15c, d. This demonstrates that the ST2 wave velocity is not sensitive to the location of the fluid annulus. Figure 4.20 shows the velocities of the ST1 and ST2 waves for different thicknesses of fluid next to the casing when the fluid completely surrounds the casing ($\theta = 360°$). The velocities of the ST waves increase as fluid thickness increases.

In poorly cemented wells, a fluid channel may exist only in a small portion of the cement. This indicates $\boldsymbol{\theta}$ in Fig. 4.14 is not 360°. Figure 4.21 shows an example, where the parameters of the fluid channel are ($\boldsymbol{r} = 122$ mm, $\boldsymbol{d} = 16$ mm, $\boldsymbol{\theta} = 90°$, and $\boldsymbol{\varphi} = 0°$). The snapshot shown in Fig. 4.21a has a mixture of the wavefields shown in Fig. 4.15a, b (or Fig. 4.15c). The first arrival is a weak casing mode. ST1 and ST2 can be observed in the wavefield. The symmetry of the wavefield is broken because the model is nonaxisymmetric. The wavefield cross section at $z = 2.5$ m in Fig. 4.21b shows that the casing modes are affected by the fluid annulus. The mode

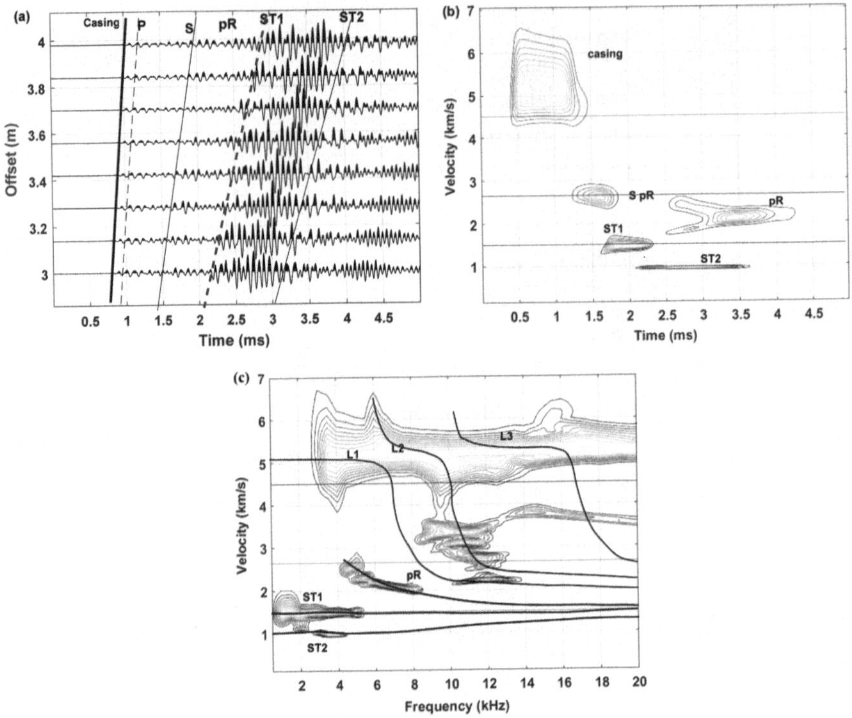

Fig. 4.18 Is the same as Fig. 4.16 except for the case where the 16 mm fluid annulus replaces the cement next to the casing. The values of parameters defined in Fig. 4.14 for the fluid annulus are ($\boldsymbol{r} = 122$ mm, $\boldsymbol{d} = 16$ mm, $\boldsymbol{\theta} = 360°$, and $\varphi = 0°$). There is a fluid annulus between the cement and casing. Fig. b and c are from Fig. 10 in Wang and Fehler (2018a)

in the borehole is not symmetrical. The time of the first arrival, marked with a blue line in Fig. 4.21c, is between that of the casing arrival ($\boldsymbol{\theta} = 360°$ case in Fig. 4.18a) and the formation P-wave, which demonstrates that the arrival time of the first arrival is not only a function of fluid thickness $\boldsymbol{d}$ but also of $\boldsymbol{\theta}$. The amplitude of the first arrival is also a function of $\boldsymbol{d}$ and $\boldsymbol{\theta}$. More detail can be seen in the experimental data in Fig. 4.27. The ST2 is weak, and it is not easy to extract the dispersion when both $\boldsymbol{d}$ and $\boldsymbol{\theta}$ are small.

4.3.2 Dipole Source

In a well-cemented cased hole, a dipole source generates a small P-wave and a strong and dispersive formation flexural wave (Fig. 4.22a). The good match, between the dispersion analysis (contour plots) and the modal dispersion curves (dashed lines) in Fig. 4.22c shows that the formation P- and S-wave velocities can be obtained from

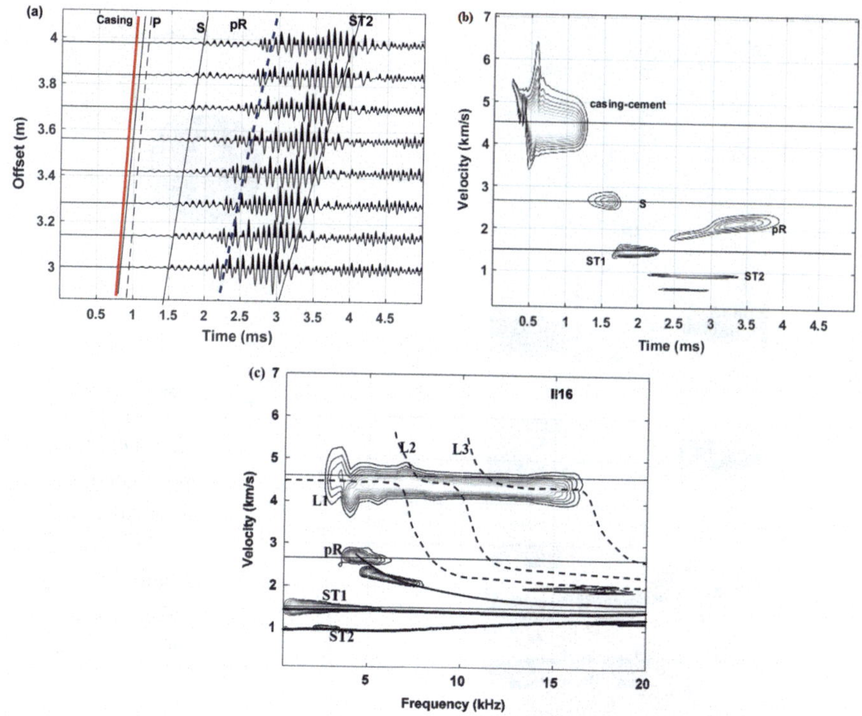

Fig. 4.19 Is the same as Fig. 4.16 except for the case where the 16 mm fluid annulus replaces the cement next to the formation. The values of parameters defined in Fig. 4.14 for the fluid channel are (***r*** = 154 mm, ***d*** = 16 mm, ***θ*** = 360°, and ***φ*** = 0°). The red line shows the casing arrival from Fig. 4.18a. Fig. b and c are from Fig. 13 in Wang and Fehler (2018a)

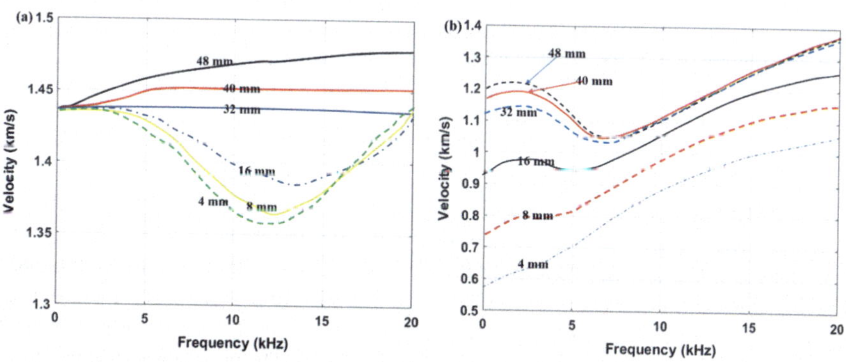

Fig. 4.20 Velocities of ST waves versus fluid thickness for r = 122 mm and θ = 360°. **a** ST1 wave in the borehole fluid. **b** ST2 wave in the fluid surrounding the casing

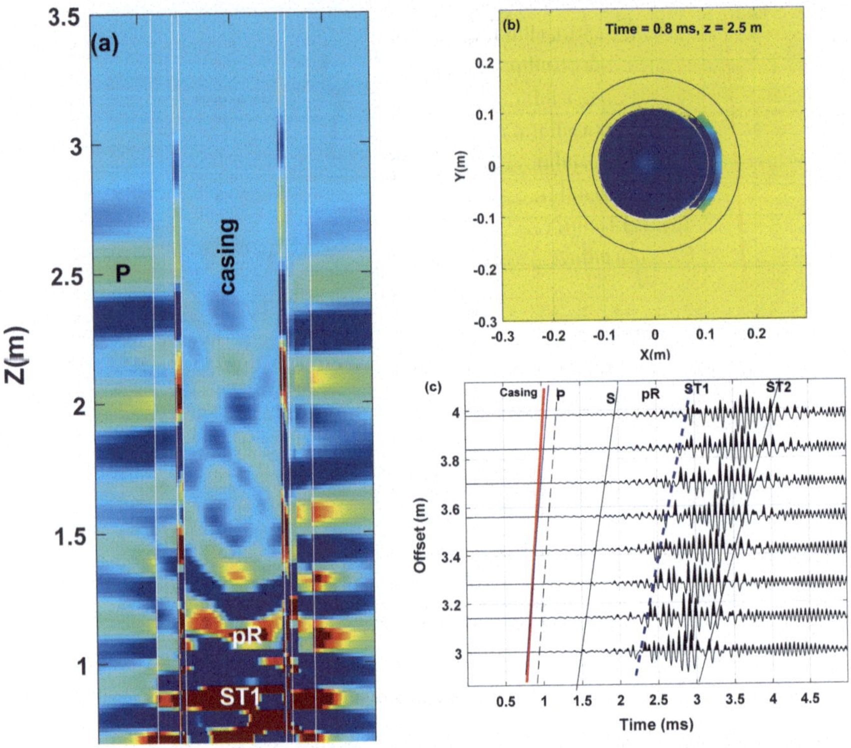

Fig. 4.21 Snapshots and waveforms for a partial fluid channel case, where the parameters are ($\boldsymbol{r}$ = 122 mm, $\boldsymbol{d}$ = 16 mm, $\boldsymbol{\theta} = 90°$, and $\boldsymbol{\varphi} = 0°$). **a** Vertical section through 60-cm-wide borehole model (including formation) with a 10-kHz monopole source centered at z = 0 m, showing acoustic waves at t = 0.8 ms. **b** Horizontal section at Z = 2.5 m. The inner two circles show the casing boundaries and the outermost circle shows the borehole wall. **c** Array waveforms, where the red line shows the casing arrival from Fig. 4.18a. and the blue line shows the first arrival of the waveform

a high-frequency dipole source in a well-cemented cased borehole. Waveforms also serve as indicators of cement quality.

The wavefields are different if there is no cement or poorly-bonded cement. Figure 4.23a shows the pressure snapshots for a free casing model where the cement in the annulus is fully replaced by fluid. The snapshot for the low source frequency (2 kHz) in the upper plot shows that there are three flexural modes, one fast flexural mode propagating along the casing (casing F2) followed by a formation flexural mode and the slowest casing flexural mode (casing F1). These modes exhibit perfect dipole characteristics. The formation flexural wave is the first arrival at the receivers in the borehole fluid because casing F2 is invisible in the borehole fluid. Casing F1 arrives later. The prominence of the formation flexural wave infers that the formation S-wave velocity can be determined from the flexural velocity at the cutoff frequency (Fig. 4.23d) when a low frequency dipole tool is used.

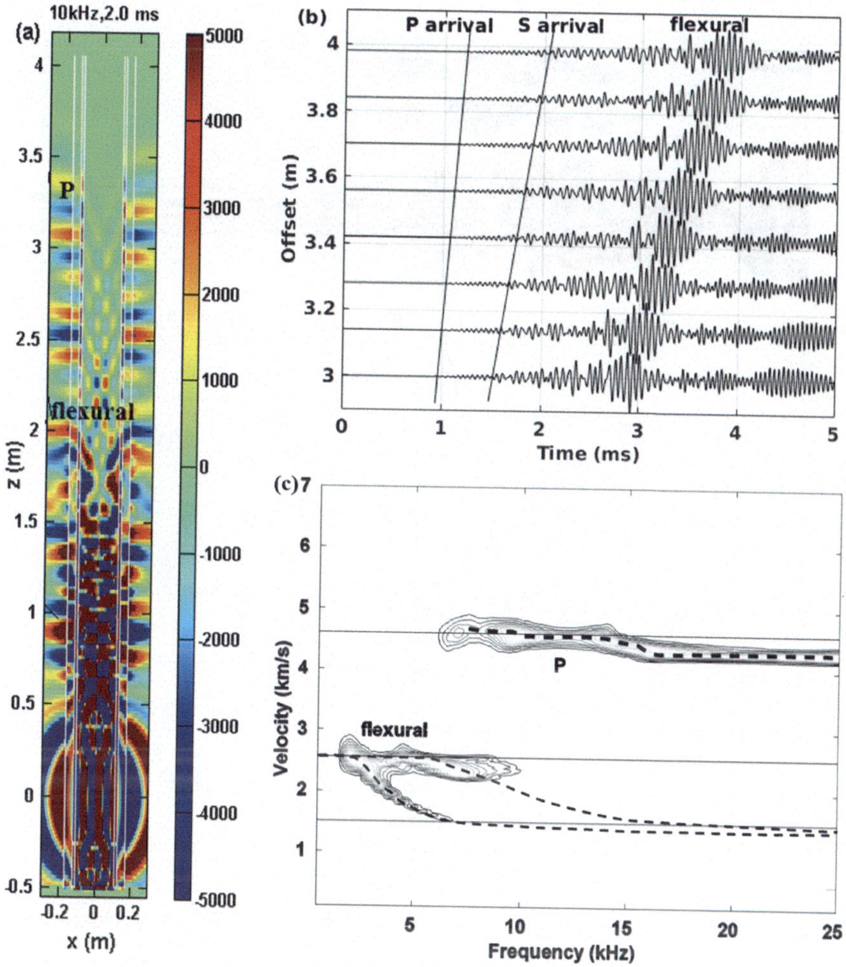

Fig. 4.22 **a** Snapshots (x–z profiles) of the wavefield at 2 ms for a cased-hole with good cement and a 10 kHz dipole source. Point sources at (0.004 m, 0, 0) and (0.004 m, 0, 0) with opposite phases are used to simulate the dipole source. **b** Array waveforms. **c** Dispersion analysis, and the modal dispersion curves are plotted with dotted lines

However, the formation flexural wave is not the first arrival when the source frequency is above the cutoff frequency of the higher-order casing modes. The snapshot in the bottom panel of Fig. 4.23a, the dense array waveforms (Fig. 4.23c) and dispersion analysis (Fig. 4.23e) show that the higher-order casing modes propagate faster than the formation flexural mode. Higher-order casing modes such as F2, F3, F4, and F5 (arrival times marked with lines in Fig. 4.23c) are faster than the formation flexural mode, and F1 is the slowest mode. The casing modes vary with source frequency, where higher frequency sources excite larger amplitude higher modes

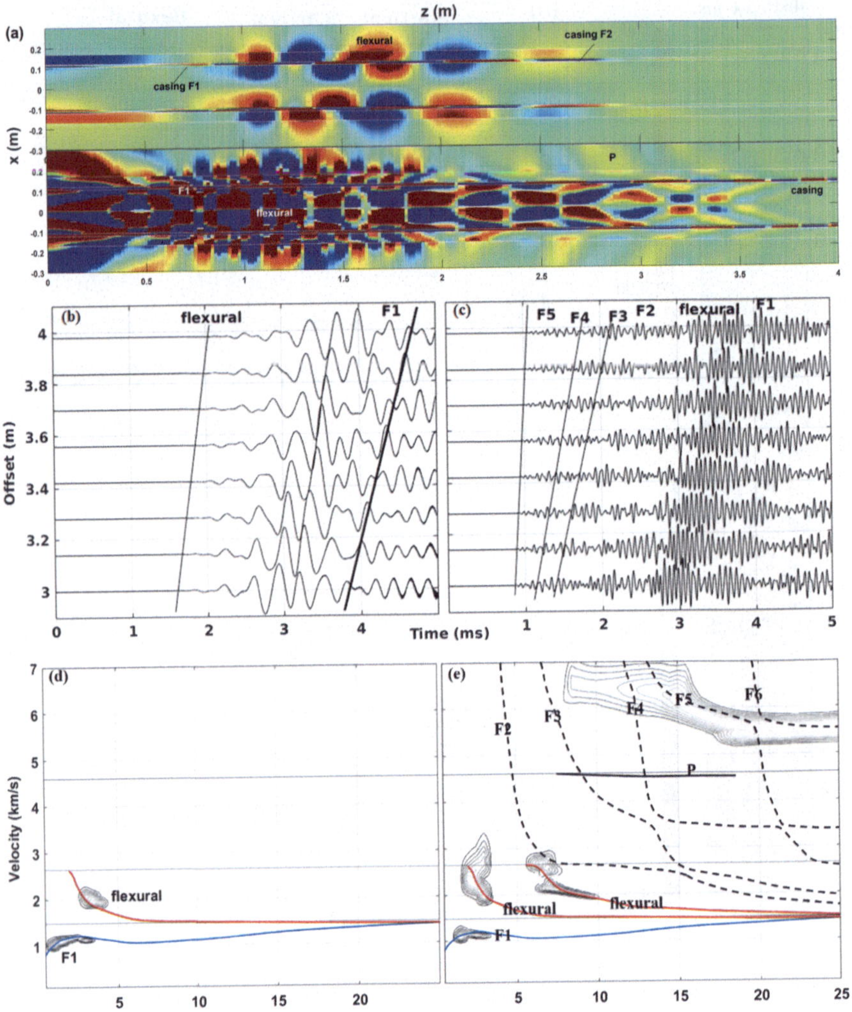

Fig. 4.23 **a** Pressure snapshots for a cased-hole with no cement (only a fluid annulus) for source frequencies of 2 kHz (top) and 10 kHz (bottom). The source position is the same as that shown in Fig. 4.22a. **b** and **c** are array waveforms for source frequencies of 2 and 10 kHz, respectively. **d** and **e** are the dispersion analyses for **b** and **c**, respectively. Modal dispersion curves are plotted as solid dashed lines. Formation P and S velocities are shown with horizontal lines

(such as F2, F3, F4, F5, and even F6). The amplitude of F1 decreases as source frequency increases.

Although a leaky P mode is excited by the high-frequency dipole source, it is difficult to identify in Fig. 4.23c. The interference from casing modes obscures this mode and hinders the formation P-wave velocity determination.

Two formation flexural modes propagate at S-wave velocity at their respective low frequency cutoffs. The dispersion analysis in Fig. 4.23e helps to identify the modes and enable the S-wave velocity measurement. A low-pass filter with a corner frequency lower than 5 kHz also helps to distinguish the formation flexural wave.

Based on the above examples and discussions, the monopole and dipole wavefields in a cased borehole can be summarized as follows:

(1) The wavefields are strongly affected by the quality of the cement bond. In a poorly cemented hole, it may be difficult to determine the formation P and S velocities from the monopole measurements.
(2) With a dipole source, there are casing flexural modes, leaky P (present only at high frequency), and formation flexural modes. Casing modes disappear if the casing is perfectly cemented. Data from a high-frequency dipole tool can be used to determine the P and S velocities from a leaky P and prominent flexural modes, respectively.
(3) When there is no cement (i.e., free pipe), the leaky P mode becomes invisible. The S-wave velocity can still be obtained from the flexural velocity at the low frequency limit.

4.3.3 Quadrupole Source

A quadrupole source in a well-cemented cased hole can be used to determine the formation P and S velocities. A low frequency source (4 kHz shown in Fig. 4.24a) excites formation screw waves, which can be used to determine the formation S-wave velocity. The formation P-wave velocity can be obtained from leaky P-waves at high frequencies, which is seen as the first arrival (Fig. 4.24b, c) and is followed by the formation screw wave. The high-frequency quadrupole tool can be used to determine both P- and S-wave velocities.

The casing modes appear when the casing is not well-cemented. Figure 4.25a shows the pressure snapshot of a 4 kHz quadrupole source in a free casing model. The formation screw wave is followed by the casing Q1 mode, which has a very slow propagation speed. The wavefield is different when the source frequency is high (such as 10 kHz in Fig. 4.25b), and the higher-order casing modes appear ahead of the formation screw wave (or formation quadrupole). The Q1 becomes weak due to the high source frequency and is overridden by formation screw waves. The formation screw wave can be used to determine the S-wave velocity (Fig. 4.25c, d).

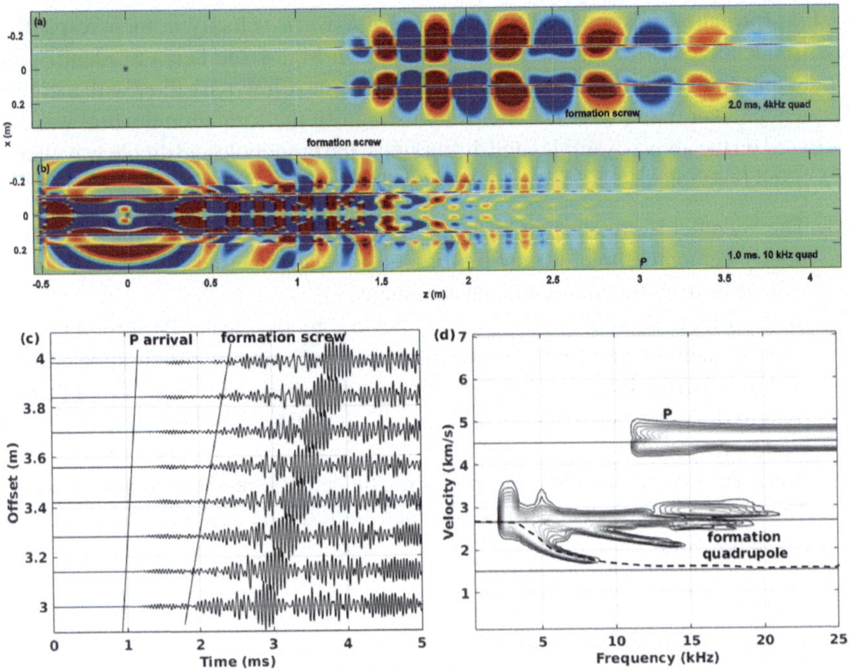

Fig. 4.24 **a** and **b** Pressure snapshot of the quadrupole wavefield in a perfectly cemented borehole model. Time and source frequency are listed on the plots. **c** Array waveforms excited by a 10 kHz quadrupole source. **d** Dispersion analysis (contour plots) of the waveforms in **c**. The modal dispersion curve is plotted as a dashed line

4.4 Cement Bond Evaluation

As a step in well completion, cement is injected into the annulus between the casing and formation. The permeability of intact cement is less than 0.1 mD (Lecampion et al. 2011), and proper cementation ensures hydraulic isolation between the formation and well and between layers and prevents fluid seepage along the casing. Proper cementing helps to avoid accidents such as the major oil spill in the Gulf of Mexico in 2010 (Deepwater Horizon Study Group 2011). For other applications, such as underground gas storage (Shahvali et al. 2014) and CO_2 storage (Lecampion et al. 2011), a good cement bond is critical for all aspects of safety.

Many technologies are utilized to evaluate the quality of cement bonds. These include temperature logging while the cement is setting, radioactive tracers, GR-density logs, thermal neutron decay time logs, "noise" logs, and acoustic logs. Acoustic methods are the major tools for cement bond evaluation. Acoustic methods can be divided into two categories: sonic (approximately 20 kHz) and ultrasonic (several hundred kHz).

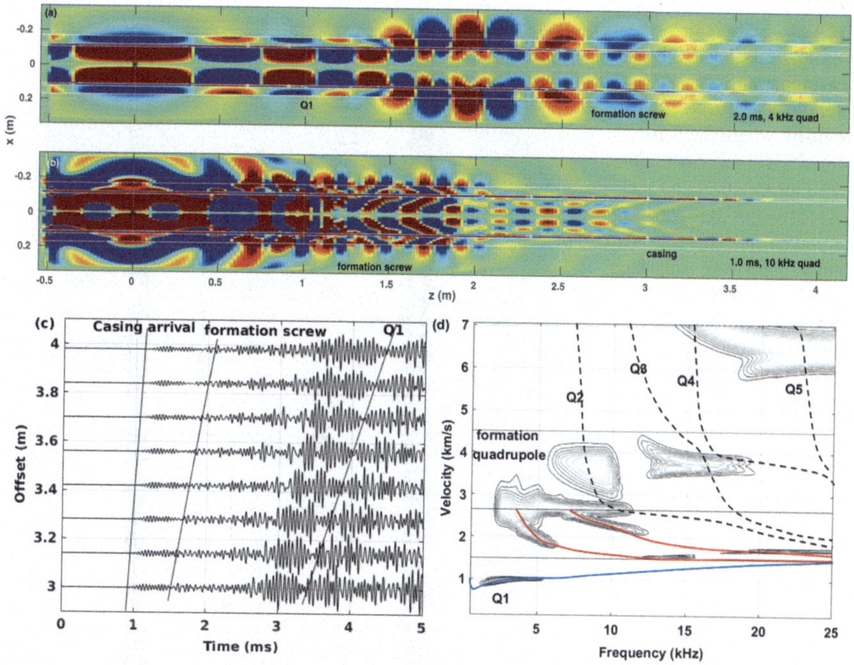

Fig. 4.25 Pressure snapshot of the quadrupole wavefield in a borehole model where the cement is replaced with fluid. Time and source frequencies are listed on the plots. **c** Array waveforms excited by a 10 kHz quadrupole source. **d** Dispersion analysis (contour plots) of the waveforms in **d**. The modal dispersion curves are plotted with different curves: blue for Q1, red for formation quadrupole modes, and dashed lines for higher casing modes

4.4.1 Sonic Methods

The sonic method, which has been extensively used in early applications (Pardue et al. 1963), is a qualitative or semiquantitative method. The attenuation and arrival time of the first arrival are measured to determine the cement conditions (Fig. 4.26). Attenuation is mainly caused by the shear coupling of the material behind the casing (Jutten and Hayman 1993). In general, high and low attenuations correspond to a good and bad cement conditions, respectively.

4.4.1.1 Cement Bond Log/Variable Density Log (CBL/VDL)

The first-generation cement evaluation method is the cement bond log (CBL). The CBL measures the attenuation of the first arrival (Pardue et al. 1963). The method is used primarily to evaluate bonding between the cement and casing (Interface I in Fig. 4.26). Several factors, such as the thickness of the cement (Walker 1968;

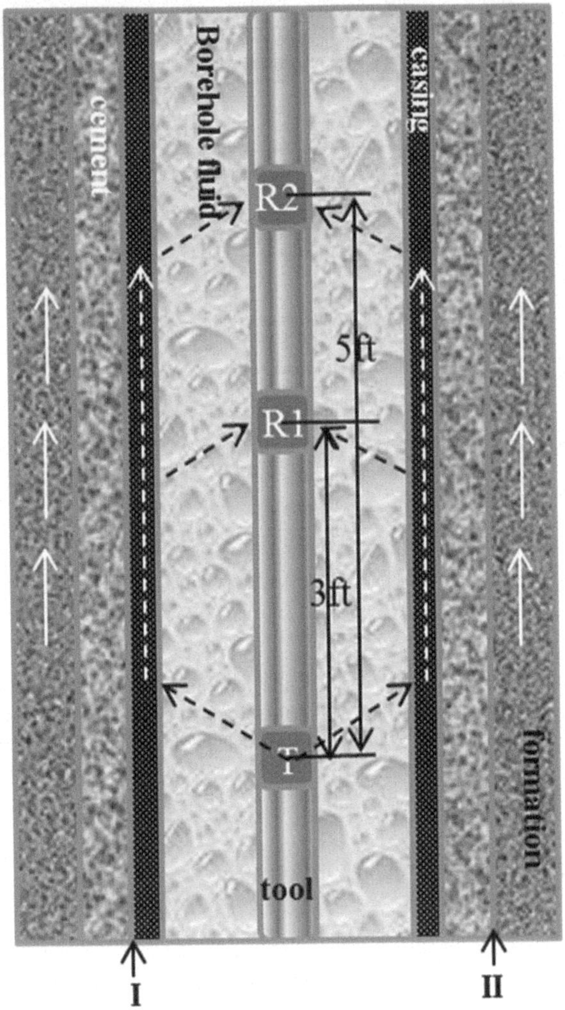

Fig. 4.26 Schematic diagram of a CBL/VDL measurement in a cased borehole. Two recordings are provided, where the 3-ft receiver (R1) records the first arrival information and the 5-ft receiver (R2) records the full waveform. There are two bonding interfaces: interface I between the casing and cement and interface II between cement and formation. T is the transmitter

Jutten and Parcevaux 1987; Jutten and Corrigall 1989), mud properties, tool eccentricity, casing eccentricity, and micro-annulus (Jutten and Hayman 1993), affect the measurement.

The introduction of full waveform recording and variable density displays (VDL) has provided a greatly enhanced capability for cement bond evaluation. The combination of the two methods CBL/VDL reinforces evaluation at both the cement-casing and cement-formation interfaces. A typical tool for cement bond logging is shown in Fig. 4.26. The industry standards of CBL/VDL are a 3-feet spacing for recording the casing wave (CBL) and a 5-feet spacing for the full waveform (VDL) acquisitions.

A monopole transducer with a source center frequency at approximately 20 kHz is used, and only the amplitude of the first wave trough is recorded in the CBL to evaluate the bonding between the cement and casing (interface I in Fig. 4.26). The mode utilized in the CBL method is one of the leaky modes of the L. The experimental results in Fig. 4.27 show that the amplitude of the first arrival is sensitive to the fluid parameters $\boldsymbol{d}$ and $\boldsymbol{\theta}$ (see definition in Fig. 4.14). In real applications, if the cement is fully replaced with fluid, the first arrival amplitude is used for calibration (or normalization) to eliminate the effect of the casing size. However, it may be difficult to find a borehole section where the cement is fully replaced by fluid. Other issues, such as the cement type, cement thickness, and tool eccenterings, also change the amplitude. To overcome these barriers, an array of transducers are employed to determine the actual spatial attenuation rate of the signal and reduce the effects of the borehole fluid, temperature, pressure and tool eccenterings (Gollwitzer and Masson 1982).

To evaluate the CBL/VDL type cement bond logging methods, we calculated synthetic seismograms for different cementing conditions (Fig. 4.28). The source is a 10 kHz monopole. The properties of formation F1, cement, casing, and fluid are listed in Table 4.1. The sonic log waveforms shown for five cases display significant variation for different cementing conditions. Free casing, seismograms are dominated by casing arrivals. For a well-cemented case, formation arrivals are the prominent modes. The presence of a fluid annulus between the casing and cement enhances the casing arrival. A fluid annulus between the cement and formation produces first arrivals controlled by the combined steel-cement properties. Formation arrivals can be extracted when the thickness of the fluid annulus is small (a few millimeters).

The calculated logs can be compared to the field data examples shown in Figs. 4.29 and 4.30 (Figs. 5 and 7 from Brown et al. 1970). Figure 4.29 shows the data collected in a borehole with free casing. The first arrival time is small and its amplitude is large. These values indicate that the first arrival is a casing mode. The VDL displayed in the

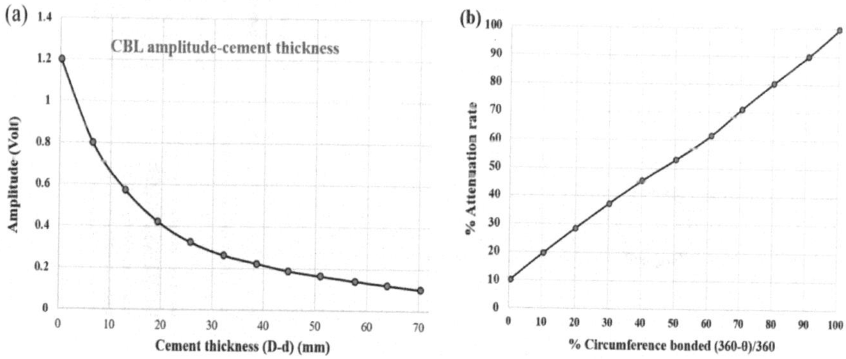

Fig. 4.27 CBL amplitude varies with fluid channel parameters $\boldsymbol{d}$ and $\boldsymbol{\theta}$. Definitions of parameters $\boldsymbol{\theta}$ and $\boldsymbol{d}$ are shown in Fig. 4.14. **a** Experimental results for various cement thicknesses (D-d) when the water ($\boldsymbol{\theta} = 360°$) appears at the casing-cement interface. Data extracted from Fig. 8 in Walker (1968). D is the annulus thickness between the casing and formation. **b** Experimental results for various $\boldsymbol{\theta}$ values when $\boldsymbol{d}$ is the annulus thickness. Data extracted from Fig. 21 of Pardue et al. (1963). Borehole diameter is 7.75 in. (20 cm). A 5.5 in. (14 cm) outer diameter industry casing was used

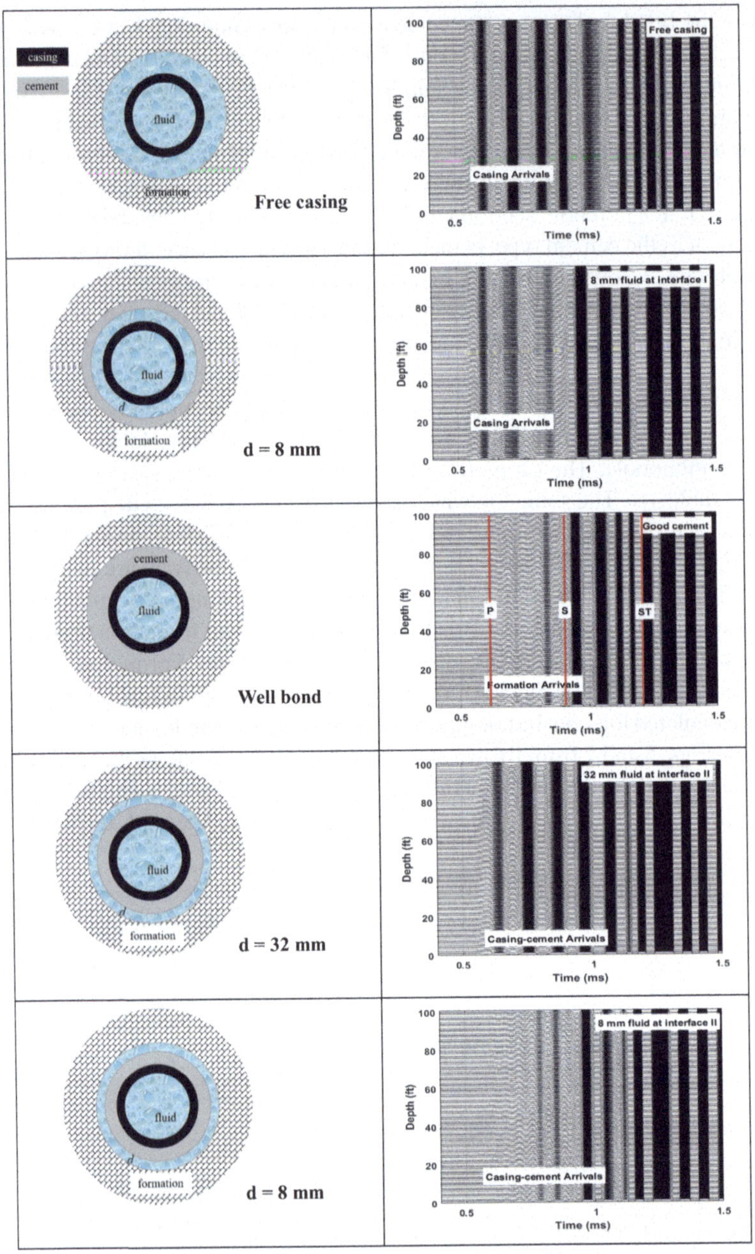

Fig. 4.28 Synthetic logs for different cementing conditions. The source-receiver offset is 1.5 m. The models are shown in the left column and the VDL logs are shown in the right column. There is one trace per foot. "I" indicates the casing-cement interface and "II" indicates the cement-formation interface. "Good" means that the cement is well bonded. "Free" means that there is no cement in the annulus between the casing and formation. The elastic parameters and geometries of the models are listed in Table 4.1. Formation F1 is used

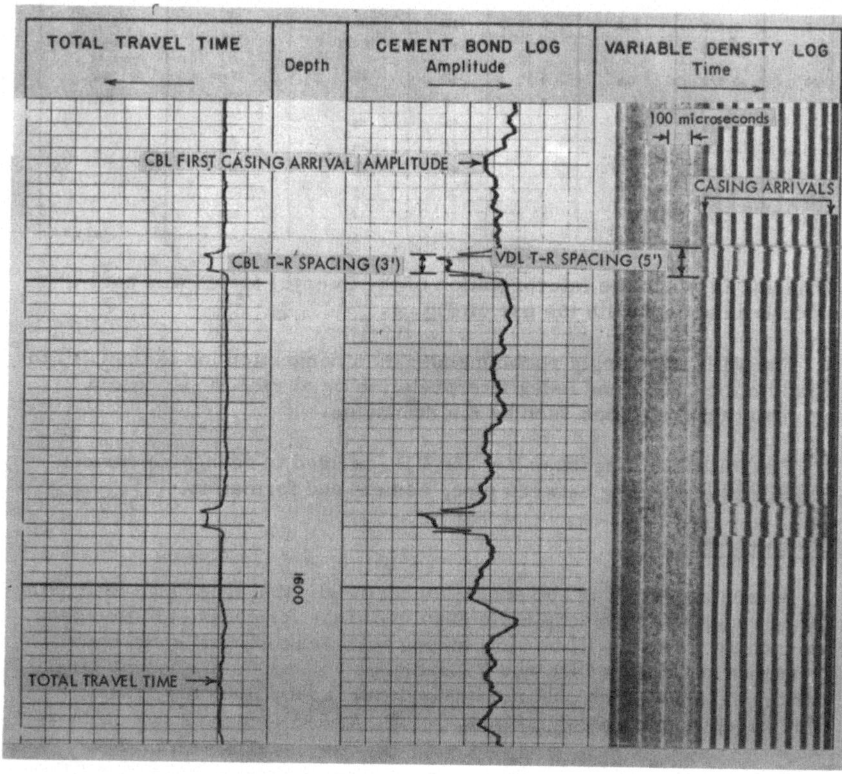

Fig. 4.29 Field example of CBL/VDL log in a free casing. The VDL log shows strong casing waves and no formation waves (Fig. 5 from Brown et al. 1970). The source-receiver offset is 1.5 m

third track shows a clear casing wave and is similar to that shown in Fig. 4.28. The casing waves are parallel due to the uniform casing transit time. The high contrast of dark and white stripes illustrates the large amplitudes of the waves.

Figure 4.30 shows another example of the CBL/VDL measurement. In this section, the casing is in a well-cemented well. A Gamma ray log in the first track shows the formation lithology, which is shale above 9590 ft and sandstone below 9590 ft. The casing amplitude in track 3 is nearly zero. The VDL shows weak casing arrivals and strong formation arrivals, similar to the well-emented case in Fig. 4.28. In the shale section, the formation waves arrive very late. In the sandstone section, we observe the P-wave and pR-wave arrivals. This section is comparable to the "Good cement" case shown in Fig. 4.28.

CBL/VDL measurements have been extensively used for cement bond evaluation for many decades. Logistically, these measurements are easy to deploy and can detect major flaws in cementing. CBL/VDL measurements are efficient and useful for an overall cement bond evaluation. CBL/VDL measurements, which have served the industry well for many decades, have some shortcomings that include a lack of angular resolution and inability to properly determine channels in cement.

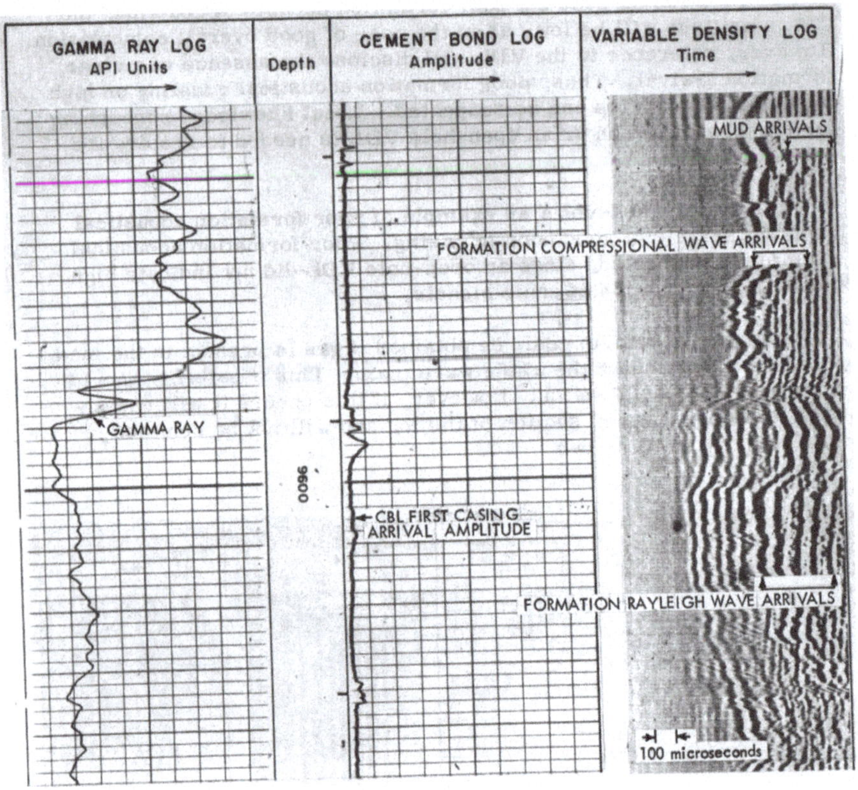

Fig. 4.30 Field example of CBL/VDL log in a well cemented borehole. CBL indicates good cement. The VDL log shows weak casing waves and strong formation waves (Fig. 7 from Brown et al. 1970). Mud arrivals indicate the borehole fluid wave velocity

A low resolution occurs in the borehole axis direction because only the amplitude of the first arrival at the 3-ft offset receiver is used. A possible solution is to obtain the attenuation between receivers at the 3 and 5-ft offsets, which improves the resolution along the borehole axis from 3 to 2 ft. Another issue is the lack of azimuthal resolution because the measurement is averaged over all azimuths. The omni-directional transducer and receivers in the CBL/VDL result in the measured bonding condition being an average of the contribution from all azimuths. The experimental results in Fig. 4.27 show that the first arrival amplitudes depend on both fluid thickness d and coverage θ. The measured value provides an estimate that is some combination of d and θ. Cement bond evaluation by sonic methods can be improved using azimuthally distributed sources and receivers. Segmented bond logs (SBT or SBL) use azimuthal measurements that can detect channels, irregularities and cement conditions. In this case, transmitters and receivers are divided into 6 or 8 segments, each covering 60° or 45° in azimuth, respectively. The sources and receivers may be held against the casing. The source frequency may be as high as 100 kHz, but only the envelope of the waveform (approximately 20 kHz) is utilized in data processing. More details on SBT or SBL will be given in Sect. 4.4.1.2.

The second challenge in the CBL/VDL is measurement in a borehole with ultralow-density cement (density less than 1.29 g/cm^3) (Ripley et al. 1981), which is widely used. The CBL/VDL measurements are affected by the combined properties of the casing and cement (Frisch et al. 2000).

In wells where the formation velocity is faster than the casing-wave velocity (such as limestone or dolomite), the first arrival would be the formation arrival and the interpretation of the CBL data to differentiate between good and poor bonds may be difficult (Smolen 1996). The VDL presentation or the frequency-time analyses of the full waveforms may help to characterize the bonding (Smolen 1996). Modeling of the waveforms for different cement properties and formation velocities can assist in the interpretation of the CBL/VDL logs.

4.4.1.2 Segmented Bond Logging Method

A segmented bond logging tool with multiple transducers and receivers at different azimuths was introduced to improve the azimuthal resolution of the CBL/VDL in the 1990 s. In a typical tool, there are 6 or 8 pads in the azimuthal direction to obtain the full azimuthal coverage (each pad covers 60° or 45°). Different tools have different designs of source and receiver arrangement, as well as different methods for coupling to the borehole sidewall. Here we show a 6-pad tool to understand the working principle of the method. Figure 4.31 shows the unwrapped configuration of a segmented bond logging tool (Smolen 1996). There is one transducer (T) and one receiver (R) on each pad and the pads, are numbered 1 to 6. Six separate measurements are made at each depth with a 100 kHz source. Signals from the firing of each transducer are received by four receivers in the adjacent four pads. For example, the

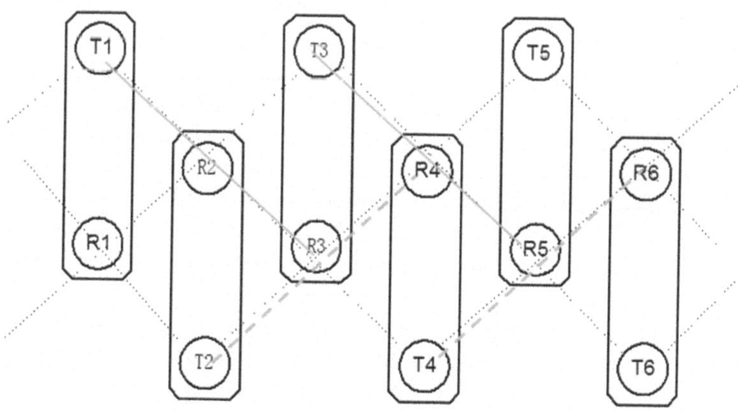

Fig. 4.31 A unwrapped configuration of a segmented bond logging tool. The interval along the propagation direction between two adjacent receivers is 15.24 cm

waveforms at receiver pairs R2 and R3, and R6 and R5 are recorded to calculate the attenuation between the two receivers (such as those between R2 and R3) when transducer 1 (T1) fires.

The pads are pressed against the casing. There are also some tools where only the transducer and receiver on the same pad are used for each measurement.

Because the 100 kHz ultrasonic frequency is used in the SBL, the excited mode is L(0, 2) or S_0. In field application, the envelope of the first 5 periods of the waveform is obtained to simulate 20 kHz waveforms to calculate the attenuation (Song et al. 2012). Thus, it is still a sonic method. From the incidence angle-frequency relationship of different modes (Fig. 4.12b), it may be possible to use relatively large incident angles to excite S_0 or L2 to L5.

In summary, a segmented log is an azimuthal CBL measurement that addresses the azimuth resolution problem. Other limitations of the CBL/VDL method still exist.

4.4.2 Ultrasonic Methods

The examples in the previous section showed that conventional acoustic logs can be used to determine whether the casing is well-cemented, very poorly cemented, or a free pipe. The channels and very thin fluid layer between cement and casing and cement and formation are difficult to detect. CBLs that operate at ultrasonic frequencies were developed for high-resolution cement bond evaluation. An ultrasonic tool, such as a pulse-echo device (see Fig. 4.32), overcomes some deficiencies of the conventional sonic logging tool. The ultrasonic methods were initially employed as caliper measurements and for wellbore imaging (Havira 1979) in uncased boreholes. Then, these methods were used for imaging of the material behind the casing (Hayman et al. 1991). The pulse-echo method utilizes the acoustic impedance contrasts at interfaces between the cement, casing, and fluid. Cement does not touch fluid unless there is a microannulus, which is either behind the casing or between the cement and formation. Measurements are affected by various factors, such as the cement and mud properties. For example, a low-density cement (density is less than 1.29 g/cm^3) (Ripley et al. 1981) reduces the impedance contrast between the cement and borehole fluid. Heavy mud used in deep water reduces the impedance contrast (Hayden et al. 2011). The low acoustic impedance contrast between the cement and borehole fluid affects the sensitivity of the pulse-echo method.

The pitch-catch method (van Kuijk et al. 2005; Herold et al. 2006; Morris et al. 2007, Froelich 2008), introduced to complement the pulse-echo, has transmitter and receivers as shown in Fig. 4.32. The waves propagating in the casing, cement, and annulus are recorded. The pulse generated by the inclined transmitter propagates along the casing and is reflected from the interfaces behind the casing. The waves are received by both receivers. By combining the pulse-echo and pitch-catch techniques, the reflection from the interfaces located behind the casing-cement interface can be identified. The combination of the pulse-echo and pitch-catch measurements provides the P-wave impedance and shear attenuation of the annulus material and

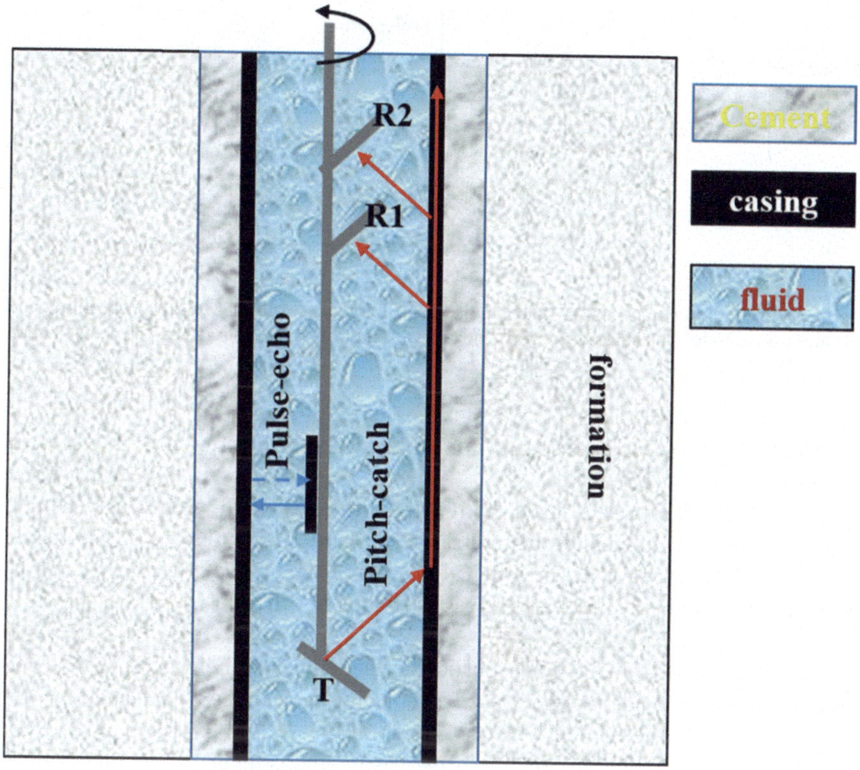

Fig. 4.32 Schematic diagram of pulse-echo and pitch-catch measurements. The pulse-echo is shown in the left part of the figure. A pulse from the transducer (a black rectangle next to the gray line in the borehole center) is fired. The echo from the borehole wall at normal incidence is recorded by the same transducer. The right part of the figure shows the pitch-catch measurement. Three inclined transducers are used in the pitch-catch measurement: one transmitter (T) and two receivers (R1 and R2). The angles of the transducers can be adjusted according to the specifics of the cased hole. More detail about the pitch-catch method can be found in Fig. 4.36 in Sect. 4.4.2.2

the reflections from various interfaces. These can be used to image and characterize the material behind the casing. Simultaneously, a rotating measurement provides azimuthal resolution and obtains images of the casing surface.

4.4.2.1 Pulse-Echo Method

Figure 4.33 shows the principle of the pulse-echo method. The received echo is the result of multiple transmissions and reflections on the inner and outer interfaces of the casing, and layers behind the casing. The time delay between different reflections from inside and outside the pipe is used to measure the pipe thickness. Amplitudes

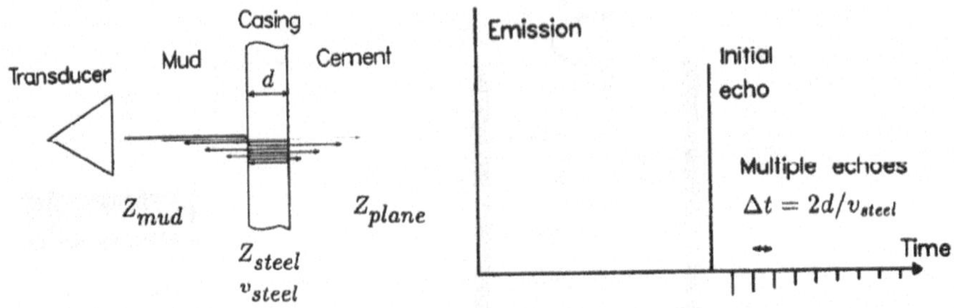

Fig. 4.33 Measurement principle of the pulse-echo technology. Left: Multiple reflections and transmissions. Right: The attenuation and time delay of multiple reflections (from Fig. 1 in Hayman et al. 1991)

of reflected echoes can be used to infer the impendence difference between the materials.

A synthetic example is used to illustrate the data and processing method (Fig. 4.34). The parameters of the model are similar to those in Fig. 4.7. The casing is perfectly bonded with ultralow-density cement. The source center frequency is 200 kHz. The total wavefield and waves attributed to different layers are shown in Fig. 4.34a.

The reflection I1 from the first acoustic interface (inner wall of casing) is modeled using a half space fluid-steel interface by replacing the cement and formation with the steel casing in the model, which is shown in Fig. 4.26. Partial reflection from the second acoustic interface (between the casing and cement), I2, is modeled by assuming an infinitely thick cement. I3 in Fig. 4.34a is obtained by subtracting I1 and I2 from the full waveform. The waveform includes the reflection from the third acoustic interface (between the cement and formation) and multiple reflections in the casing and cement layer. The different reflections can be observed in the full

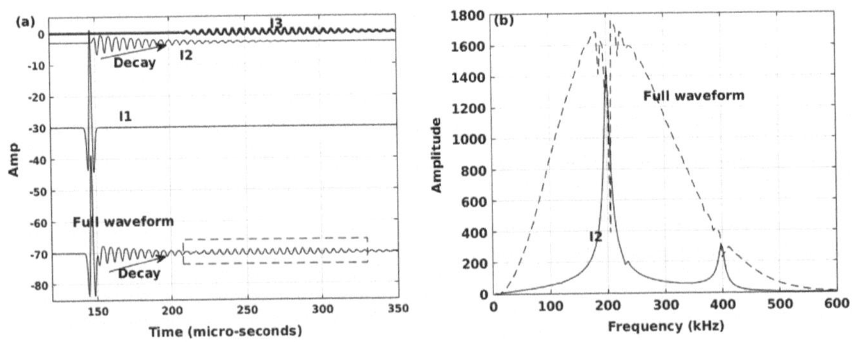

Fig. 4.34 **a** Full waveform and reflections from different acoustic interfaces in the pulse-echo technology. **b** Spectrum of waveforms. Labels I1, I2, and I3 are the reflections at the first (casing inner interface), second (casing-cement), and third (cement-formation) interfaces, respectively

waveform. The waveforms marked with arrows can be used for determination of the material impedance behind the casing.

It has been suggested (e.g. Zeroug and Froelich 2003; Sinha and Zeroug 1999; Miller and Stanka 1999) that the waveform in the dashed rectangle in Fig. 4.34a can be used for determining the bonding condition of the third acoustic interface (bonding interface *II*). The small amplitude of this wave train includes the reflection from the third acoustic interface and multiple reflections in the casing. The wave train is affected by the bonding condition of both the second and third acoustic interfaces. An elaborate deconvolution method may help identify the different echoes. It is reasonable to state that the ultrasonic pulse-echo faces challenges in the evaluation of bonding conditions at different interfaces.

The determination of casing thickness is useful for old wells or for determining corrosion. The frequency spectra of the full waveform and of the I2 reflector are shown in Fig. 4.34b. In a real measurement, the "notch" in the amplitude spectra may not be significant in the variable downhole environmental conditions because the amplitude spectra are highly dependent on transducer variations and borehole fluid attenuation. The group delay (GD) of the spectra is used: $GD = -d\varphi(\omega)/d\omega$, where $\varphi(\omega)$ is the phase spectra. The notch frequency position in the GD is used to determine the casing thickness. The depth and width of the notch are used to determine the annulus impedance (Hayman et al. 1991; Thierry et al. 2016) because the deep notch indicates a strong impedance contrast.

The inversion for the impedance of annulus material was initially based on a simple one dimensional (1D) three-layered model where the borehole fluid impedance (Zmud) is the input parameter and the casing thickness (d) and annulus material impedance (Zcmt) are the two output parameters. Plane wave propagation (only P-wave) is assumed in the model. One model used in industry is $B = k_2 Zcmt + k_3 Zmud$, where B is the width of the notch in the GD. k_2 and k_3 are constants (Kuijk et al. 2006). The inversion scheme is very sensitive to the accuracy of Zmud because the errors in the Zmud are amplified 5 times in the estimate of Zcmt (Thierry et al. 2016). Several methods are commonly used in industry, such as free-casing calibration and dedicated mud measurement cells.

A calibration in the free casing (no cement behind the casing) is a reliable method for estimating Zmud, but it is not easy to find a free casing section in actual measurement situations and the mud property may vary with depth and azimuth.

Figure 4.35 shows an example of the dedicated mud measurement, where two measurements are conducted with a pulse-echo tool. First, the transducer faces a built-in target steel plate to measure the in-situ mud wave velocity because the downhole temperature and pressure change the properties of the mud. Second, the reflections from the casing and material behind the casing are obtained. For the material impedance calculations, the properties of both the casing and mud are used. Azimuthal imaging is archived by rotating the transducer 360° inside the casing. Moving the tool along the borehole axis provides the depth coverage.

Wave propagation in the pulse-echo method can be modeled by a simple layered model due to the high frequency. The multiple reflections complicate the waveform.

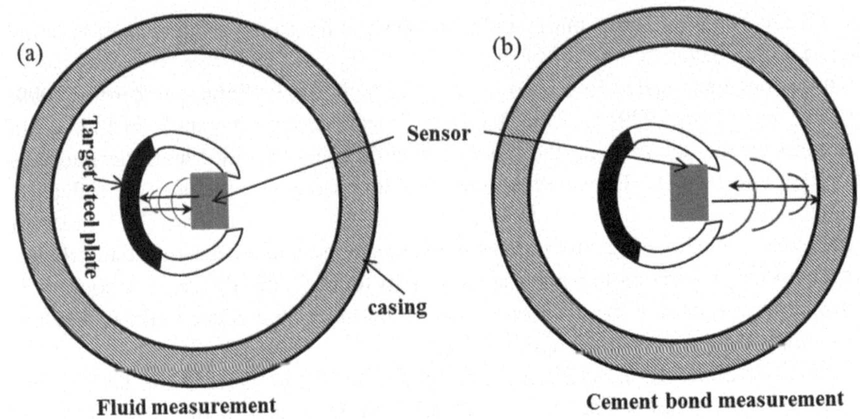

Fig. 4.35 Two measurement modes of pulse-echo technology (modified from Hayman et al. 1991). **a** Down hole fluid wave velocity measurement. **b** Cement bond measurement

Using the first pulse as the source wavelet, other reflections can be identified by deconvolution.

Challenges for the application of pulse-echo measurements originate from the difficult downhole environment caused by the necessity for safety and environmental protection because the cement bond is evaluated after well construction in addition to over the entire life of the well. Additionally, in heavy mud, a thick, large diameter casing is common. In these cases, lower source frequencies are used and the curvature of the casing cannot be ignored. Simultaneously, casing eccentering should also be considered in the inversion. An advanced full waveform inversion method based on a 3D model has been proposed (Zeroug et al. 2016). In the inversion method, the synthetic full waveform from 3D modeling is used to match the recorded waveforms, where both P-and S-waves are included, and Zmud, d, and Zcmt are the output parameters. To improve the efficiency of the pulse-echo method in the heavy mud environment, an advanced ultrasonic transducer needs to be developed that has improved sensitivity and a broader frequency bandwidth. For example, medical ultrasound transducers may be reengineered for downhole measurements (Thierry et al. 2016, 2017).

4.4.2.2 Pitch-Catch Method

The pitch-catch method (see schematics in Fig. 4.32) employs mode A_0 (see Fig. 4.12a) to obtain the shear wave attenuation of the material (cement or others) behind the casing and the reflection from the cement-casing interface (Zeroug and Froelich 2003).

The effective excitation of mode A_0 and suppression of other modes can be archived by choosing the source frequency and incidence angle. Knowledge of the group velocity dispersion can help in choosing the source frequency. The group velocity dispersion curve of A_0 has a very slight dispersion when the frequency is higher than a certain value (f_g) (Fig. 4.36). The f_g is a function of pipe thickness and increases with decreasing thickness. It is approximately 100 kHz for a 14-mm-thick pipe and approximately 200 kHz for a 5-mm-thick pipe. Once the source frequency is selected, the incidence angle can be determined from the relationship between the frequency and incidence angle (shown in Fig. 4.12b) to effectively excite a pure A_0 mode.

To demonstrate the interaction of a wavefield with the casing, borehole fluid, cement, and formation in the pitch-catch method, the simulated results are shown in Fig. 4.37. The P and S velocities of steel pipe are 5860 and 3130 m/s, respectively. The density is 7850 kg/m^3. The Vp, Vs, and density of ultralow-density cement are 1600 m/s, 1000 m/s, and 1100 kg/m^3, respectively. The outer radii of borehole fluid, casing, and cement are 100 mm, 105 mm, and 120 mm, respectively. The source frequency and incidence angle are 240 kHz and 38°, respectively.

There are two modes in the casing: S_0, the fastest mode visible from z = 85 to 90 cm and A_0 following S_0. The difference between those two modes can be seen in the snapshot. There is only one color at a given offset indicating symmetrical motion about the casing (white and black colors are positive and negative pressures, respectively) for S_0, and there are two colors indicating antisymmetrical motion about the casing for the A_0 mode. The leaky A_0 is clear on both sides of the casing. The leaky A_0 in the borehole fluid is the first arrival received by the receivers.

The A_0 in the annulus (between the casing and formation) is reflected from the third interface and transmitted into the casing and then leaks into the borehole fluid as the secondary leaky mode, marked as TIE (third interface echo) in Fig. 4.37. The energy in the annulus is different in the two models. There is more leakage into the borehole when the casing is perfectly cemented.

The waveforms at two receivers, 30 and 36 cm from the source, are shown in Fig. 4.38. The first distinct wave (approximately 140 ms) is the primary leaky A_0 mode from the casing. The second wave is the reflected wave from the third acoustic

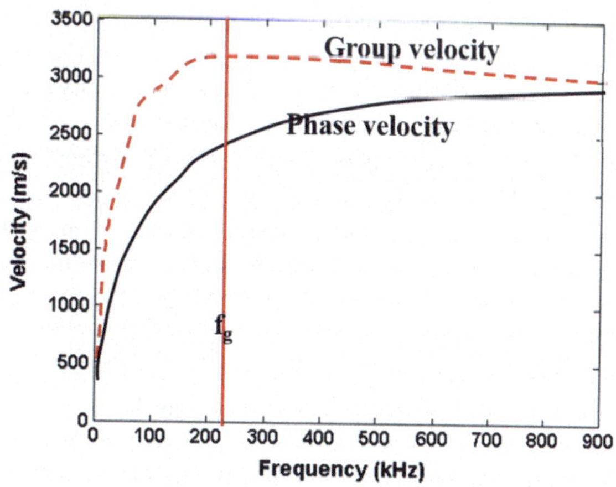

Fig. 4.36 Dispersion curves of mode A_0 in a 5-mm-thick pipe immersed in water determined from Eq. 4.1b and shown in Fig. 4.12a

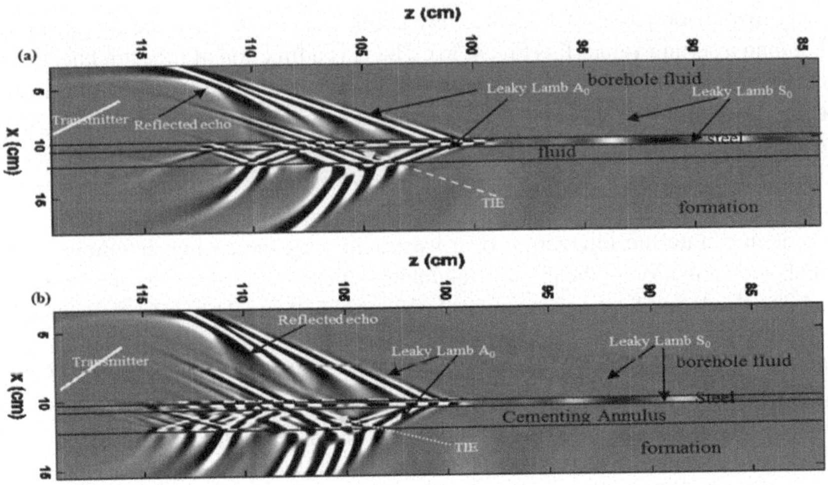

Fig. 4.37 Pressure wavefield snapshots at 75 us for two models with the pitch-catch method (Fig. 11 in Wang et al. [2016a]). An illustration of the pitch-catch method is shown in Fig. 4.32. **a** Cement fully replaced with fluid (free casing) model. **b** Cased and cemented (ultralow density) model. The source frequency is 240 kHz, and the incidence angle is 38°. Layered plain models are used to approximate the cased hole models at ultrasonic frequencies

interface (TIE). The amplitude of the TIE is relatively large due to constructively interfering reflections at the third acoustic interface (cement-formation interface). The shape and amplitude of the TIE depend on the cementing material. The TIE in the slow cement case (Fig. 4.38b) is larger and less dispersive than that in the fluid annulus case (Fig. 4.38a). Other factors such as attenuation, roughness of the third interface, and casing eccenterings, also affect the TIE amplitude.

Because a high frequency is utilized in the pitch-catch method, the ray theory can be used for interpreting the wavefield. The guided A_0 Lamb mode in the casing leaks energy into both sides of the casing. The direct leakage into the borehole fluid provides the first arrival received by the receivers in the borehole fluid (Primary A_0 in Fig. 4.39). The leakage of A_0 in the annulus (between the casing and formation) depends on the contrast between the group velocity (V_A) of the A_0 mode at the source frequency and the cement P-wave velocity (V_{pc}). If V_{pc} is larger than V_A, with the exception of leaky waves, there are no P-related waves (P, P-P, and P-S) in the annulus. Additionally, the S-P waves are not transmitted into the casing. In this case, the cement is defined as fast cement and other cements (with V_{pc} below V_A) are slow cement (van Kuijk et al. 2005). In slow cement, the waves in the annulus include P and S related modes (P, S, P-P, P-S, S-P, and S-S). If the cement is fully replaced with fluid, there is only the fluid P-wave in the annulus. The transmission angle is less than that of the S-wave in Fig. 4.39. Part or all of the reflected waves (P-P, P-S, S-P, and S-S) from the third acoustic interface (interface next to the formation) are transmitted into the casing as Lamb waves. They leak into the borehole as the secondary leaky modes (marked as TIE in Fig. 4.39). The primary A_0 mode in the borehole (marked

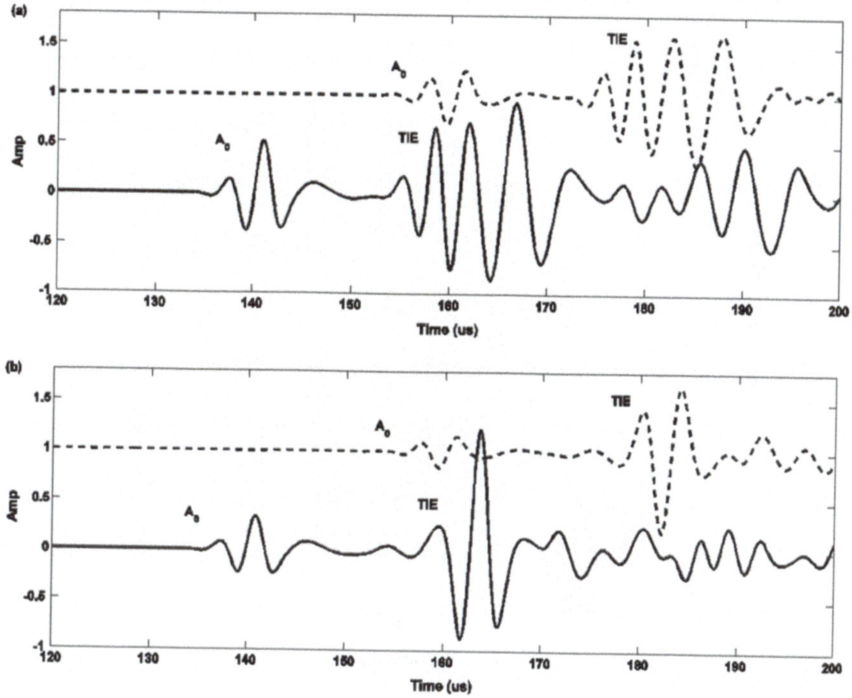

Fig. 4.38 Synthetic waveforms of a pitch-catch measurement (Fig. 12 in Wang et al. [2016a]). **a** Free casing model. **b** Cased and cemented model. Solid and dashed lines are the waveforms at the near (30 cm source-receiver distance) and far (36 cm source-receiver distance) receivers

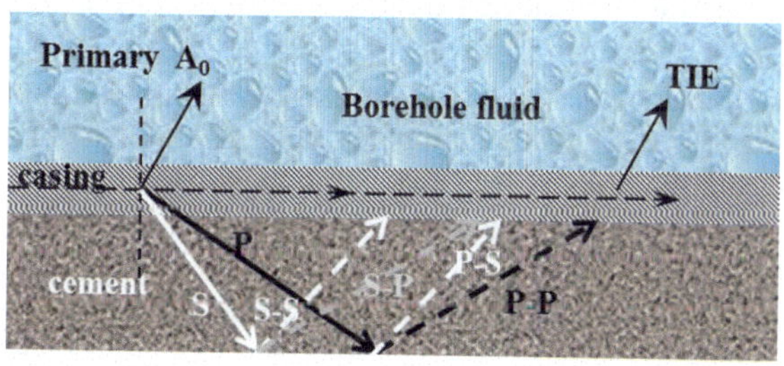

Fig. 4.39 Ray path of the wave in a perfectly cemented cased-hole using the pitch-catch method

as A_0 in Fig. 4.37) is only affected by the material in the annulus. The attenuation of the primary A_0 mode is related to the shear attenuation of the material in the annulus. A cross-plot (see the example in Fig. 4.40) between the shear attenuation and P-wave impedance (from the pulse-echo method) can be utilized to distinguish low-density cement from fluid to avoid ambiguity from using a single property.

Shear attenuation can be defined by $\alpha = 20\log_{10}(P_1/P_2)/\Delta L$ (dB/cm), where P_1 and P_2 are the amplitudes of the waves at two receivers separated by an interval ΔL.

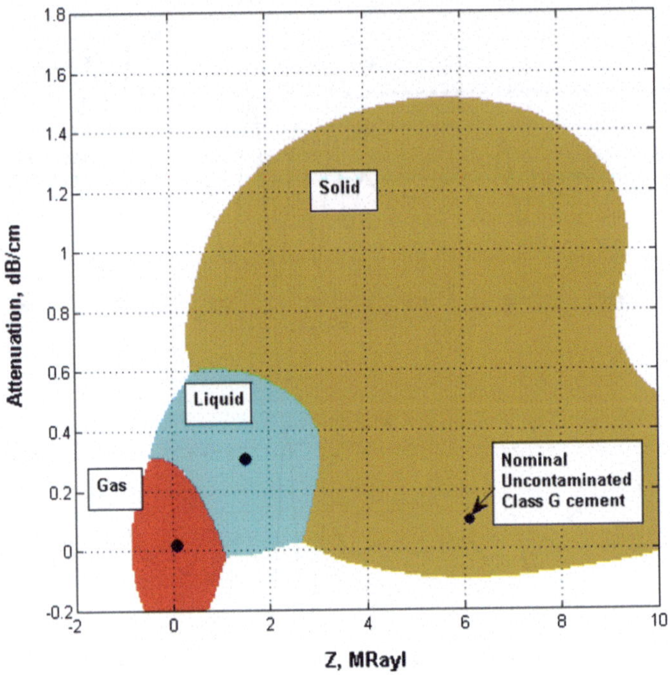

Fig. 4.40 An example of Solid-Liquid-Gas (SLG) mapping using combined pulse-echo and pitch-catch measurements (Fig. 8 in Van Kuijk et al. 2005). Z is the cement impedance from the pulse-echo measurement and attenuation is obtained from the pitch-catch measurement

The shear attenuation from the waveforms in Fig. 4.38 are 0.864 and 1.165 dB/cm for the free casing and for the ultralow-density cement, respectively.

In Sect. 4.4.2.1, the influence of casing eccentering on impedance determination was stated. The attenuation determination is also affected by the casing eccentering where the primary A0 mode is contaminated by the S0 mode or others. Efforts are devoted to extract the clear A0 mode used to calculate attenuation (Le Calvez and Brill 2018).

4.4.2.3 Data Examples

Ultrasonic measurements were conducted in a full-scale experimental well shown in Fig. 4.41. In the experiment, the inner radii of the casing, cement, and formation are 88.9, 99.3, and 113 mm, respectively. The depth of the well is more than 14 m. There are 6 segments in the model. Each 1.5-m-thick segment has different cementation conditions. Segment 1: the cement is eccentered by 12 mm as shown in Fig. 4.41. Segments 2 to 6 are the centralized models with a 3-mm fluid channel at the interface between the casing and cement. The channeling angles (definition see θ in Fig. 4.14)

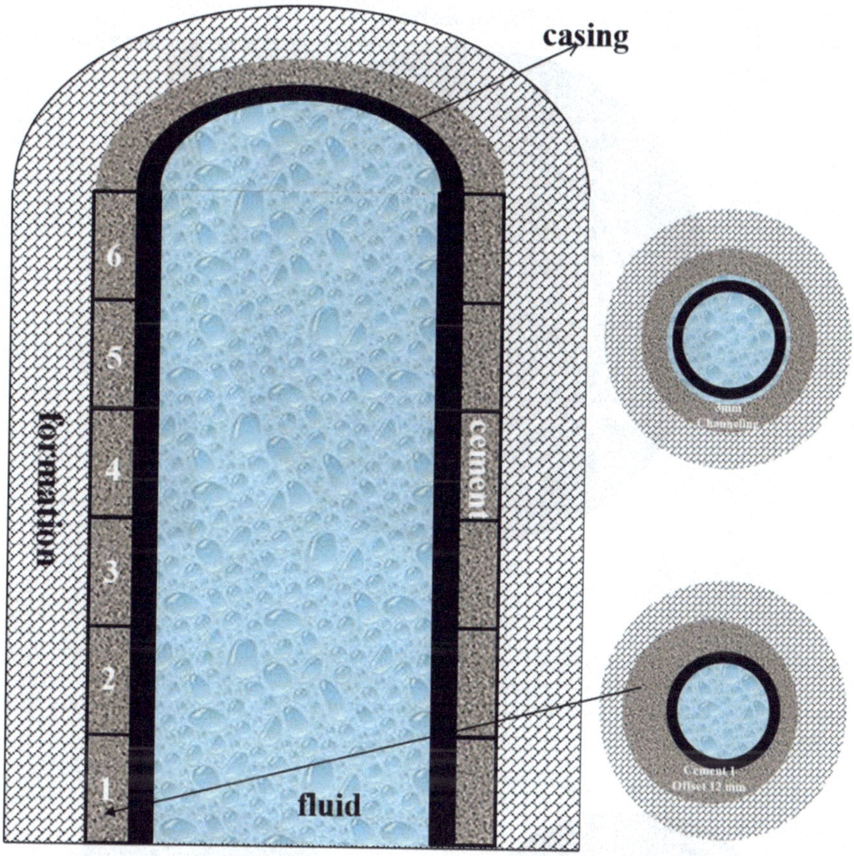

Fig. 4.41 Full-scale experimental well. The cements are separated into 6 segments each having a height of 1.5 m. Numbers identify different segments with different cementing conditions. Segment 1 has cement with eccentering of 12 mm to model an eccentered casing. Segments 2 to 6 have centralized casing with a 3-mm-thick fluid channel at the interface between the casing and cement. The channeling angles (definition see θ in Fig. 4.14) for cements 2 to 6 are 30, 60, 90, 180, and 360°

for cements 2 to 6 are 30, 60, 90, 180, and 360°. The P-wave velocity and density of the formation are 3900 m/s and 2100 kg/m^3, respectively. The cement density is 1200 kg/m^3. A standard industry 7-in. casing is used in the model. The P-wave velocity and density of the borehole fluid are 1460 m/s and 1150 kg/m^3, respectively.

In the experiments, the pulse-echo and pitch-catch measurements were conducted together on a single tool. The configuration of the tool is similar to the schematic diagram shown in Fig. 4.32. The distances from the source transmitter of the pitch-catch to the pulse-echo transmitter, near and far receivers are 140, 250, and 350 mm, respectively.

Figures 4.42, 4.43, 4.44 and 4.45 show the data collected in segment 1 (eccentered cement case), where the distance from the measurement location to the bottom of the

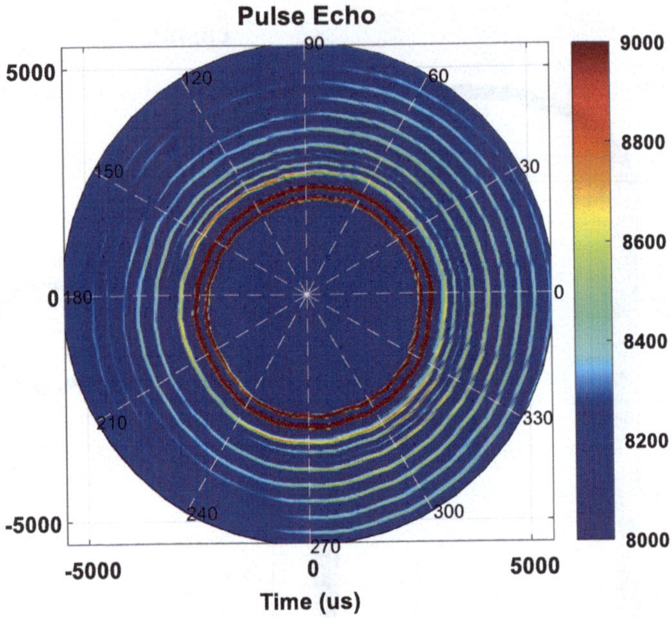

Fig. 4.42 Polar plot of the pulse-echo measurement in segment 1, 0.847 m from the bottom of the well in Fig. 4.41. The radius of the polar plot is time (unit, us). The azimuthal direction is the azimuthal angle. Colors show the amplitudes of the waveforms. The horizontal and vertical axes are time (unit: us)

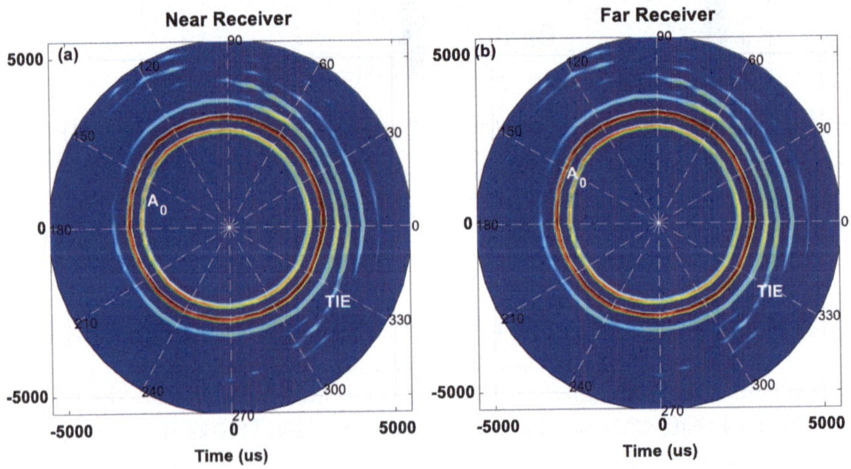

Fig. 4.43 Polar plots of the pitch-catch measurements in segment 1 of Fig. 4.41. **a** Near receiver. **b** Far receiver. The measurement was conducted in segment 1, 0.847 m from the well bottom. The radius of the polar plot is time (unit, us). The azimuthal direction is the azimuthal angle. Colors show the amplitudes of the waveforms

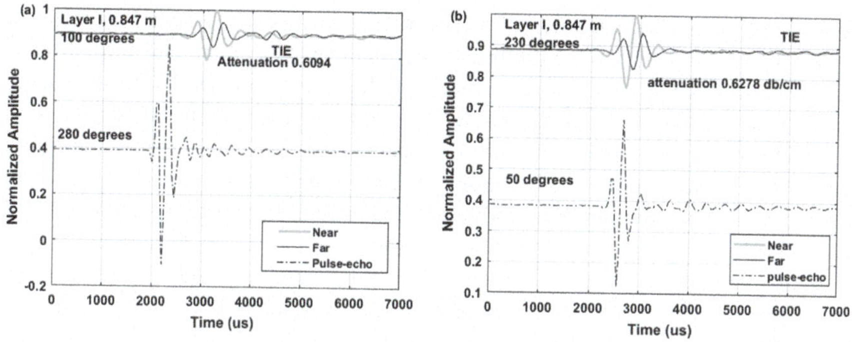

Fig. 4.44 Pitch-catch and pulse-echo waveforms at selected azimuthal angles in eccentered cement section (segment 1 in Fig. 4.41)

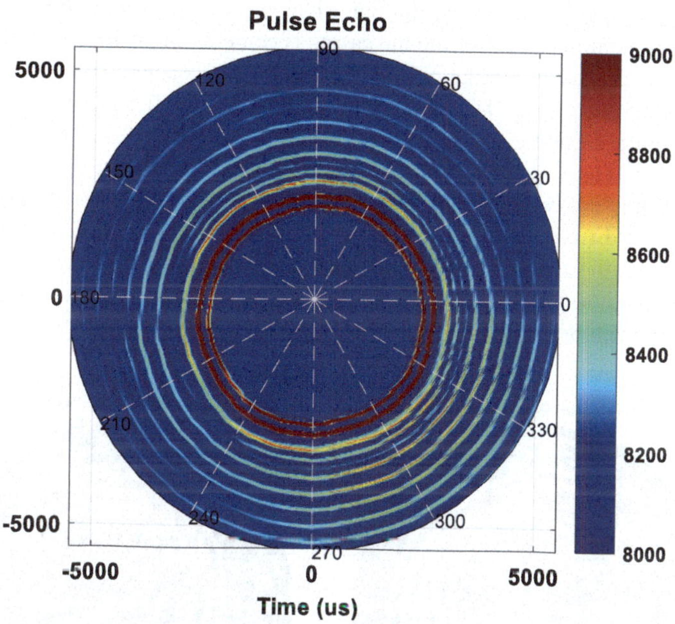

Fig. 4.45 Polar plot of the pulse-echo measurements at segment 5 in Fig. 4.41. The radius of the polar plot is time (unit, us). Colors show the amplitudes of the waveforms. The tool is eccentered in the casing but the casing is centered in the borehole

well is 0.847 m. The measurement was conducted at all azimuthal angles. The data are shown as a polar display. The radius is time (0 to 8000 us), and the polar angle is the azimuthal direction. The amplitudes of the waveforms are shown with different colors. Azimuth 0° is the direction where the casing is closest to the formation.

Figure 4.42 shows the pulse-echo measurement result. There is a strong event at approximately 2000 us, corresponding to the initial echo from the casing inner wall.

Subsequently, there are a number of weak circles that are echoes from the outer wall of the casing (interface I) and the interfaces between the cement and formation as well as multiples in the casing and cement. Echoes are attenuated with increasing time. By measuring the attenuation, the properties of the cement bonding can be determined (see additional explanation in Sect. 4.4.2). We can see from the figure that the tool is not centralized because of the asymmetry in the arrival in different echoes, where the azimuth 300° arrivals are later than the those at azimuth 120°. Waveforms at azimuths of 280° and 50° are shown in Fig. 4.44a, b, respectively.

Figure 4.43 shows the measurements of the pitch-catch method. These measurements are different from the waveforms obtained from the pulse-echo method. The first event (circle approximately 2500 us) is the A0 mode shown in Fig. 4.39. The reflection from the cement-formation interface is at approximately 4500 us. The waveforms at azimuths of 100 and 230° are shown in Fig. 4.44a, b, respectively.

Figures 4.45, 4.46, and 4.47 show the data collected from segment 5, where one-half (180°) is well-cemented but the other half has a 3-mm-thick fluid channel at the casing-cement interface. While the casing is centered in the borehole, the tool is eccentered in the casing. Similar to Fig. 4.42, the tool is eccentered because the initial echoes, approximately 2000 us, are not azimuthally symmetric. The TIE in Fig. 4.46 (marked with a dashed circle) is clearer than that in Fig. 4.43 because the casing is centralized. The interface between the cement and formation can be observed although it may not be clear at some azimuths (from 90 to 180° in Fig. 4.45). The changing time difference at different azimuths between the A_0 and TIE illustrates the changes in the third reflection interface due to the 180° fluid channel behind the casing outer wall. The waveforms from the pulse-echo near and far receivers of the pitch-catch are shown in Fig. 4.47 for the same azimuths as those shown in

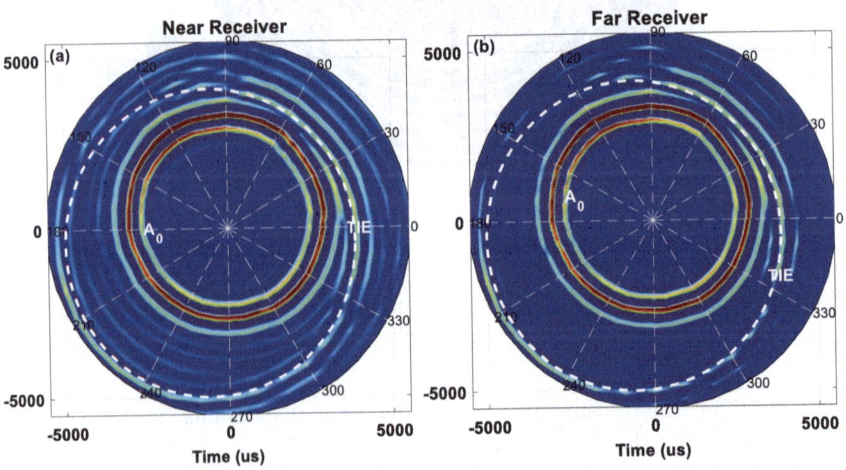

Fig. 4.46 Polar plots for the pitch-catch measurements. **a** Near receiver. **b** Far receiver. The measurement was made at segment 5. The radius of the polar plot is time (unit, us). Colors indicate the amplitudes of the waveforms

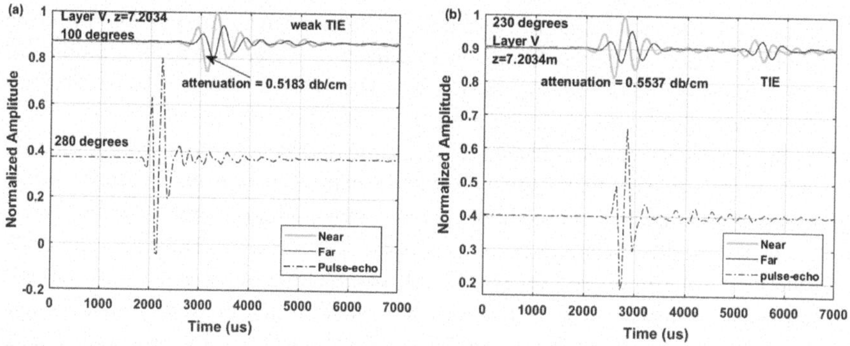

Fig. 4.47 Waveform at two azimuthal angles in segment 5 in Fig. 4.41

Fig. 4.44. The initial echoes in the pulse-echo are different at the two azimuths due to the different cement conditions. In the data collected by the pitch-catch method, the TIE amplitude is obvious at the azimuth angle of 230°, as shown in Fig. 4.47b. However, it is not clear at 100° in Fig. 4.47a.

As presented in Sect. 4.4.2.2, the A0 mode at the near and far receivers can be used to determine the shear attenuation and characterize the material behind the casing. The attenuations obtained from the waveforms shown in Fig. 4.44 are 0.6094 and 0.6278 dB/cm for azimuth angles of 100° and 230°, respectively. The attenuations are 0.5183 and 0.5537 dB/cm for azimuth angles of 100° and 230°, respectively, in Fig. 4.47. The P-wave impedance can be obtained from the pulse-echo measurement, and then, the material behind the casing can be determined.

4.5 Summary

In this chapter, wavefields in a cased hole have been discussed. Starting from the case of a pipe immersed in fluid, coverage extends to the free casing, well-bonded casing, and a series of poorly bonded cases. The cement bond evaluations using both sonic and ultrasonic waves (pulse-echo and pitch-catch, respectively) were discussed. A summary of the results is as follows:

(1) In monopole logging, there are casing modes, and P, S, pR, and ST waves in the wavefields. The formation properties can be determined when casing is well bonded. However, when the casing is not cemented, the P-wave becomes submerged in the casing modes and the S-wave loses coherence. An additional ST wave appears due to the fluid channel in the annulus. The formation velocity determinations in poorly bonded cased holes are difficult and require extensive modeling for the interpretation of waveforms.

(2) A dipole source induces a P mode (present only at high frequencies), flexural mode, and some casing modes. Casing modes disappear if the casing is perfectly

cemented. Data from a high-frequency dipole tool can be used to determine the P- and S- wave velocities. If there is no cement (i.e., free pipe), the amplitude of the refracted P-wave becomes very small. The S-wave velocity can be obtained from the flexural-wave velocity at the low frequency limit.

(3) In the quadrupole wavefields, there are casing quadrupole modes and formation quadruple modes, which can be used to determine the S-wave velocity.

(4) Cement bond evaluation methods include sonic and ultrasonic methods. Sonic methods provide a general evaluation of cement integrity. Their azimuth resolutions can be improved by the use of azimuthal sources and receivers. The ultrasonic methods (e.g., pulse-echo and pitch-catch) provide the means for high-resolution cement bond evaluation. These methods are slow and require extensive effort for data interpretation.

Chapter 5
Acoustic Logging-While-Drilling

Acoustic logging-while-drilling (ALWD) is an advanced technology developed to determine the formation properties in real-time while drilling. Its advantages include obtaining formation properties while conserving drill-rig time, true formation information without invasion, acquisition of information for drilling safety such as over-pressured formations, and for geo-steering.

A multitude of technical problems need to be solved for successful application of ALWD. These include the presence of the drill collar, harsh environment, data transmission and noise associated with drilling.

In the early years, ALWD measurements followed the idea of the wireline sonic time delay measurement (Aron et al. 1994) and recorded the times of the first arrival. Transducers were installed in the collar approximately 12 m above the drill bit (Minear et al. 1995). The schematics for four different tools, shown in Fig. 5.1, illustrate the evolution of the ALWD measurements. From left to right, the tools correspond to interval transit time, borehole compensation, monopole/dipole, and multipole measurements. Following the developments in wireline logging, ALWD introduced full waveform data acquisition, azimuthal coverage, and advanced transducers (Varsamis et al. 1999; Joyce et al. 2001). Now, ALWD tools include monopole, dipole, and quadrupole sources and receivers (Leggett et al. 2001; Tang et al. 2002a).

In a multipole tool, the source may consist of several groups of transducers to obtain different radiation patterns. Each group consists of several (usually 8) PZTs (Piezoelectric transducers). Different source modes can be realized by controlling the excitation of the transducers. Excitation of multiple sources with the same phase generates a monopole source. The dipole source can be generated by positive excitation on one half of the total PZTs and negative phases on others. For the quadrupole source, the PZTs are divided into 4 parts. The amplitude of excitation and the source pattern can be controlled to generate different source patterns. The source frequency ranges from 2 to 20 kHz. The source-receiver distance usually is half of that of a wireline tool due to the length limitation of the collar, approximately 10 ft (1 ft = 0.3048 m), and the receiver interval is between 1 and 2 ft.

H. Wang et al., *Borehole Acoustic Logging—Theory and Methods*, Petroleum Engineering, https://doi.org/10.1007/978-3-030-51423-5_5

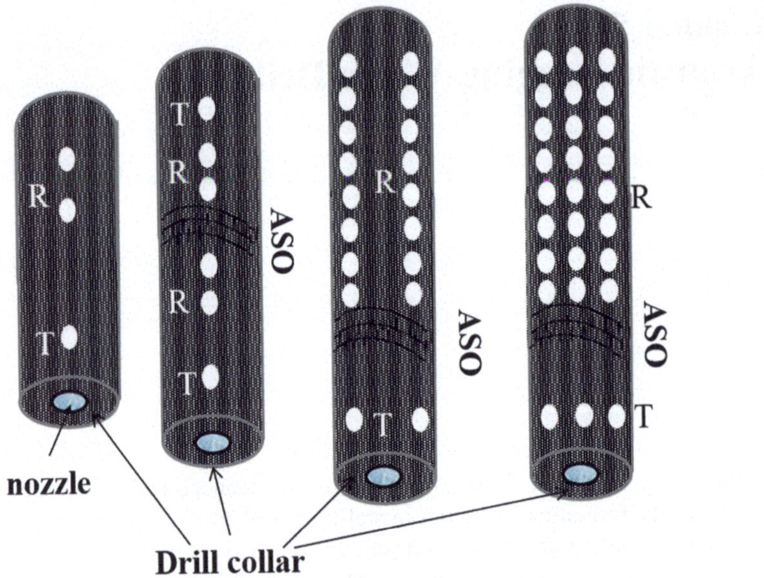

Fig. 5.1 Configurations for different tools (T, R and ASO are source, receivers and isolation, respectively)

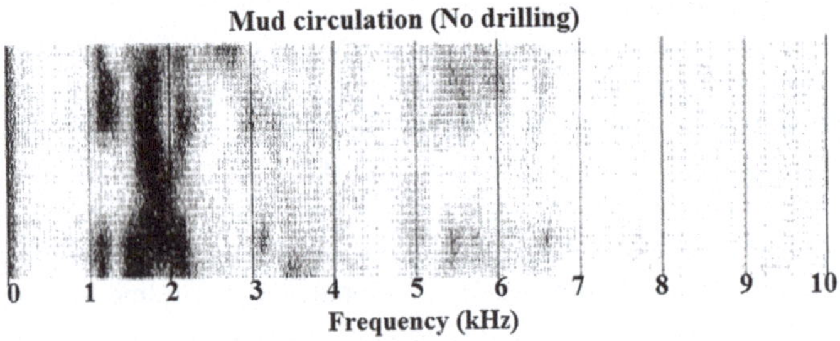

Fig. 5.2 Spectrum of mud circulation signal recorded by an accelerometer when the drill bit is not moving (Fig. 6 in Joyce et al. 2001). The vertical axis is time. The darker area indicates the larger amplitude from the VDL display. The frequency of noise ranges from 1 to 6.5 kHz and the noise is strongest between 1 and 2.3 kHz. The maximum is at 1.7 kHz

Drilling noise has a major effect on ALWD. The noise comes from mechanical vibrations and fluid movements. Interaction of drill pipe, drill collar and the drill bit with the formation, and mud circulation contribute to the noise. Examples of noise spectra due to fluid circulation and drilling are shown in Figs. 5.2 and 5.3, respectively (Joyce et al. 2001). Drilling noise has higher frequency than that of fluid circulation. Fortunately, the noise decreases above 4 kHz.

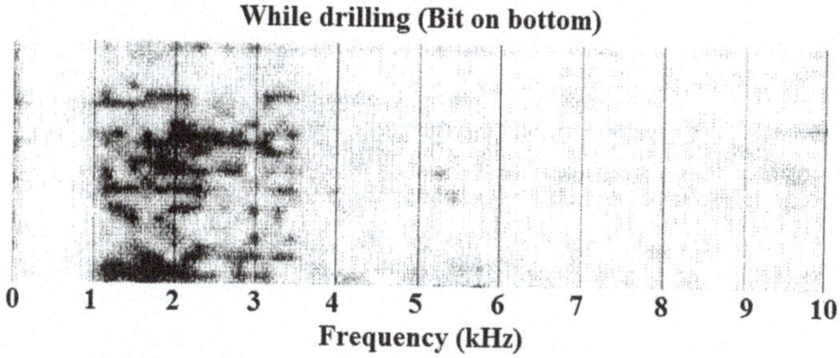

Fig. 5.3 Spectrum of drilling noise recorded by an accelerometer during drilling (Fig. 6 in Joyce, et al. 2001). The vertical axis is time. The darker area indicates the larger amplitude from the VDL display. A 6.75-in. drill pipe accompanied with a polycrystalline diamond compact (PDC) drill bit is in an $8^{5/8}$-in. borehole. The surrounding formation is a fast formation with a P-wave velocity of approximately 4000 m/s. Drilling noise ranges from 1 to 3.5 kHz. Signals below 1 kHz have been removed

5.1 Wave Modes in the Borehole with a Cylindrical Collar

The ALWD system consists of the drill collar, the fluid-filled borehole, and acoustic sources and receivers, embedded on the drill collar. The influence of the massive drill collar, inside the borehole, has a significant impact on the wavefields.

As shown in Fig. 5.4, there are four zones in the ALWD model cross section: formation, outer fluid annulus, collar (or drill string), and inner fluid column. The sources and azimuthal array receivers are embedded on the outer surface of the drill collar. The source can be monopole, dipole, or quadrupole, as shown in Fig. 5.4a–c.

The expressions of the multipole source in the frequency-wavenumber domain are,

Monopole:

$$\sigma^s(k_z, \omega) = p_0 F(\omega) U(k_z), \tag{5.1a}$$

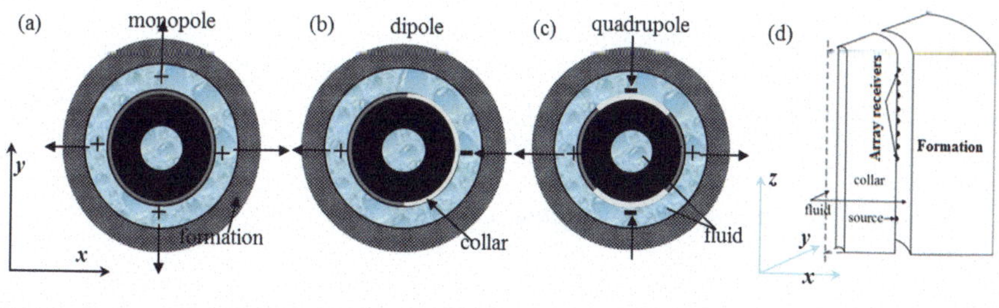

Fig. 5.4 Schematic diagram of ALWD model cross section. **a–c** Top-down view of the model showing multipole ALWD sources and **d** side view of the model. Directions x, y, and z are marked in the figure. The origin is the center of the innermost circle in **a–c**. The gray and white portions of the ring source embedded on the outer surface of the drill collar are the positive and negative phases of the source, respectively

Dipole:

$$\sigma^s(k_z,\omega,\theta) = p_0 F(\omega) U(k_z) \frac{2}{\pi} \sum_{m=0}^{\infty} \frac{\sin(2m+1)\theta_0}{2m+1} \cos(2m+1)\theta, 0 \le \theta_0 \le \pi/2, \tag{5.1b}$$

Quadrupole:

$$\sigma^s(k_z,\omega,\theta) = p_0 F(\omega) U(k_z) \frac{2}{\pi} \sum_{m=0}^{\infty} \frac{\sin 2(2m+1)\theta_0}{2(2m+1)} \cos 2(2m+1)\theta, 0 \le \theta_0 \le \pi/4, \tag{5.1c}$$

where σ^s is source in the frequency (ω)-wavenumber (k_z) domain. θ is azimuth angle of the source. p_0 is source amplitude. $F(\omega)$ is the Fourier transform of the source time history as described in Sect. 2.1.3.3. We use a Ricker wavelet. $U(k_z)$ is a distribution function of a source with a height of h. $U(k_z)$ can be obtained from Eq. 2.27. θ_0 is the half distribution angle of the transducers. m is a number to specify the mode.

With the boundary condition between fluid and solid, described in Sect. 2.1.3, the characteristic equations for ALWD are,

$$[m_{ij}]_{12\times12} \cdot [a_i]_{12\times1} = [b_i]_{12\times1}, \tag{5.2}$$

where the elements of a_i are $[a_i] = [A^i_n, A_n, B_n, C_n, D_n, E_n, F_n, A^o_n, B^o_n, B^t_n, D^t_n, F^t_n]$. Superscripts i and o of the coefficients denote the inner fluid and outer fluid of the collar, and t denotes the formation related elastic terms. $[b_i]_{12\times1} = [0, 0, 0, 0, 0, \sigma^s_n, 0, 0, 0, 0, 0, 0]$ designates the source radiation. The elements of $[m_{ij}]_{12\times12}$ are given in Appendix A. By taking the determinant of $[m_{ij}]_{12\times12}$ as zero, the analytical dispersion curves are obtained and used in this chapter.

The displacement (5.3a) and stress (5.3b) potentials in the outer fluid column are (Wang and Tao 2011),

$$u^o_f(r,\theta,k,\omega) = \sum_{n=0}^{\infty} A^o_n[\frac{n}{r} I_n(fr) + f I_{n+1}(fr)] + B^o_n[\frac{n}{r} K_n(fr) - f K_{n+1}(fr)] \cos n\theta, \tag{5.3a}$$

$$P^o_f(r,\theta,k,\omega) = \sum_{n=0}^{\infty} \rho_f \omega^2 [A^o_n I_n(fr) + B^o_n K_n(fr)] \cos n\theta. \tag{5.3b}$$

The wavefield in the space-time domain is (Tang and Cheng 2004),

$$\varphi^o_f(r,\theta,z,t) = \frac{1}{2\pi} \int_{-\infty}^{\infty} \int_{-\infty}^{\infty} \varphi^o_f(r,\theta,k,\omega) S(\omega) e^{-j\omega t} U(k) dk. \tag{5.4}$$

For acoustic logging-while-drilling (ALWD) modeling, two fast formations F1 and F2, and two slow formations S1 and S2 shown in Table 5.1 are used. The

Table 5.1 Formation classification in wireline and LWD cases. V_p and V_s are the velocities of the P- and S-waves, respectively; V_f is the P-wave velocity of the borehole fluid; V_{fl} is collar flexural-wave velocity at the source frequency range

Wireline formation classification	LWD formation classification
Fast formations ($V_f < V_s$)	(1) Fast-fast formation (F1) ($V_f < V_s$ and $V_{fl} \ll V_p$)
	(2) Slow-fast formation (F2) ($V_f < V_s$ and $V_p \approx$ or $< V_{fl}$)
Slow formations ($V_s < V_f < V_p$)	(3) Conventional slow formation (S1) ($V_s < V_f < V_p$)
	(4) Very slow formation (S2) ($V_p \approx$ or $< V_f$, $V_p/V_s > 2$)

Table 5.2 Parameters for the LWD models. V_p and V_s are the velocities of the P- and S-waves; OR: outer radius, IF: inner fluid, OF: outer fluid, D: density; C12, C22, C32 are various collars used for modeling

	V_p (m/s)	V_s (m/s)	D (g/cm^3)	OR (mm)
IF	1500	–	1.00	27
OF	1500	–	1.00	117
C12	10,000	6000	7.85	90
C22	8000	5000	7.85	90
C32	5860	3300	7.85	90
Formation F1	4500	2650	2.40	
Formation F2	3000	1800	2.00	
Formation S1	2300	1000	2.00	
Formation S2	2000	500	2.00	

slow formations are also classified into slow (S1) and very slow (S2) formations, as described in Sect. 2.3. Formations S1 and S2 in Table 5.2 are the slow and very slow formations, respectively. The formation properties are the same as those used in Chap. 2.

5.1.1 Wavefields in a Fast Formation

The ALWD wavefield is similar to that in the uncemented cased hole that was covered in Chap. 4. Waves in ALWD are the longitudinal (L), flexural (F), and quadrupole (Q) modes for the monopole, dipole, and quadrupole sources, respectively. In addition, there are two Stoneley (ST) waves.

5.1.1.1 Monopole Source

The dispersion curves for ALWD with a monopole source are shown in Fig. 5.5. These are different from those in the cased hole (Fig. 4.18c) due to the presence of a thick collar and small inner radius of the collar. In the monopole case, the L modes move to a higher frequency range than those in the cased hole. Thus, there are only L1 and L2 modes in the 0–25 kHz frequency range. In Fig. 5.5, the dashed curves are for a collar immersed in an infinite fluid, and solid curves for a borehole model with surrounding formation F1. Red curves show modes that are present only when there is a formation. The dispersion curves of collar (L1 and L2 are shown in Fig. 5.5) and inner ST modes are independent of formation properties because they are the same for the two models with formation and formation replaced by fluid. This indicates the velocity of L and inner ST modes are not affected by the formation within the frequency range of interest. However, the excitation amplitudes of the L and ST modes change with formation velocity and with frequency.

In the complex contour or branch-cut integration method (see Sect. 2.1.3.1), the waveform can be calculated using the residue of different poles in Eq. 2.17. Those poles are modes in the waveforms. The spectral amplitude of each mode is called the wave mode excitation function. Figure 5.6 shows the excitation function (excitation amplitude with frequency) of collar L1 and L2 modes for formations given in Table 5.2. There are distinct differences between the excitation curves within an infinite fluid (no formation in the model) and models with various formations. If there is no formation in the model, the amplitude of the collar L1 (dashed curves in

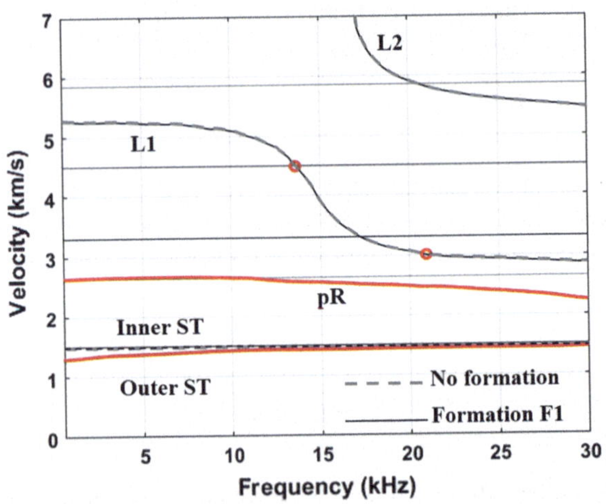

Fig. 5.5 Dispersion curves of the monopole LWD measurements. The collar is C32 and the formation is F1 (Table 5.2). Dashed curves are for dispersion when the collar is immersed in infinite fluid (without a formation). Solid curves are for collar in a borehole embedded in a formation. The red curves show modes (pR and Outer ST) that are present only when there is a formation. Red dots are critical frequencies for formation F1 and F2 (see text)

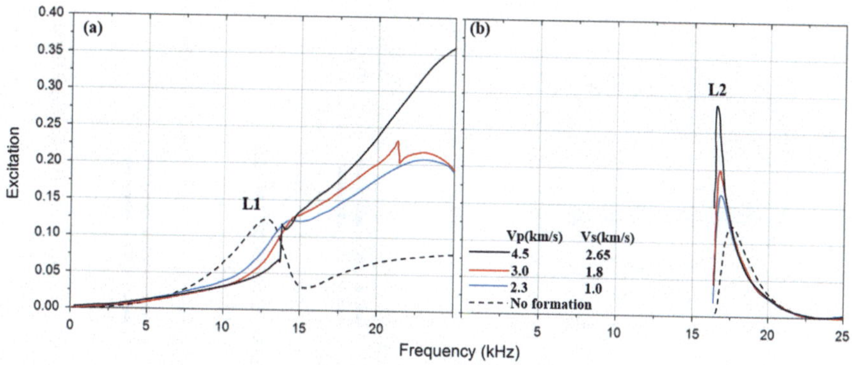

Fig. 5.6 Excitation function of L modes in the ALWD model in various formations. Collar is C32. Formation velocities for different curves are listed in **b** and Table 5.2

Fig. 5.6) increases with frequency when the frequency is less than a critical frequency (approximately 12 kHz). It reaches a peak at the critical frequency and then decreases with frequency until 15 kHz. Although it increases with frequency above 15 kHz, the amplitude is small. When the formation is included in the model, the situation is different. For the dispersion curve of the collar L1 mode in various formations, there is also a critical frequency. This is due to the existence of a pole close to the formation P-wave branch point (Yu 1984; Kurkjian 1985). At the critical frequency, the velocity of L1 is equal to formation P-wave velocity (Zheng and Hu 2017b; He et al. 2017). The critical frequency varies with formation P-wave velocity. The critical frequencies for formations F1 and F2 are 13.5 and 21 kHz, shown as red circles in Fig. 5.5. At the critical frequency, the L1 mode resonates with the formation P-wave. This results in a large amplitude P-wave. Below the critical frequency, the amplitude of L1 decreases when the formation velocity increases because collar L1 acts as a leaky mode. More energy leaks into the formation when the formation velocity is low. The collar L2 (second order collar longitudinal mode) is excited only at high frequency. Most energy of the leaky L2 is reflected by the borehole wall. A formation with a larger velocity (larger impedance) results in a larger amplitude of the collar L2 mode, as shown in Fig. 5.6b. As indicated in Fig. 5.6, there is a stop-band (frequency range where L1 excitation is small) between modes L1 and L2 when the collar is in an infinite fluid without a formation. However, the L1 mode that is reflected from the borehole wall at high frequency makes the natural stop band between L1 and L2 no longer distinct when the formation is included in the LWD model.

Figure 5.7 shows side views of the wavefield pressure snapshots, calculated by the 3DFD method described in Appendix B, for the monopole measurements within formations F1 and F2. In the simulation, 36-point sources shown in Fig. 5.7a are used to approximate the ring source. The source is located at $z = 0$ m and the receivers at eight different offsets are located from $z = 3$ m to $z = 4.05$ m with 0.15 m separation along the borehole. The source function is a 10 kHz Ricker wavelet. All 36-point sources with identical phases are simultaneously fired for the monopole simulation. The 36-point receivers are located around the collar at each receiver ring (8 receiver

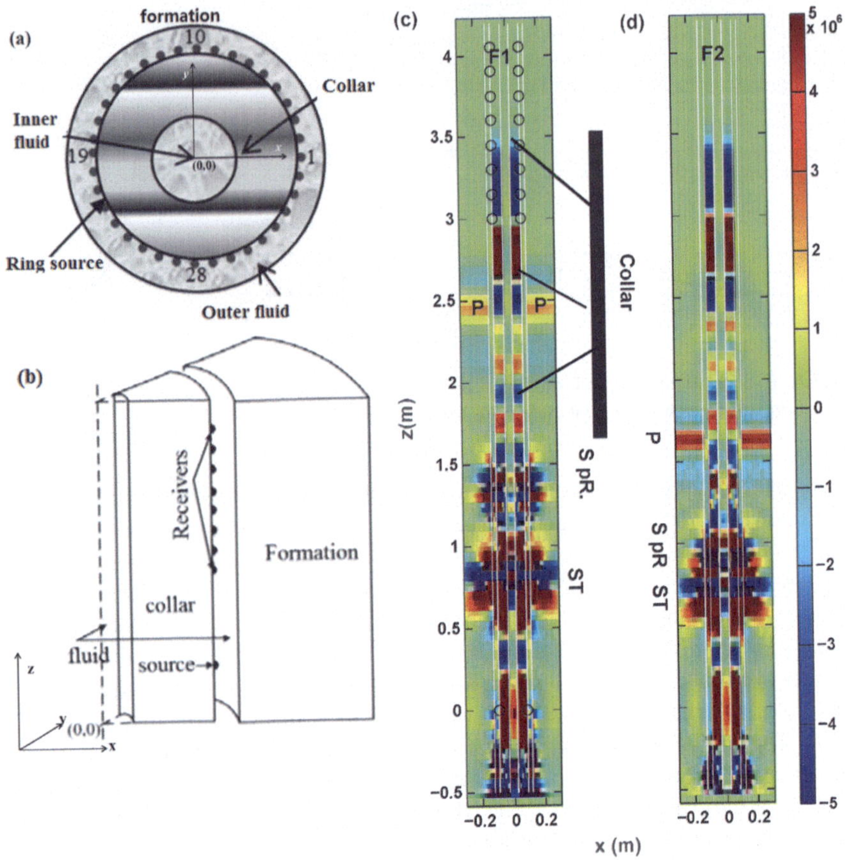

Fig. 5.7 Model used for the ALWD simulation and pressure snapshots for a monopole source (Fig. 1 in Wang et al. [2017]): **a** horizontal cross section of the model; **b** side view. Pressure snapshots at 0.7 ms in the x-z profile for a centralized tool in fast formations F1 (**c**) and F2 (**d**). The source is at $z = 0$ and receivers are marked by circles in **c**. There are six white solid lines from $z = -0.45$ m to $z = 4.1$ m. The two innermost lines are the boundaries of the inner fluid and the two outermost lines are the borehole wall, and the other two lines are the outer boundary of the collar. Collar C32 is used

rings). Receiver azimuths are measured relative to receiver 1, where receiver 1 has azimuth angle 0°, receiver 10 is at 90°, receiver 19 is at 180° and receiver 28 is at 270°. The dimensions of the simulation model are 0.6 m, 0.6 m and 4.55 m in x, y, and z, respectively. The borehole center is (0, 0) in the x-y plane in Fig. 5.7a, which indicates the coordinates for receivers 1, 10, 19 and 28 are (0.09, 0), (0, 0.09), (−0.09, 0), and (0, −0.09), respectively.

The collar longitudinal (L) modes propagate as the fastest waves along the z direction. The amplitude distribution of collar mode at $z = 3.2$ m and time 0.7 ms is shown in Fig. 5.8. Note that energy is proportional to the absolute value of pressure

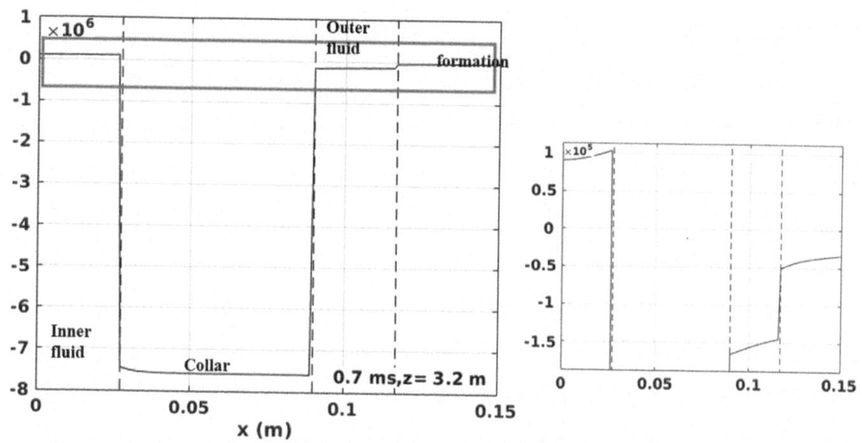

Fig. 5.8 Distribution of pressure amplitude of collar mode in Fig. 5.7c at Z = 3.2 m at 0.7 ms. The solid rectangle region in the left part is zoomed out and displayed in the right part of the figure. The vertical axis is the magnitude of the pressure on a linear scale. Dashed lines from left to right are inner and outer walls of the collar and borehole wall

amplitude. Most of the energy is trapped in the collar and it dramatically decreases outside.

The formation P-wave, S and pR, and ST waves (marked in Fig. 5.7) follow in sequence. The ST mode is the slowest mode in the wavefield. The P-wave motion can be observed in formation F1 (Fig. 5.7c), even though it is submerged within the collar L waves. The simulated array waveforms in Fig. 5.9a show the collar, S and pR, and ST waves. Because the source and receivers are placed in the fluid rather than in the collar, the amplitude of the observed collar wave is small. The first 1.2 ms of the waveforms in Fig. 5.9a are amplified and displayed above the main traces. The velocities are obtained from the array waveform using the time

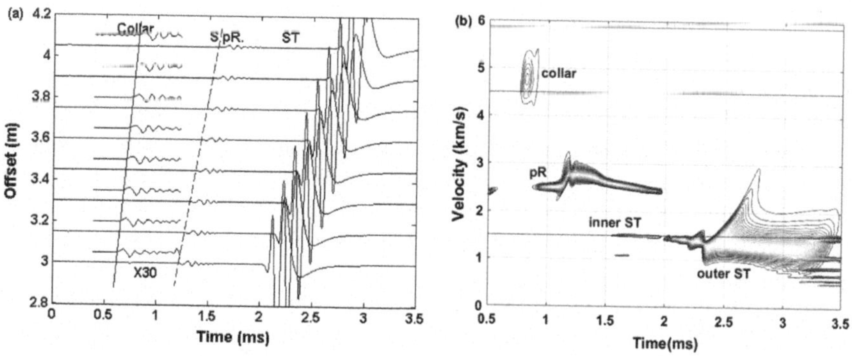

Fig. 5.9 Waveforms for a monopole ALWD tool in formation F1. Parameters of the collar and formation are C32 and F1 in Table 5.2. **a** The array waveforms. **b** Velocity-time semblance plot. The waveforms of the first 1.2 ms are amplified and displayed above the complete traces in **a**

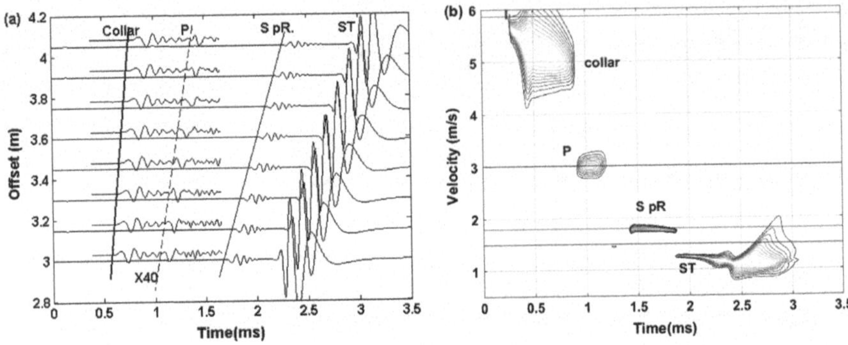

Fig. 5.10 Waveforms for a monopole ALWD tool in formation F2. Parameters of the collar and formation are C32 and F2 in Table 5.2. **a** The array waveforms. **b** Velocity-time semblance plot. The waveforms of the first 1.5 ms are amplified and displayed above the complete traces in **a**

semblance method, and are shown in Fig. 5.9b. Collar, S, pR, and ST waves stand out. However, it is difficult to determine the P-wave velocity because of the interference of high-frequency collar L waves.

The situation is different when the formation P-wave velocity is much lower than the collar L-wave velocity. In the case of formation F2, shown in Fig. 5.7d, the P-wave is clearly observed in the outer fluid column in the borehole. The calculated waveforms for formation F2 are shown in Fig. 5.10a. The arrivals of collar, S, pR, and ST waves are clearly observed. The velocity-time semblance analysis (Fig. 5.10b) shows clear peaks for all modes.

In ALWD, the massive drill collar and associated waves dominate over formation modes much more than in the case of wireline tools. It is thus tempting to seek approaches to reduce the ALWD tool effects by obtaining the wavefields of the tool without a formation and subtracting these from measurements in the borehole. This was tested using the synthetic wavefields calculated for scenarios with and without formation. Wavefield snapshots calculated for the ALWD model in formations F1 and F2 are shown in Fig. 5.7c, d.

The wavefield snapshots at 0.7 ms of the ALWD in fluid (without formation) are shown in Fig. 5.11a. The two difference wavefields, between with and without formations F1 and F2, are shown in Fig. 5.11b and c, respectively. A monopole 10 kHz source was used for all calculations. In the wavefield for the model with a collar in fluid without formation (Fig. 5.11a), the collar L modes dominate between 1.5 and 3.5 m. The inner ST mode is located approximately 1 m. The strong outer ST modes are visible from 0.8 to 1 m. By subtracting Fig. 5.11a from Fig. 5.7c, we obtain the wavefield shown in Fig. 5.11b. Although the P-wave becomes stronger in Fig. 5.11b than in Fig. 5.7c, there are still residual collar waves that affect the formation P-wave determination. The same phenomenon is also observed in Fig. 5.11c, obtained by subtracting Fig. 5.11a from Fig. 5.7d. The residual collar wave is clear. This illustrates that the collar waves are affected by the formation surrounding the borehole.

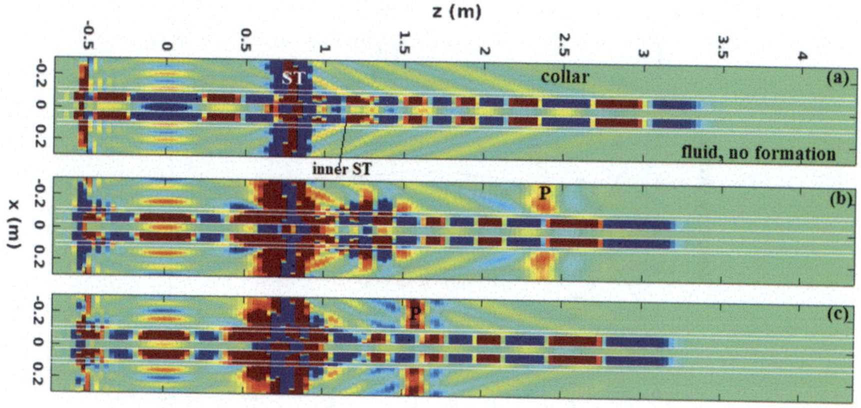

Fig. 5.11 Wavefield snapshots (0.7 ms) for **a** collar in fluid (without formation), **b** subtraction between wavefields with formation F1 (Fig. 5.7c) and (**a**), and **c** subtraction between wavefields with formation F2 (Fig. 5.7d) and (**a**). Color scale is the same as Fig. 5.7d. The source frequency is 10 kHz

Subtracting the free tool wavefield from those obtained in the borehole, the P-wave velocity determination in a fast formation is still a challenge, especially when the velocity difference between the collar and formation P-waves is small.

5.1.1.2 Dipole Source

Figure 5.12 shows the dispersion curves for the dipole LWD measurement in models with formation F1 (solid curves) and without formation (dashed curves). The red

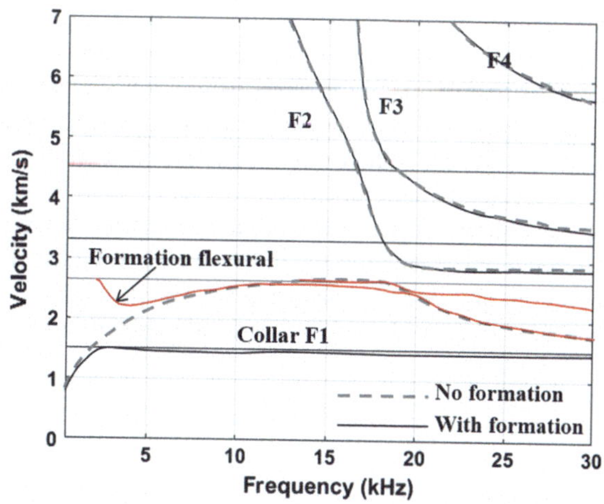

Fig. 5.12 Dispersion curves of the dipole LWD measurements. C32 is the collar and formation is F1. The dashed curves are for dispersion when the collar immersed in an infinite fluid (without a formation). Solid curves are for collar in a borehole in formation F1. The red curves show modes that are present only when there is a formation

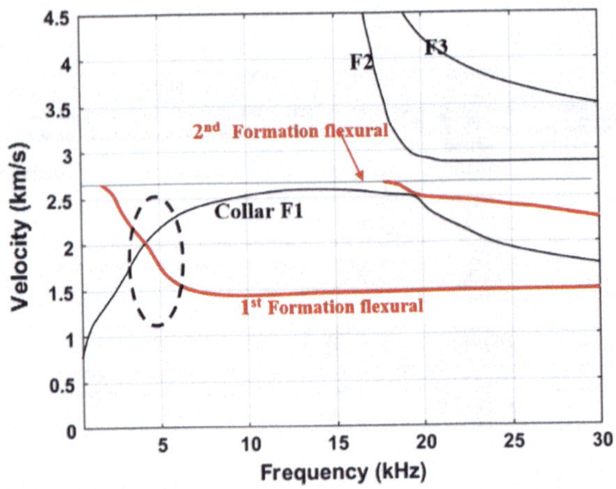

Fig. 5.13 Dispersion curves for the interaction between formation flexural and collar F1 modes. Formation F2 is used here. The red curves are the modes which only exist in the borehole model with a formation

curves show modes that are present only when there is a formation. Collar flexural waves, except for the fundamental mode (collar flexural F1), have large cutoff frequencies.

The dispersion curve of the collar flexural wave (shown in Fig. 5.12) is only affected by the properties of the collar when it is immersed in an infinite fluid. However, the fundamental collar flexural wave (F1) interacts with the formation waves when the collar and formation phase velocities are close. The details of this interaction can be seen in Fig. 5.13, where formation F2 is used. The 1st and 2nd order of formation flexural waves are marked by red curves. The cutoff frequencies of these two modes are 2 kHz and 18 kHz. The velocity of 1st collar flexural mode (F1) approaches that of the 1st formation flexural mode approximately 4 kHz (the region marked by a dashed circle). It also approaches the 2nd formation flexural mode at approximately 20 kHz. The strong interaction affects the determination of formation S-wave velocity if a low frequency source is used. Identifying the frequency range with least interference between modes is important for reliable S-wave velocity determination.

Waveforms for a 10 kHz dipole source are shown in Fig. 5.14. The two modes, the collar and formation flexural waves, can be identified in the seismograms and in the velocity-time semblance plots. The formation flexural wave velocity is nearly constant above 5 kHz and prominently shown in the velocity-time semblance plot (Fig. 5.14b).

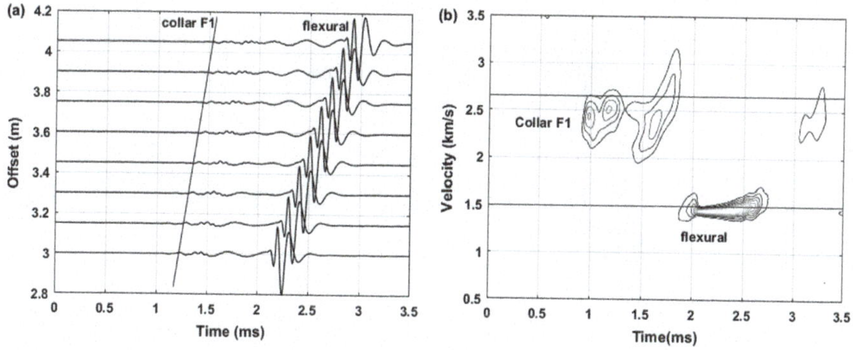

Fig. 5.14 High-frequency dipole ALWD measurement (10 kHz) in formation F2 with collar C32. **a** Array waveforms. **b** Velocity-time semblance. The mode identifications are marked based on the dispersion curves in Fig. 5.13

5.1.1.3 Quadrupole Source

Figure 5.15 shows the dispersion curves for the quadrupole LWD measurement in models with formation F1 (solid curves) and without formation (dashed curves). There are two formation screw modes in the frequency range from 0 to 30 kHz, with cutoff frequencies at 4 and 12 kHz, respectively. The collar screw mode Q1 disappears when the collar is put in a borehole surrounded with a formation. The lowest order collar quadrupole mode is Q2 with a cutoff frequency approximately 10 kHz.

If the source frequency is above the cutoff frequency of the collar quadrupole (or screw) mode, there are two modes as shown in Fig. 5.16. At high frequency, the collar quadrupole mode as the first arrival with a small amplitude. It is followed by

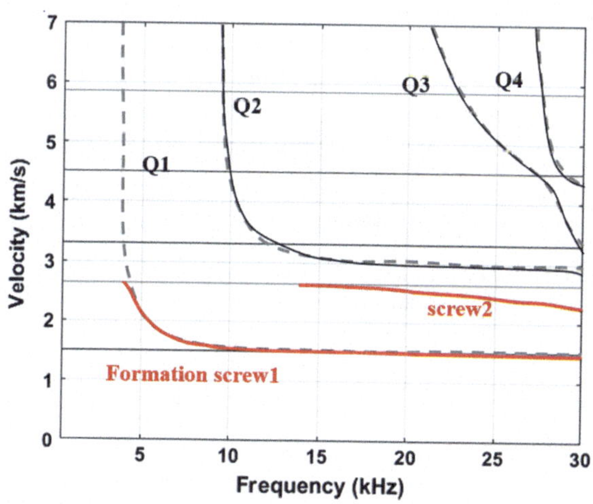

Fig. 5.15 Dispersion curves of the quadrupole LWD measurements for Collar C32 and formation F1. The dashed curves are for dispersion when the collar is immersed in an infinite fluid (without a formation). Dispersion curves in red only exist when the model has a formation

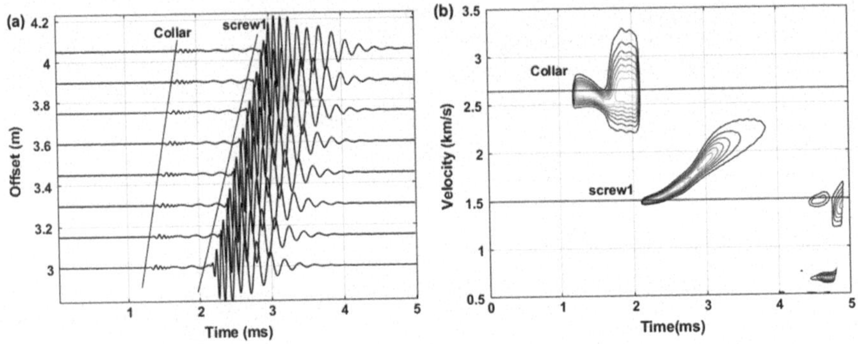

Fig. 5.16 High-frequency (10 kHz) ALWD quadrupole source in formation F1 (Vp = 4500 m/s, Vs = 2650 m/s). **a** Array waveforms. **b** Velocity-time semblance

the dispersive 1st order formation screw mode with a relatively large amplitude. The formation S-wave velocity cannot be determined.

There is only a 1st order formation "screw" mode when the frequency is low, such as shown for the 4 kHz central frequency waveforms in Fig. 5.17a. The fundamental formation "screw" wave has a strong dispersion at low frequency. Thus, there are two peaks in the velocity-time semblance plot in Fig. 5.17b. The S-wave velocity can be determined from the first peak (2 ms, 2.65 km/s). In summary, S-wave velocity may be determined from both the 2nd order screw mode at high frequency and the 1st order screw at low frequency.

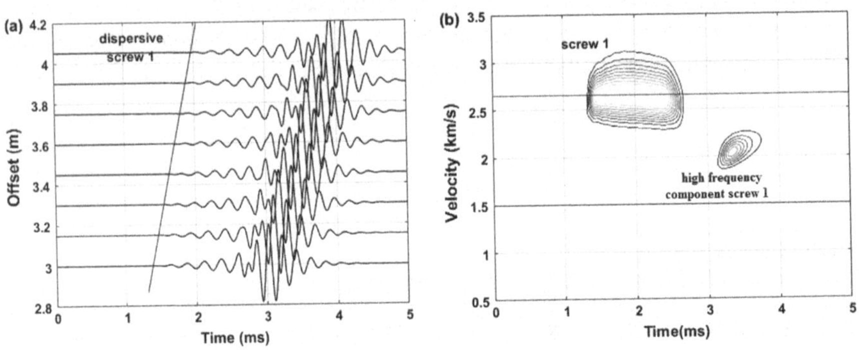

Fig. 5.17 Low frequency (4 kHz) quadrupole ALWD source in formation F1. **a** Array waveforms. **b** Velocity-time semblance

5.1.2 Wavefields in a Slow Formation

5.1.2.1 Monopole Source

In this section, the monopole ALWD wavefields in the slow and very slow formations (S1 and S2) are presented.

The pressure wavefield snapshot at 1 ms for the 10 kHz monopole source in slow formation S1 is shown in Fig. 5.18. There are only collar L, leaky P, and ST modes in the wavefield. The ST modes are visible from 0.5 to 1 m. The leaky P-waves are visible from 1 to 1.8 m. The wavefront at 1.4 m is oblique which indicates that there is energy leakage into the formation. The collar L modes are from 1.8 m to the top of the model. The leaky P-wave can be clearly observed due to the large velocity difference between the collar L modes and formation P-wave velocities. The array waveforms in Fig. 5.19a show the collar, P, and ST arrivals. The dispersion curves in Fig. 5.19c include both the contour plot computed from the array waveforms and analytical curves. The analytical curves (solid) show that there is no overlap between collar L1 and leaky P-waves. The leaky P-wave in the array waveform is visible and can be used for formation P-wave velocity determination (Fig. 5.19b). The inner ST wave is very small and controlled by the property of inner fluid column and the collar S-wave velocity. The outer ST wave is sensitive to formation S-wave velocity and propagates with a velocity lower than the formation S-wave velocity. Although the S-wave velocity may be determined from the outer ST wave dispersion curve, the S-wave velocity may be determined directly from dipole and quadrupole measurements.

The leakage of the P-wave becomes greater in the very slow formation. Dispersion curves for formation S2 are shown in Fig. 5.20. The leaky P-wave velocity is close to the formation P-wave velocity at low frequencies and to the borehole fluid-wave velocity at high frequencies. For this reason, the source frequency is very critical for accurately determining the formation P-wave velocity in the very slow formations. A low frequency source is helpful (Tang et al. 2005; Wang and Tao 2011).

Another factor affecting the waveform is the borehole radius. In the very slow formation, there may be borehole enlargement during drilling in the weak formation. The effects of the source frequency and borehole radius are discussed below.

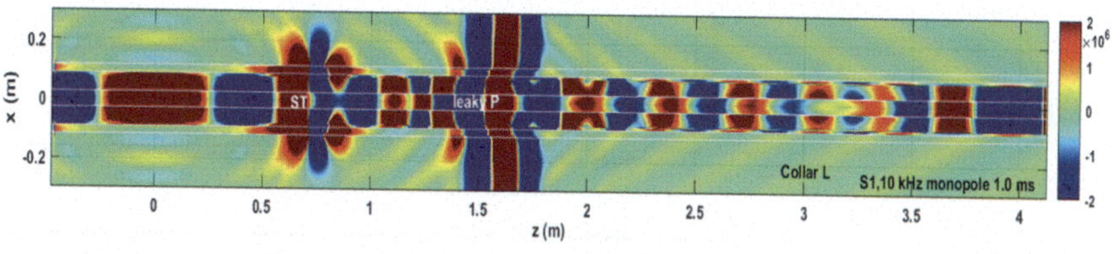

Fig. 5.18 Pressure x-z profile snapshots at 1 ms for a centered monopole tool in slow formation S1. The two innermost lines around x = 0 are the boundaries of the inner fluid. The two outermost lines are boundaries of the borehole wall, and the other two lines are the outer boundary of the collar

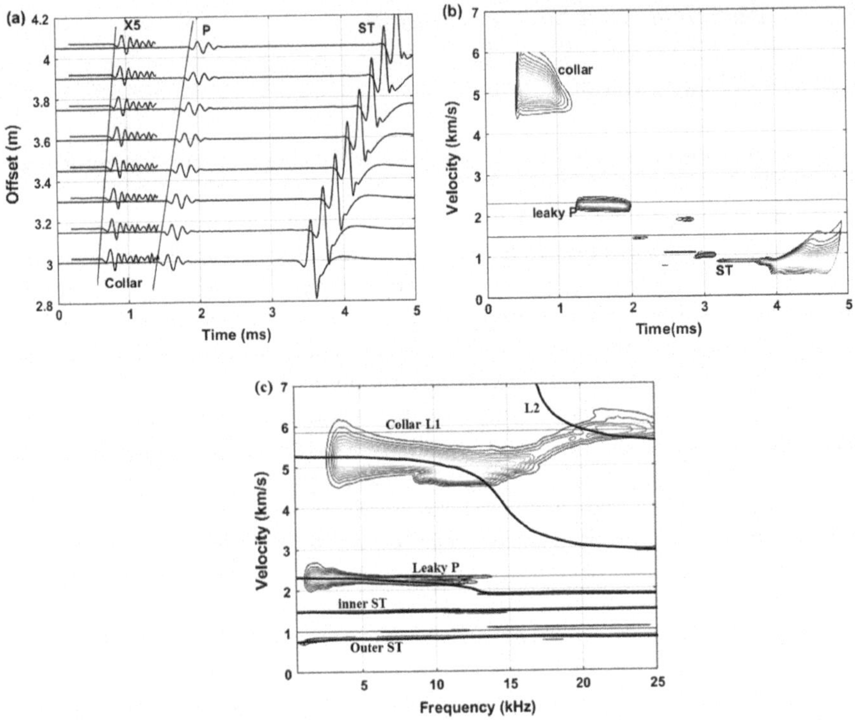

Fig. 5.19 High-frequency monopole ALWD measurement (10 kHz) in slow formation S1 in Table 5.2 with collar C32. **a** Array waveforms. **b** Velocity-time semblance. **c** Dispersion analysis of array waveforms. The contour plot was computed from the array waveforms and the solid lines are the analytical dispersion curves. Light gray horizontal lines show model velocities (Collar P 5860 m/s; Formation P 2300 m/s; Formation S: 1000 m/s; Borehole fluid: 1500 m/s)

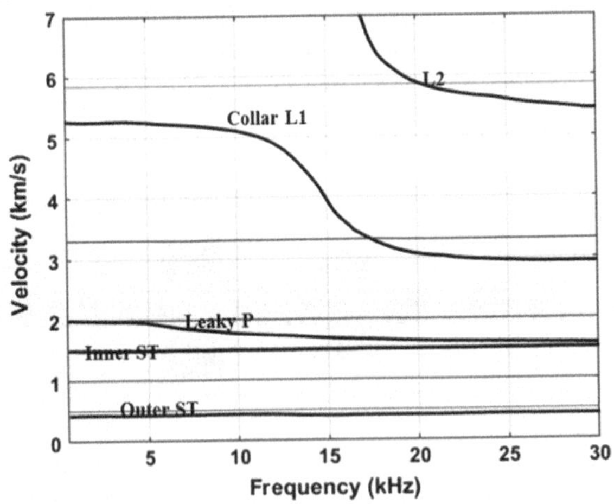

Fig. 5.20 Analytical dispersion curves of the monopole ALWD measurement in the slow formation S2, collar C32 (see Table 5.2)

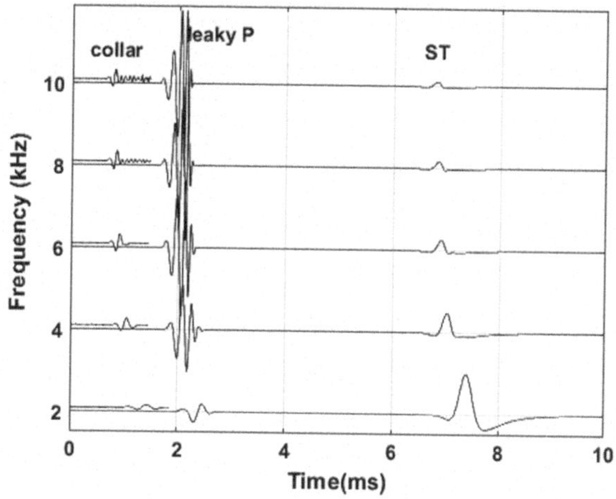

Fig. 5.21 Monopole ALWD waveforms (3 m offset) for different source frequencies in a very slow formation (S2). All waveform amplitudes are normalized by the waveform at 2 kHz. The collar wave is amplified 20 times

Monopole waveforms (3 m offset) at five different source frequencies (2, 4, 6, 8, and 10 kHz) in a very slow formation S2 are shown in Fig. 5.21. The waveforms show three events: the early arrivals, including the collar and leaky P-waves, and ST wave. In the real tool, when the source and receivers are placed on the collar body, the amplitude of collar wave is strong. In this calculation, because the source and receivers are placed in the fluid and not on the collar body, the amplitude of the collar mode is very small so the waveform is amplified 20 times for display here. The large amplitude leaky P-wave dominates the full waveforms at all source frequencies. The ST wave amplitude rapidly decreases with the increasing source frequency. This is similar to the case of wireline acoustic logging as described in Chap. 2.

The ST-wave velocity is lower than the formation S-wave velocity and the coherence of the ST wave becomes weaker and the leaky P-wave velocity decreases as the source frequency increases (Fig. 5.22).

Borehole radius also affects the waveforms. Figure 5.23a shows the collar wave (a 3-kHz source frequency) at 3.0 m offset with various borehole radii. The amplitudes of the collar waves decrease with the increasing borehole radius. The leaky P-wave, as shown in Fig. 5.23b, becomes larger with larger borehole radius. The ST wave amplitude decreases when borehole radius increases as shown in Fig. 5.23c. The relationship between the ST amplitude and distance to the borehole wall is similar to that shown in Fig. 2.28d. The dispersion curve of leaky P-wave moves toward the lower frequency when the borehole radius increases, as shown in Fig. 5.23d. P-wave velocity can still be determined from the leaky P-wave.

The impacts of the source frequency and borehole radius on the measurements suggest that a low frequency monopole source (lower than 3 kHz) can be used to measure the P-wave velocity in slow formations.

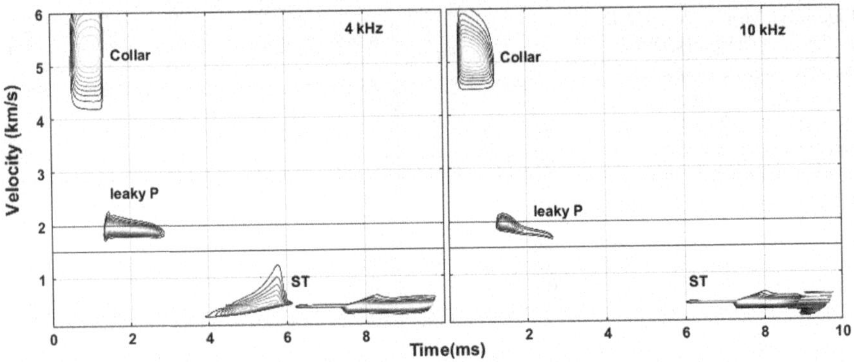

Fig. 5.22 Velocity-time semblance contour plots for the array waveforms (offsets from 3 to 4.05 m with a 0.15 m interval) from the monopole ALWD measurements in a very slow formation (S2) at two source frequencies

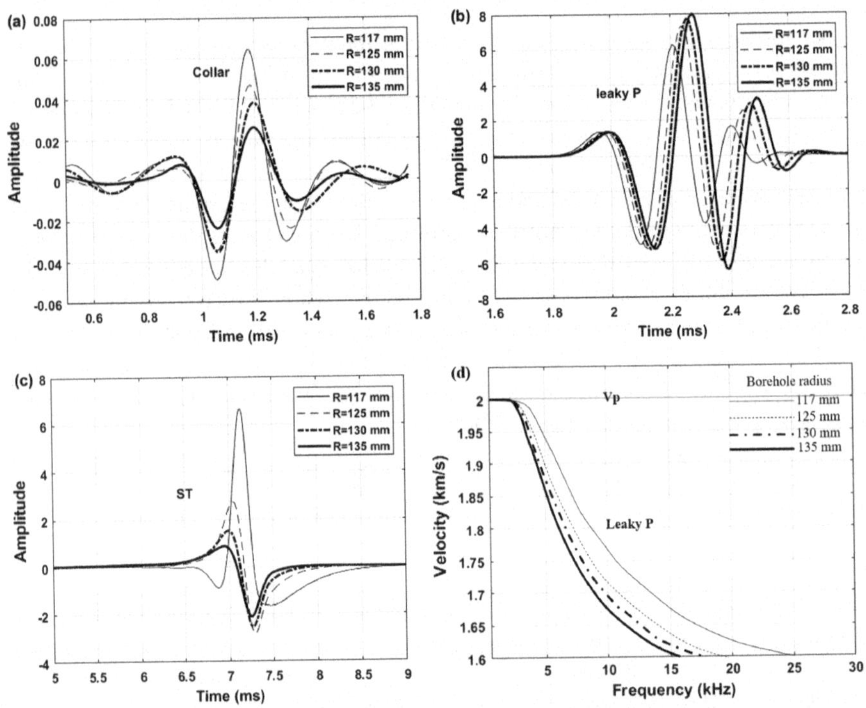

Fig. 5.23 Effect of borehole radius on the monopole LWD measurement in a very slow formation (formation S2 in Table 5.2. **a** 3 m offset waveforms (collar wave) with various borehole radii. **b** 3 m offset waveforms (leaky P-wave) with various borehole radii. **c** 3 m offset waveforms (ST wave) with various borehole radii. **d** Dispersion curves of leaky P-wave. Formation P-wave velocity is shown by horizontal line

5.1.2.2 Dipole Source

In the wireline case, direct measurement of the S-wave velocity in slow formations can be done using a low frequency dipole tool. The flexural-wave velocity at the cutoff frequency is equal to the formation S-wave velocity. Similar measurements of S-wave velocity can be made by a dipole source in the ALWD case. However, some complications arise because of interference by collar flexural waves (Tang et al. 2002; Wang et al. 2009).

Dispersion curves for the dipole AWLD wavefields for formations S1 and S2, are shown in Fig. 5.24. The collar F1 and formation flexural modes cross at very low frequency (marked with a red ellipse). The influence of the interference is shown by wavefield snapshots at 2 ms (2 kHz dipole source) in formation S1 in Figs. 5.25 and 5.26. There are only two modes in the wavefield snapshots (Fig. 5.25), where the highly dispersive collar flexural mode extends from 0.5 to 4 m and the low dispersive formation flexural mode extends from 0.5 to 1.3 m. There is strong interference between the collar flexural and formation flexural waves. The interference can also

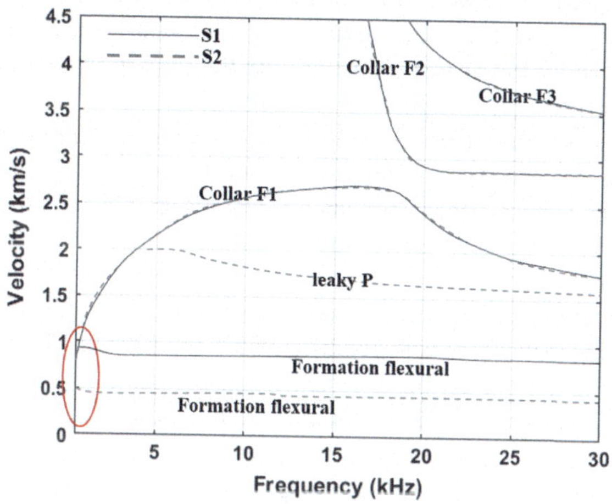

Fig. 5.24 Analytical dispersion curves of the dipole ALWD measurement in the slow formations S1 (solid curves) and S2 (dashed curves). Geometrical and elastic parameters are listed in Table 5.2. Collar C32 in Table 5.2 is used

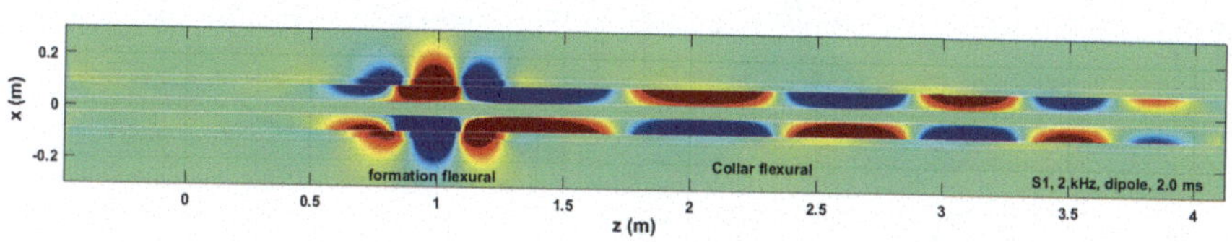

Fig. 5.25 Pressure x-z profile snapshots at 2 ms for a centralized dipole tool (2 kHz) in slow formation S1. Sources are at (0, 0.09) and (0, −0.09). Horizontal white lines are described in Fig. 5.18

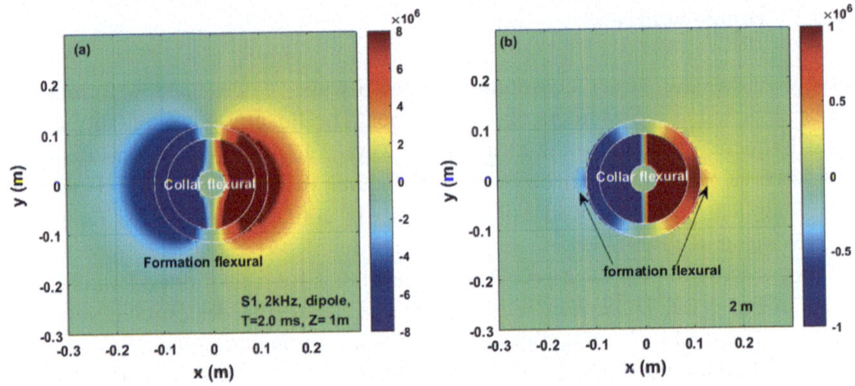

Fig. 5.26 Pressure x-y (horizontal) profile snapshots at 2 ms for a centralized dipole tool in slow formation S1. **a** Z = 1 m. **b** Z = 2 m. The two innermost circles are the inner and outer boundaries of the collar. The outermost circle is the borehole wall

be found by comparing the top-down view (x-y profiles) snapshots at different depths (z locations) in Fig. 5.26. The collar flexural wave dominates at z = 2 m (Fig. 5.26b). The interference of the collar flexural wave with the formation flexural wave makes it difficult to determine the formation S-wave velocity.

In the array waveforms (Fig. 5.27a), the first arrival is the strongly dispersive collar flexural wave, and its coda interferes with the arrival time of the formation flexural wave. The velocity-time semblance contour plot in Fig. 5.27b also shows the interference between the later part of the collar wave and formation flexural wave from 2 to 4 ms.

There is no interference between the collar flexural and formation flexural waves at high frequency due to the large velocity difference between them. However, the high-frequency measurements must be corrected for dispersion effects to obtain the true formation S-wave velocity. Figure 5.28a shows synthetic array waveforms generated

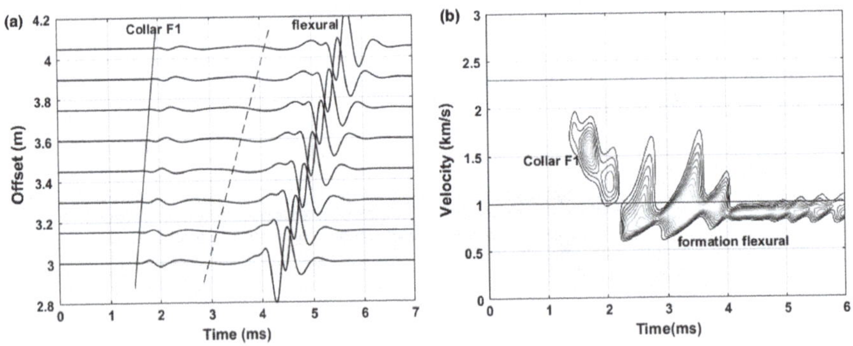

Fig. 5.27 Low frequency (2 kHz) dipole ALWD wavefield in slow formation S1. **a** Array waveforms. **b** Corresponding velocity-time semblance contour plot of the waveforms in **a**

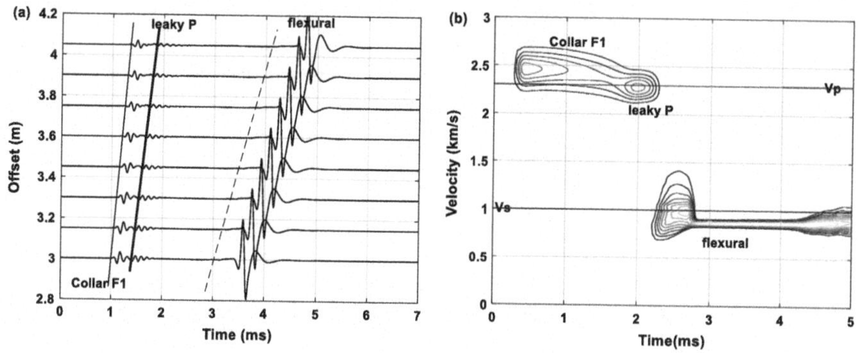

Fig. 5.28 High-frequency (10 kHz) dipole ALWD wavefield in slow formation S1. **a** Array waveforms. **b** Corresponding velocity-time semblance contour plot of the waveforms in **a**

by a 10 kHz dipole source. The collar F1 mode is well separated from the formation flexural wave. Their velocities can be obtained from the velocity-time semblance analysis, as shown in Fig. 5.28b. The measured formation flexural-wave velocity is 865 m/s, which is lower than the S-wave velocity (1000 m/s) by approximately 13.5%. Correcting the measured flexural-wave velocity to obtain the true S-wave velocity requires calculating the dispersion curve for each case.

As discussed in Chap. 2, there is a dipole induced leaky P-wave in a slow formation. Leaky P is difficult to identify at low frequency. Using high source frequency measurements, it can be identified, following the collar F1 as shown in Fig. 5.28a. Its amplitude is small. The obtained velocity is very close to formation P-wave velocity (Fig. 5.28b).

In summary, for slow formations, the S-wave velocity may be obtained from the dipole ALWD flexural wave with a proper dispersion correction. The high-frequency measurement may be used for P-wave velocity determination from dipole induced leaky P-waves.

In very slow formation S2, both the flexural and leaky P-waves become prominent. The separation between collar flexural wave and formation flexural wave increases. The S wave velocity of formation S2 (500 m/s) can be determined from the velocity of the formation flexural wave at the cutoff frequency (approximately 2 kHz). Figure 5.29 shows synthetic waveforms at different source frequencies in formation S2. There is no interference between the collar wave and formation flexural wave. The velocity-time semblance of the array waveforms (Fig. 5.30) shows good coherence of the formation flexural wave, the velocity of which is equal to the formation S-wave velocity (500 m/s).

The leaky P mode, is more dispersive in formation S2 than that in formation S1. The velocity of the leaky P wave reaches the formation P-wave velocity at the cutoff frequency (approximately 5 kHz) and approaches the borehole fluid-wave velocity at high frequency. The interference between the leaky P and collar flexural waves in the low frequencies (Fig. 5.29) affects the P-wave velocity determination from the leaky P-wave. The leaky P travels faster than the collar flexural wave at low frequencies. At

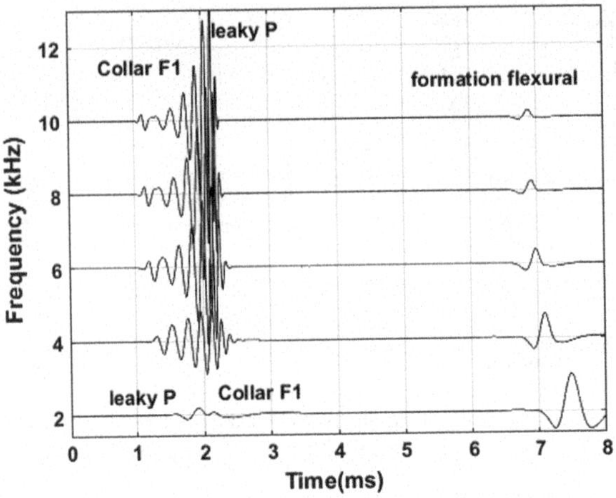

Fig. 5.29 Dipole ALWD waveforms at 3 m offset in slow formation S2 with various source frequencies

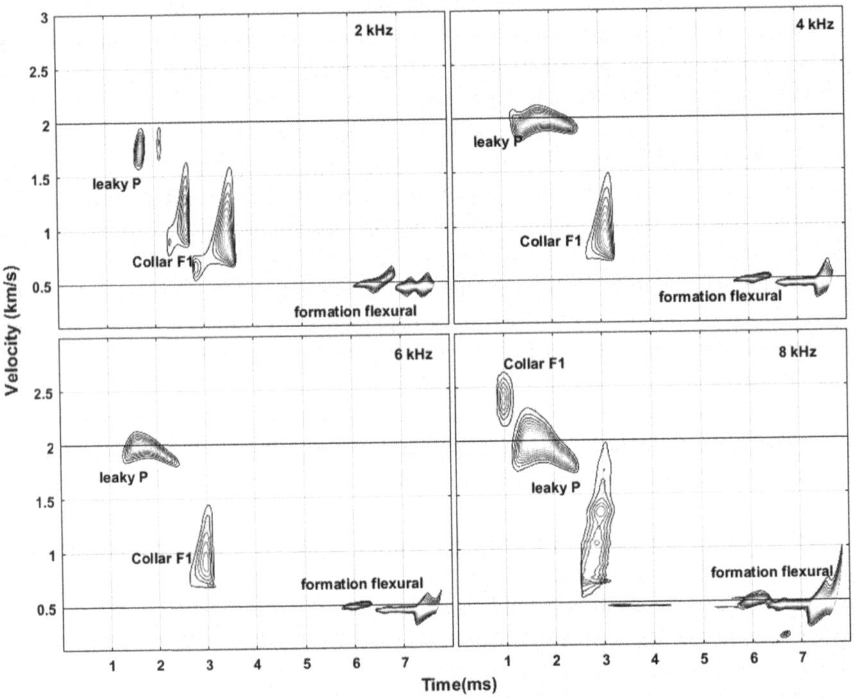

Fig. 5.30 Velocity-time semblances of the dipole ALWD array waveforms with various source frequencies for formation S2. Offsets of the array waveforms are from 3 to 4.05 m with a receiver interval of 0.15 m. Formation P and S velocities are 2000 and 500 m/s, respectively

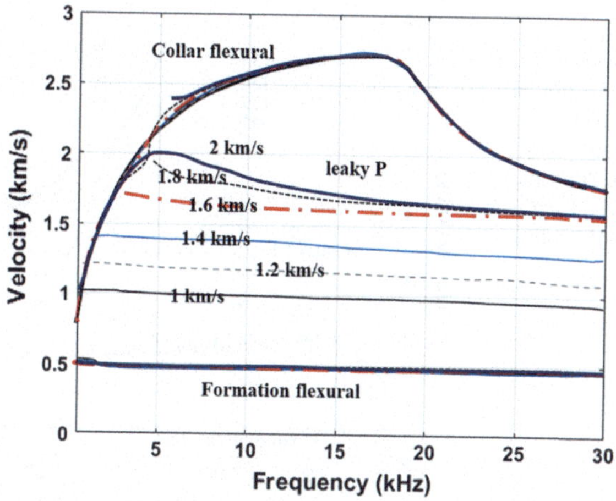

Fig. 5.31 Dispersion curves with various formation P-wave velocities and a fixed S-wave velocity ($V_S = 500$ m/s). Different lines designate different P velocities: thick solid black for 2000 m/s, thin dot black for 1800 m/s, red dot-dashed for 1600 m/s, thin blue for 1400 m/s, grey dash for 1200 m/s, and thin black for 1000 m/s. Dispersion curve for formation flexural and collar flexural waves do not change with formation P-wave velocity

higher frequency (f ≥ 8 kHz), the collar flexural wave arrives earlier than the leaky P-wave. The velocity-time semblance plots in Fig. 5.30 also show the trend of the velocities of the collar flexural and leaky P-waves at different source frequencies. The leaky P-wave overlaps with the collar wave when the source frequency is 2 kHz. Some leaky P is generated by the 2 kHz source because the source band extends above the cutoff frequency of leaky P-wave. The dipole leaky P-wave separates from the collar flexural wave at higher frequencies (f > 4 kHz), and yields the formation P-wave velocity.

The dispersion curves for formations having common S-wave velocity (500 m/s) and various P-wave velocities (between 2000 and 1000 m/s), shown in Fig. 5.31, illustrate the effect of the P-wave velocity and Poisson's ratio. The dispersion curve of the formation flexural wave is controlled by the formation S-wave velocity and does not change with P-wave velocity. The collar flexural and formation flexural waves do not interfere at low frequency and the formation S-wave velocity can be determined from the formation flexural-wave velocity at the cutoff frequency. The leaky P-wave, becomes more dispersive when P-wave velocity increases and interferes with the collar flexural wave at low frequency. The P-wave velocity can be determined from the leaky P when the P-wave velocity is lower than the collar flexural-wave velocity. Dispersion curves help identify the best frequency range for P-wave velocity determination.

As mentioned in Sect. 5.1.2.1, borehole enlargement is commonly encountered in very slow formations. Changing borehole radius has little effect on formation velocity measurements. Figures 5.32 and 5.33 show the dispersion curves and waveforms for different borehole radii. The waveforms for a 3-kHz source frequency are shown

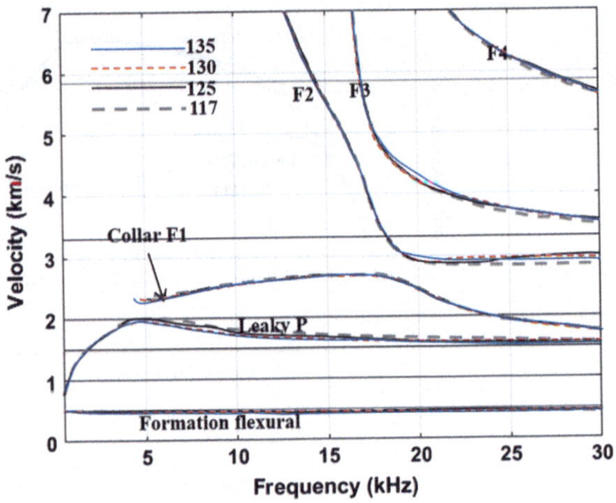

Fig. 5.32 Impact of borehole radius on the dispersion curves of different modes in dipole ALWD measurement. Formation P and S velocities are 2000 and 500 m/s, respectively. Lines are coded by borehole radius in mm

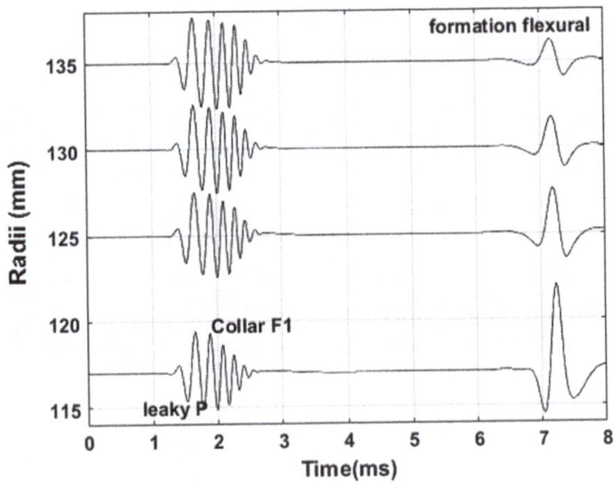

Fig. 5.33 Impact of radius on the waveforms in dipole ALWD measurement. Formation P and S velocities are 2000 and 500 m/s

in Fig. 5.33. For formation S2, the amplitude of the flexural wave decreases when borehole radius increases.

In summary, ALWD measurements made using a dipole source in a slow or very slow formation include the collar flexural, leaky P, and formation flexural waves. The leaky P-wave is affected by the formation Poisson's ratio and it is more dispersive in formations with high Poisson's ratios. The formation flexural wave is most sensitive to formation S-wave velocity, and the formation flexural-wave velocity reaches the formation S-wave velocity at the cutoff frequency. However, the two waves interfere at very low frequency (see the ellipse region in Fig. 5.24), which makes the S-wave velocity determination difficult. The P-wave velocity can best be determined from the leaky P-wave at high frequency. In very slow formation S2, dispersion analysis is an effective method to distinguish leaky P and collar flexural waves. Using a low frequency dipole source is the best approach for determining both P and S velocities in very slow formations.

5.1.2.3 Quadrupole Source

In Sect. 2.3.3, it was shown that there are leaky P and formation screw waves in a quadrupole wavefield in a slow formation. In addition to these, there are collar waves in ALWD measurements. Figure 5.34 shows the dispersion curves of different quadrupole ALWD modes in both formations S1 and S2. The dispersion characteristic of the collar quadrupole (or "screw") wave is not affected much by formation properties. This is similar to the case of fast formations. There is a cutoff frequency for each collar screw mode and it is influenced most by the outer and inner radii of the collar. The cutoff frequency of the fundamental collar screw wave is approximately 10 kHz for collar (C32). The formation screw wave is strongly affected by the formation S-wave velocity. The formation screw-wave dispersion curve reaches the formation S-wave velocity at the cutoff frequency (approximately 2 kHz) and the formation screw-wave velocity decreases when frequency increases. The dispersion of the leaky P wave is controlled by Poisson's ratio as described in Sect. 2.3.3.

Synthetic waveforms and their velocity-time semblance contour plots, for a quadrupole source with 10 kHz center frequency, are shown in Fig. 5.35. There are collar and formation screw waves in the wavefields. The collar screw mode is the first arrival and the formation quadrupole (or "screw") mode is the last one. The leaky P-wave is indiscernible due to its small amplitude and the interference of the collar mode. The corresponding velocity-time semblance contour plot is shown in

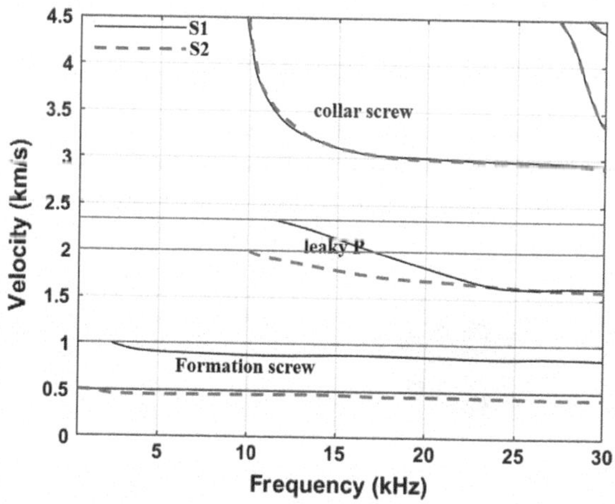

Fig. 5.34 Modal dispersion curves of quadrupole ALWD measurements in slow formations S1 and S2. Modes in different formations are marked with different lines and listed on the plot. Gray horizontal lines show material velocities (see Table 5.2). Collar is C32

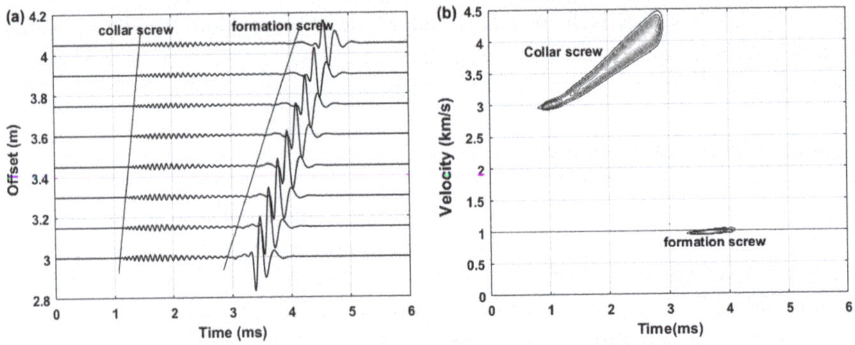

Fig. 5.35 Wavefield of high-frequency quadrupole ALWD measurement (a 10-kHz source frequency) in the slow formation S1 with collar C32. **a** Synthetic waveforms. **b** Velocity-time semblance contour plot

Fig. 5.35b. The region from 1 to 3 ms corresponds to the collar screw wave. It is dispersive. The peak from 3 to 4 ms corresponds to the formation screw wave. The velocity obtained by semblance (980 m/s) is slightly lower than the real S-wave velocity (1000 m/s).

If the source frequency is lowered to 2 kHz, the collar screw wave is not easy to excite because of its high cutoff frequency (approximately 10 kHz for the 1st order collar screw wave) as shown in Fig. 5.34. Figures 5.36 and 5.37 show the wavefield snapshots of the 2 kHz quadrupole source in formation S1 with collar C32. From the x-z profile shown in Fig. 5.36, there is only the formation screw mode and no collar mode. Figure 5.37 shows the x-y profile at z = 1 m. It is different from that of the dipole case shown in Fig. 5.26. There is no collar mode in the quadrupole case.

In the array waveforms shown in Fig. 5.38a, the weak leaky P can be observed when amplified 100 times. The formation screw wave is the later arrival (approximately 3 to 6 ms). The velocities of the leaky P and formation screw waves, obtained from the velocity-time semblance contour plot in Fig. 5.38b, are 2255 and 1000 m/s. The formation S-wave velocity is directly obtained from the low frequency quadrupole ALWD measurement without any dispersion correction. In addition, the P-wave velocity may be determined from the leaky P-wave at low frequency although

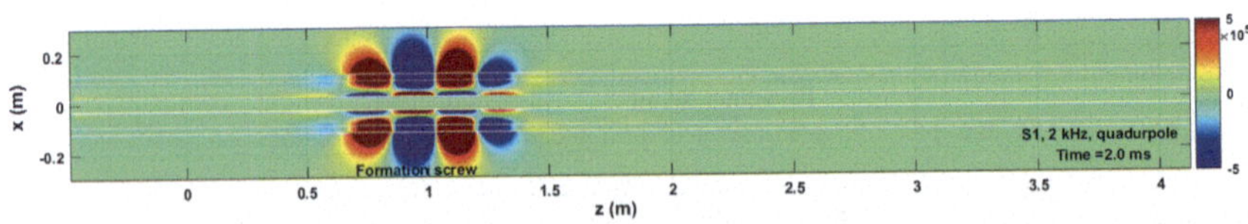

Fig. 5.36 Pressure snapshot at 2 ms for the x-z profile for a centralized quadrupole tool (2 kHz) in slow formation S1 with collar C32. Source positions are at (0, 0.09) and (0, −0.09). The two innermost lines are the boundaries for the inner fluid and the two outermost lines are boundaries of the borehole wall, and the other two lines are the outer boundary of the collar

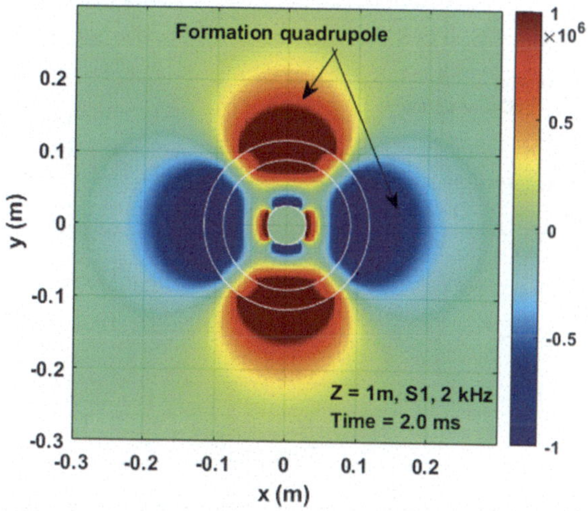

Fig. 5.37 Pressure snapshot at 1 ms for the x-z profile for a centralized quadrupole tool for slow formation S1 with collar C32. The two innermost circles are the inner and outer boundaries of the collar. The outermost circle is the borehole wall

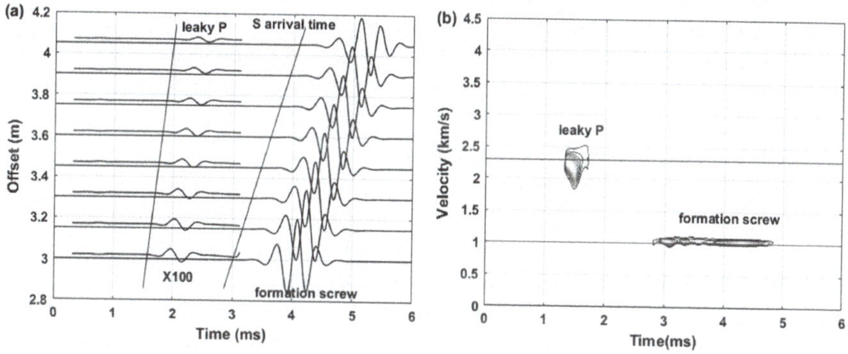

Fig. 5.38 Seismograms of quadrupole ALWD measurement (a 2-kHz source frequency) in slow formation S1 with collar C32. **a** Array waveforms. **b** Velocity-time semblance contour plot made from the array waveforms

the amplitude is very small and signal to noise ratio (SNR) in the field data could affect the accuracy of measured velocity.

The quadrupole ALWD waveforms in the very slow formation S2 are shown for different source frequencies in Fig. 5.39. The waveforms are normalized by the maximum amplitude at 2 kHz. The waveforms include the early arrivals (collar screw and leaky P-wave) and later arrival (formation screw wave).

The arrival time of the formation screw waves does not change with the source frequency. However, its amplitude decreases as the source frequency increases. The

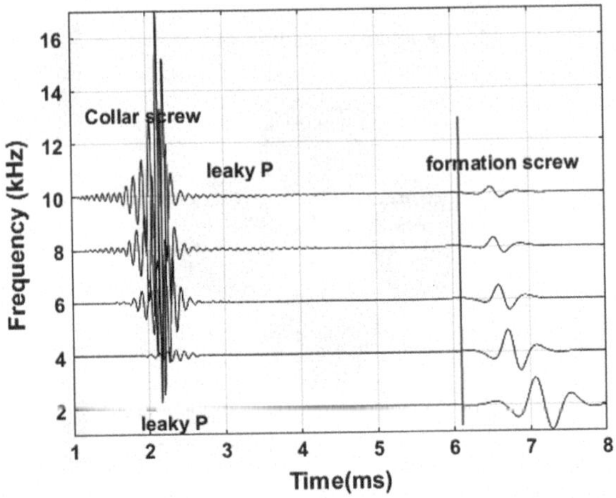

Fig. 5.39 Quadrupole ALWD waveforms at the 3 m offset at different source frequencies in formation S2. Vertical line shows arrival time of formation screw wave, which does not change with frequency

formation S-wave velocity in the very slow formation can be measured best by a low-frequency quadrupole tool.

In the earlier arrivals (approximately 2 ms in Fig. 5.39), there is the leaky P-wave. The amplitude of leaky P increases with the increasing source frequency. When the source frequency increases above the cutoff frequency, the collar screw mode is excited. The amplitude of the collar wave increases with the increasing source frequency. The weak leaky P-wave is submerged within the coda of the collar screw wave.

The leaky P-wave is affected by the formation Poisson's ratio. Figure 5.40 shows the effect of the formation S-wave velocity on the dispersion curves when the formation P-wave velocity is fixed at 2000 m/s and the formation S-wave velocity changes from 500 to 1000 m/s. The dispersion curves of the collar screw modes are not affected by the changing formation S-wave velocity. The dispersion curve of the leaky P-wave moves to higher frequency when the formation S-wave velocity is increased.

Commonly encountered borehole enlargement in very slow formations moves the leaky P and collar screw modes to a lower frequency range. Examples of the effects of borehole radii on the modes in formation S2 are shown in Figs. 5.41 and 5.42. The amplitude of the leaky P-wave increases in the enlarged borehole. The enlarged borehole only decreases the amplitude of the formation screw mode. The enlarged borehole radius (140 mm) makes the dispersion curve of the second order leaky P-wave visible below 30 kHz in Fig. 5.41, and it can also be observed at approximately 3 ms in the waveforms shown in Fig. 5.42 after 10,000 times amplification. Although the leaky P at low frequency can be used for formation P-wave velocity measurement, the drilling noise (Figs. 5.2 and 5.3) and its very small amplitude make it impractical.

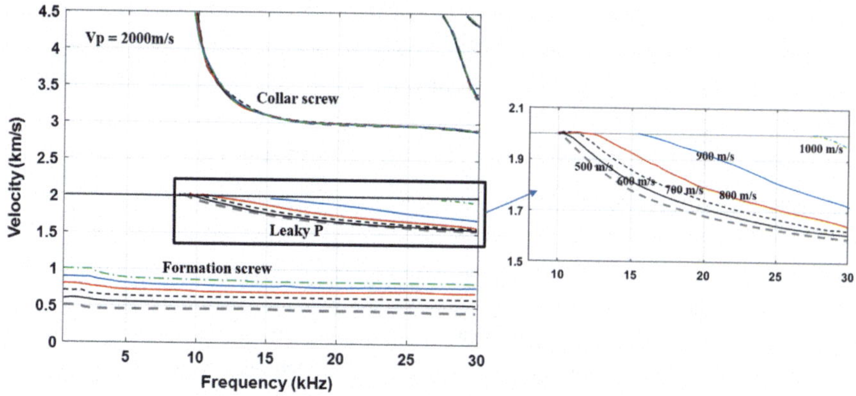

Fig. 5.40 The impact of formation S-wave velocity on the dispersion curves in the quadrupole ALWD measurement in very slow formations. The P-wave velocity of the formation is 2000 m/s and the S-wave velocity changes from 500 to 1000 m/s. Lines for different S-wave velocities are listed in the right plot

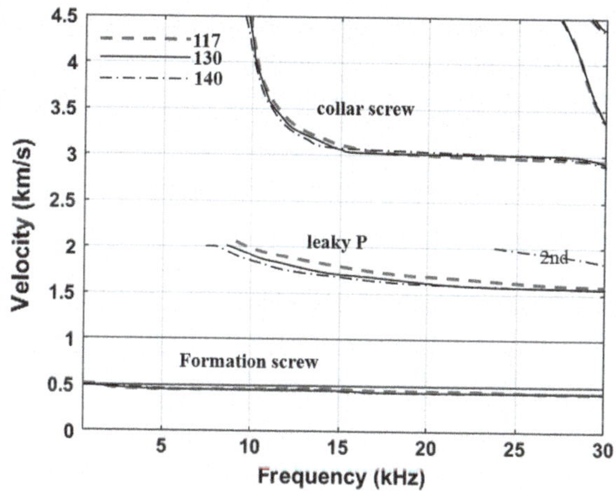

Fig. 5.41 The impact of borehole radius (listed in mm) on the dispersion curves of different modes in the quadrupole LWD measurement in a very slow formation (formation S2 in Table 5.2)

5.2 Improvements for Velocity Measurements in ALWD

5.2.1 Formation P-Wave Velocity Measurement in Fast Formations

As discussed in Sect. 5.1.1.1, the longitudinal modes propagating along the collar affect the formation P-wave measurements in fast formations (Aron et al. 1994; Leggett et al. 2001). To determine the P-wave velocity, the influence of the collar

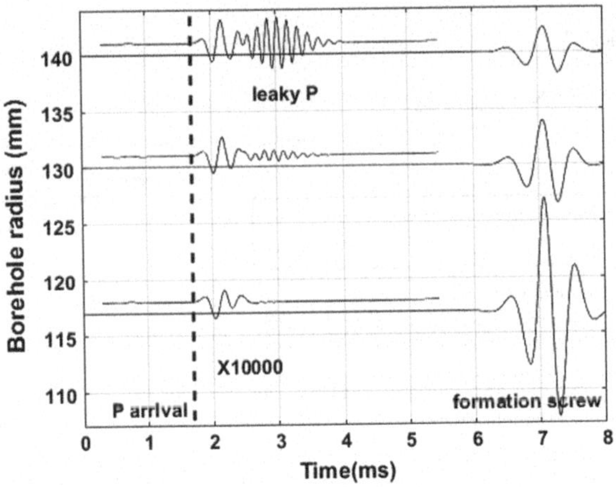

Fig. 5.42 The impact of borehole radius on the waveforms at 3 m offset (2 kHz quadrupole source) in a very slow formation (formation S2 in Table 5.2)

L modes must be suppressed. Decoupling the transducers from the collar body dramatically reduces the collar wave amplitude.

In this section, different solutions to reduce the influence of the collar L waves on the velocity measurements of P- and S-waves are discussed in detail.

5.2.1.1 Slotted Collar

A grooved/or slotted collar along with absorbing material between transducers and the collar have been used to suppress longitudinal collar waves. A collar wave may be attenuated 60 dB with slotted isolators (Aron et al. 1994; Varsamis et al. 1999; Leggett et al. 2001; Joyce et al. 2001; Kinoshita et al. 2010; Su et al. 2015). Figure 5.43 shows a schematic diagram of the grooves on both inner and outer boundaries of the

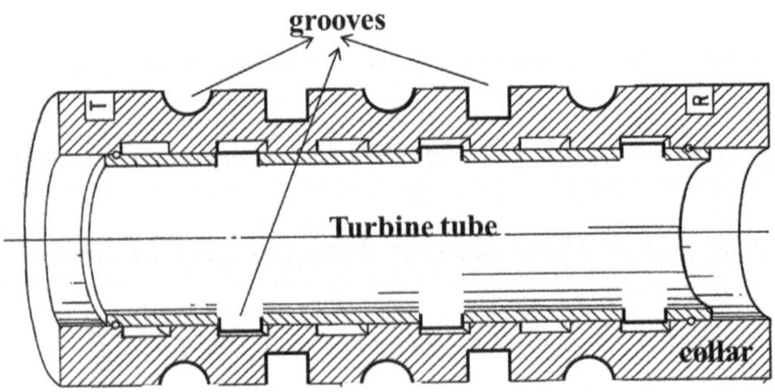

Fig. 5.43 Schematic diagram of the grooves on both inner and outer boundaries of the collar

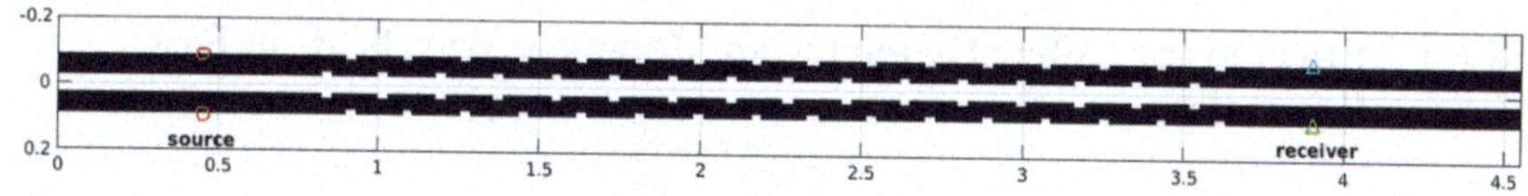

Fig. 5.44 Side view of a grooved collar used in finite difference simulations (scale in meters). Source and receiver positions are marked. The cuttings are axisymmetric. There are 16 grooves on the outer and inner surfaces of the collar. The length of each groove is 24 mm. The depth is 15 mm. The separation between two grooves is 180 mm

collar. The depth and width of the grooves, and spacing between the grooves, can be optimized through the use of numerical simulations.

We simulated the effects of grooves on collar waves with the tool in fluid and in borehole in formation F2. Figure 5.44 shows a side view of a grooved collar used in the 3D finite difference simulations (see Appendix B). Source and receiver positions are marked with circles and triangles, respectively. Both the inner and outer surfaces of the collar have grooves as shown in Fig. 5.44. The length of each groove is 24 mm. Their depth is 15 mm. The separation between two grooves on one surface is 180 mm. Figure 5.45 shows the synthetic waveforms for an un-grooved collar (thin blue curve) and grooved collar (thick curve). Collar properties for C32 in Table 5.2 are used. For the tool in fluid, there is no surrounding formation. The collar wave is attenuated by the grooves and arrival time is delayed as shown in Fig. 5.45a. When the collar is put in a borehole surrounded with formation F2, the amplitude of the collar waves in both ungrooved and grooved case are smaller than the fluid case. The collar wave is attenuated and delayed. The formation P-wave at approximately 1.5 ms is stronger for the grooved collar. The reduction of collar wave amplitude can be optimized by design of the length, depth, and separation distance of the slots.

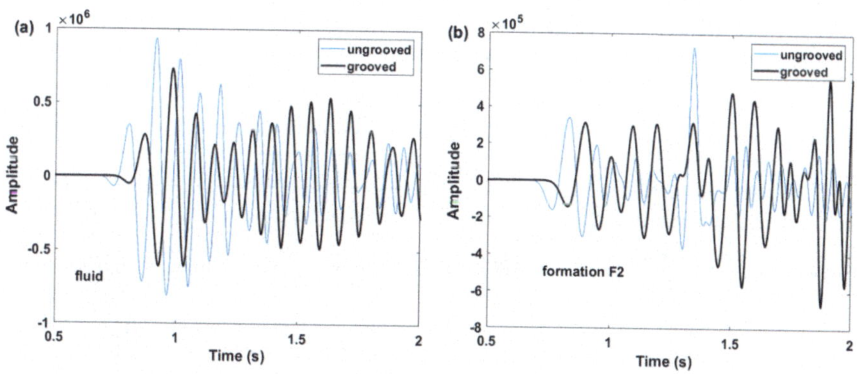

Fig. 5.45 Synthetic waveforms for an un-grooved collar (thin blue curve) and grooved collar (thick curve). **a** Tool in fluid. **b** Tool in formation F2 (parameters see Table 5.2). Collar properties are as for C32

5.2.1.2 Effect of the Collar Geometry on Monopole Wavefields in Fast Formations

In Sect. 4.3.1, we showed the effects of casing properties on logs. In ALWD, the drill collar is much bigger and thicker than the casing. The geometry of the drill collar, its thickness, radii, and material affect the wavefields. Here, we simulate array traces to show the effects of the drill collar properties in different formations.

Figure 5.46 shows array waveforms for varying collar outer radius with fixed inner radius (27 mm). Decreasing outer radius increases the separation between collar and formation P-waves. Increasing the outer radius decreases the thickness of fluid annulus, and increases the amplitude of the collar and ST waves. The amplitudes of pR waves decrease with decreasing thickness of the fluid annulus because of higher cutoff frequency. The explanation for why the large outer radius makes the formation waves obscure is shown in Fig. 5.47, where the dispersion curves of different modes with various collar outer radii are shown. A larger collar radius moves the collar modes to a lower frequency range.

The effect of the collar inner radius on the waveforms is shown in Fig. 5.48. A larger collar inner radius moves the collar waves to a lower frequency range, resulting

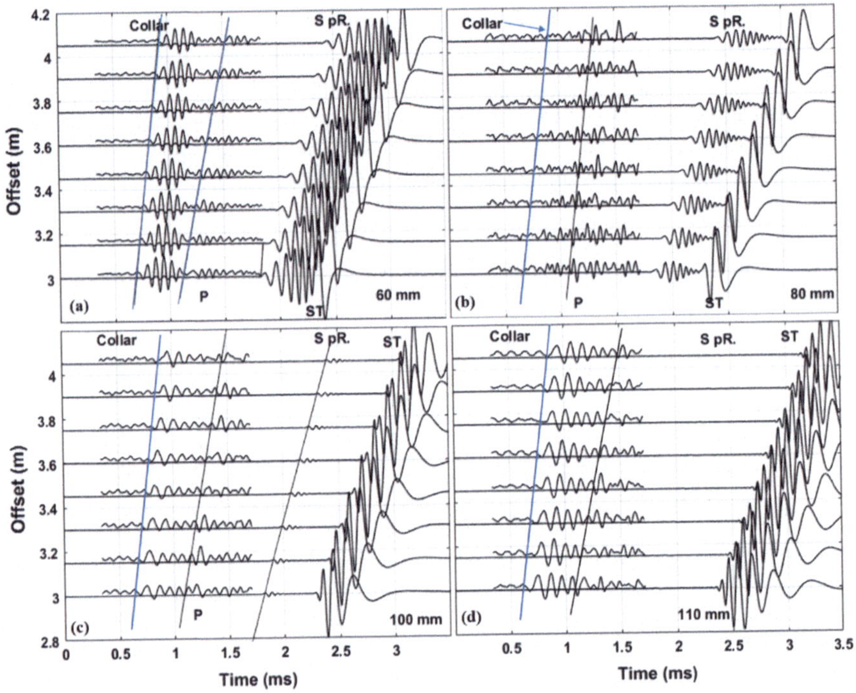

Fig. 5.46 The effect of collar outer radius on different monopole modes. Collar inner radius is fixed at 27 mm and the outer radii are listed on each plot. Formation is F2 in Table 5.2 and the source frequency is 10 kHz. Collar C32 is used

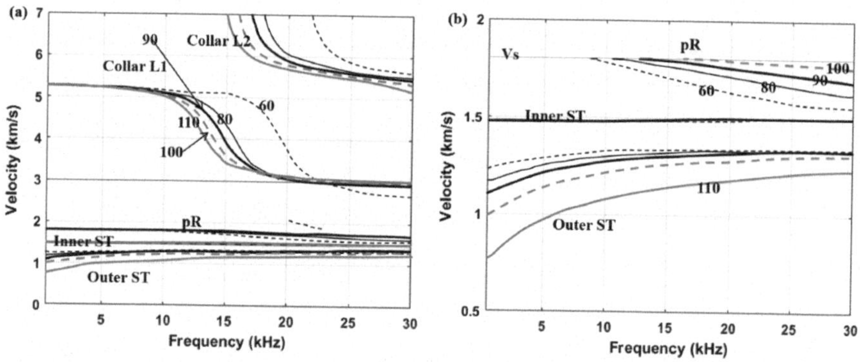

Fig. 5.47 The effect of collar outer radius on dispersion curves of different modes for a monopole source. **a** All modes. **b** Modes in a velocity range from 500 to 2000 m/s. Collar inner radius is fixed at 27 mm. Formation F2 is used. Borehole radius is 117 mm. Outer radii (in mm) of the collars are marked on the lines. Collar C32 is used

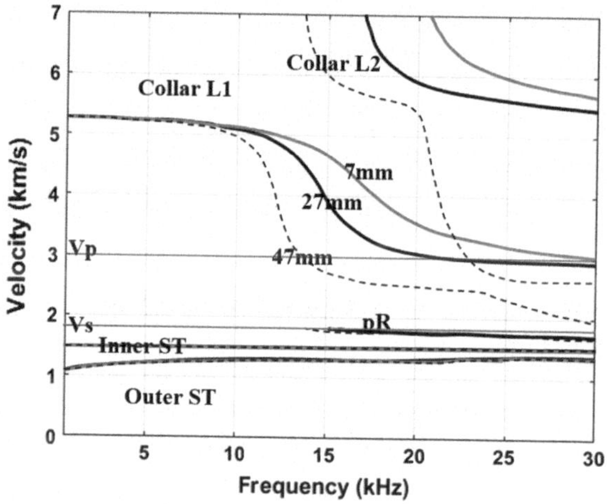

Fig. 5.48 The effect of various collar inner radii on dispersion curves of different modes. Collar outer radius is fixed as 90 mm. Formation F2 is used. Borehole radius is 117 mm. The inner radii of the collar are marked on the lines. Collar C32 is used

in longer duration of the collar waves that overwhelm the formation P-waves. For example, in the right panel of the waveforms shown in Fig. 5.49, the P-wave is difficult to see for the collar inner radius of 47 mm. The P arrivals can be seen when the collar inner radius is 7 mm. Table 5.3 gives the P-wave amplitudes with different collar inner and outer radii.

By reducing the collar outer and inner radii, the amplitude of the collar wave can be reduced and the formation P-wave might be visible.

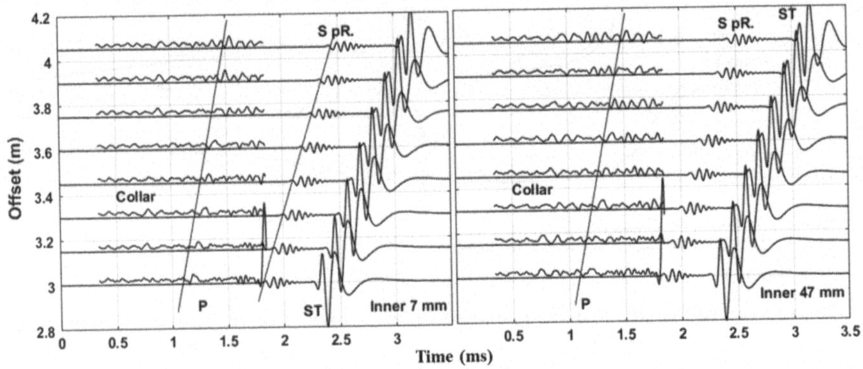

Fig. 5.49 The effect of collar inner radius on different modes. Collar outer radius is fixed as 117 mm and the inner radii are listed on each plot. Formation F2 in Table 5.2 and a 10 kHz monopole are used. Collar modes have been amplified. Collar C32 is used

Table 5.3 Maximum amplitude (**MA**) of P-wave (10 kHz monopole) with various collar radii. OR: outer radii, IR: inner radii. Maximum amplitude of the source is set as 10,000

NO.	1	2	3	4	5	6	7
IR (mm)	27	27	27	27	7	47	67
OR (mm)	50	70	90	100	90	90	90
MA	47.5	83.3	162.0	226.7	162.4	154.3	127.0

5.2.1.3 Effects of Collar Material Properties on P-Wave Velocity Determination

Here the effects of the elastic properties of the collar on wavefields in formation F1 (see Table 5.2) are presented. The collar used in the simulations in this section does not have slots.

Figure 5.50 shows the dispersion of the monopole ALWD wavefields in formation F1 for different properties of the collar material. The S-wave velocity of the collar material in subplots are 3.3 (Fig. 5.50a), 4 (Fig. 5.50b), 5 (Fig. 5.50c), and 6 km/s (Fig. 5.50d). The effect of the collar P-wave velocity on the dispersion curves of different modes is small. The influence of the collar P-wave velocity on collar L1 wave dispersion curves becomes obvious when the material S-wave velocity changes. The material density has little impact on the dispersion. Collar material properties do not affect the dispersion of pR and ST waves. However, a mode (denoted as P in the figures) propagating with nearly the formation P-wave velocity appears when the collar S-wave velocity is equal to or above 4000 m/s. This dispersive mode is marked as "P" in Fig. 5.50. It is a combination of the formation P-wave and the leaky collar L modes (see explanation in Fig. 5.7, where the collar longitudinal mode leaks energy into the formation).

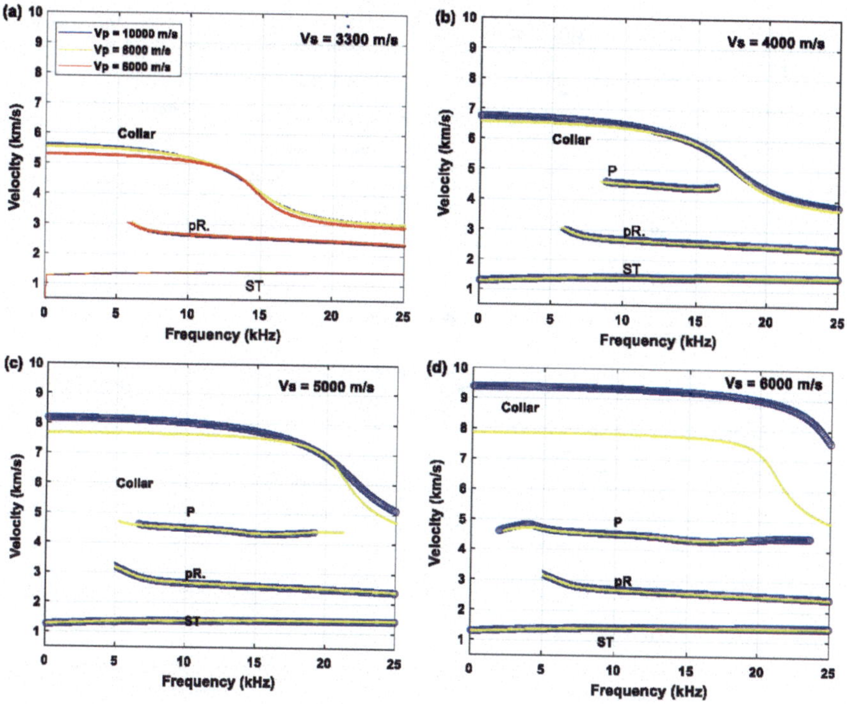

Fig. 5.50 Dispersion curves for monopole ALWD wavefield with various collar material P and S velocities. Material S-wave velocity is listed on each plot. Color legend for different material P-wave velocity is labeled in **a**. Formation is F1 (Vp = 4500 m/s, Vs = 2650 m/s, and density = 2400 kg/m^3)

Figure 5.51 shows dispersion curves with various collar S-wave velocities when collar P-wave velocity is 10 km/s (Fig. 5.51a), 9 km/s (Fig. 5.51b), 8 km/s (Fig. 5.51c) and 7 km/s (Fig. 5.51d), respectively. The phase velocity of the collar L1 decreases with decreasing collar S-wave velocity. The formation P-wave disappears when the collar S-wave velocity is at or below 4 km/s. The pR and ST waves are only slightly affected by the collar material S-wave velocity.

In conclusion, the S-wave velocity of the collar is a key parameter that controls the phase velocities of the collar L modes.

The effect of collar density on waveforms is demonstrated by simulated waveform in Fig. 5.52. The collar velocities are those of C12 and two densities are extreme values: 9.0 and 1.60 g/cm^3. Figure 5.52 shows the waveforms with collars of two different densities. Increasing collar material density reduces the amplitudes of different modes. The collar L waves are most strongly influenced by the collar density. The amplitude and arrival time of ST wave are slightly affected by collar density. Because the amplitude of collar L modes is significantly lowered by the increased collar material density, the formation P-wave becomes more visible and easier to identify. This can be verified from the velocity-time semblance analysis in Fig. 5.52c, where the formation P-wave has a stronger coherence for the high collar density.

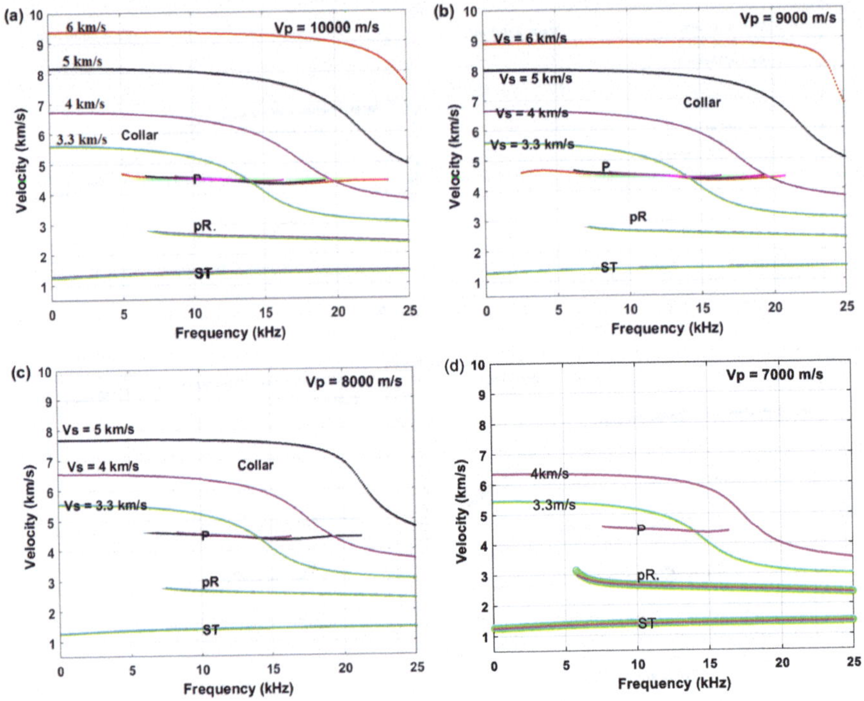

Fig. 5.51 Dispersion curves for monopole ALWD wavefield with various collar material S-wave velocities for different material P-wave velocities. Collar material S-wave velocities are listed above each line. Formation is F1 in Table 5.2

The influence of density of collar C22 on the waveforms (in Fig. 5.53) is similar to that of collar C12. However, the reduced velocities of the collar material (compared with C12) makes the collar L and P waves arrive more closely in time making the determination of the formation P-wave velocity difficult.

For the same collar density, the waveforms with various collar velocities in fast formation F1 are shown in Fig. 5.54 (All collars have density of 7.85 g/cm^3). Waveforms of the first 1.2 ms (Fig. 5.54a) give details about the collar waves from the simulations for three different collars. The collar wave frequency increases when velocity of the collar material increases. The formation P-wave appears when collar velocities increase from C32 to C22, and the formation P-wave separates much more from the collar wave when the collar is C12. Compared to C32, the relatively low frequency and large amplitude of the collar wave makes the P-wave apparent in C12 and C22. The increasing collar velocity enhances the amplitude of pR and ST waves (Fig. 5.54b, c).

From above examples, we conclude that a collar with greater material density and S-wave velocity makes the formation P-wave more obvious and easier to pick. This point may be considered for choosing the material for the collar and lead to the use of an advanced composite material for the collar (Wang et al. 2016c).

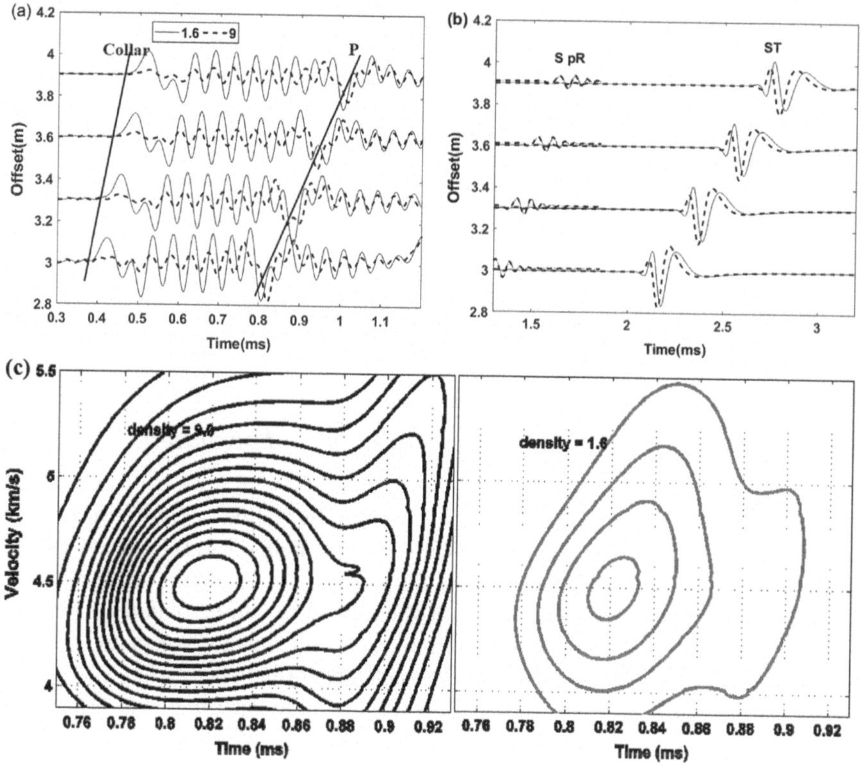

Fig. 5.52 Effect of collar density on waveforms for C12 in formation F1 (parameter see Table 5.2). The solid and dash lines indicate collar density of 1.60 and 9.00 g/cm^3, respectively. **a** Collar and P waves; **b** S, pR and ST waves; **c** Contour plots showing velocity analysis using velocity-time semblance method for the P-wave portion of the array waveforms

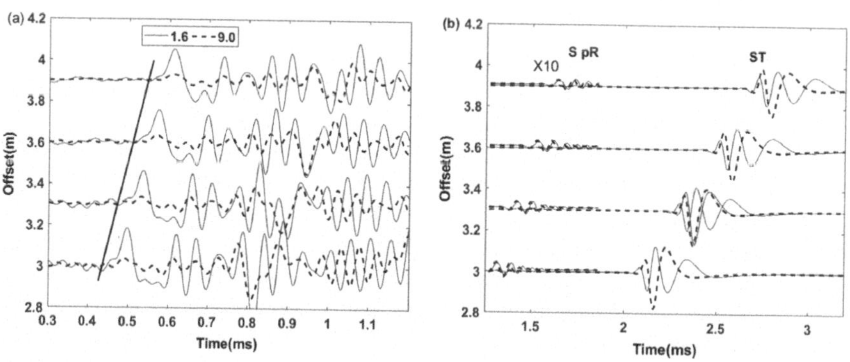

Fig. 5.53 Effect of collar density on waveforms for C22 in formation F1. Density of 1.60 and 9.00 g/cm^3 cases are labeled as 1.6 and 9.0, respectively. **a** Collar and P waves; **b** S, pR and ST waves

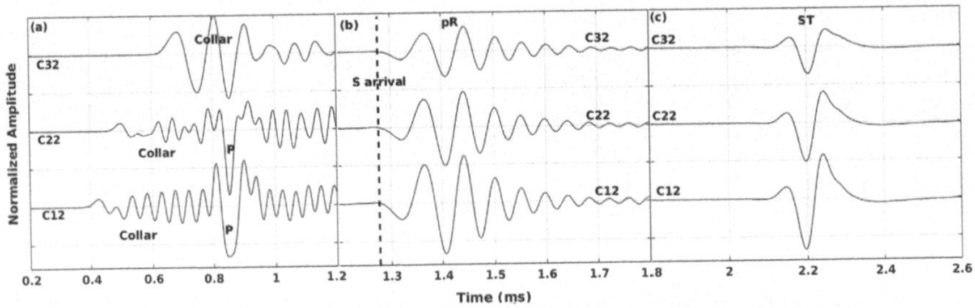

Fig. 5.54 ALWD waveforms in formation F1 for collars C12, C22, and C32. **a** First 1.2 ms waveforms. **b** pR waves. **c** ST waves. Amplitudes are scaled independently in each time window. All collars have density of 7.85g/cm^3

5.2.2 S-Wave Velocity Measurement in Slow Formations

The measurement of the S-wave velocity in slow formations is more difficult in ALWD than in the wireline case. In ALWD, the strong collar flexural wave interferes with the formation flexural wave at low frequency. This hinders S-wave velocity determination. However, the formation S-wave velocity can be obtained from the formation flexural mode generated by a dipole source using the dispersion curves at high frequency and applying a frequency dependent correction.

The quadrupole source is also used for S-wave velocity measurements. The interference of the collar modes with the formation waves is reduced because of the high cutoff frequency (approximately 10 kHz) of the collar screw wave.

5.2.2.1 Measuring Formation Wave Velocities with the Dipole Measurement

As mentioned in Sect. 5.1.2.2, it is possible to determine both the P- and S-wave velocities from a dipole ALWD tool in a very slow formation, due to the large velocity difference between the formation and collar flexural-waves.

In this section, the effects of the collar properties, including radii, velocities, and density are investigated. Figure 5.55 shows the effects of collar radii on the dispersion curves, where the collar C32 and slow formation S1 (Table 5.2) are used. The borehole radius is fixed at 117 mm. Increasing inner or outer radii lowers the cutoff frequency of the higher-order collar flexural modes such as F2. The collar F1 mode is slower in the thin collar case. Changing collar outer radius affects collar F1 more than the inner radius. Only the high frequency part of the collar F1 mode is affected by the increased inner radius and the interference between collar F1 and leaky P becomes stronger. The leaky P and formation flexural waves are not affected by changing the inner radius. However, both the leaky P and formation flexural waves are affected by the changing outer radius, where the modes become more dispersive when the collar outer radius increases. Because increasing outer radius of the collar increases the velocity of the collar F1 mode, the interference between the collar F1 and leaky P

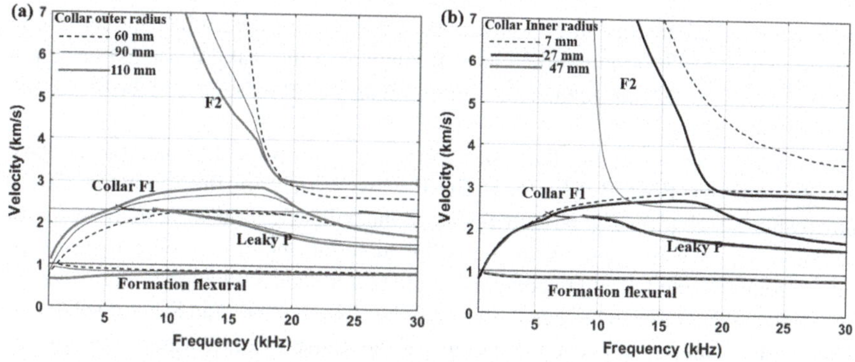

Fig. 5.55 Effect of various collar radii on dispersion curves of different modes (dipole measurement in formation S1). **a** Effect of collar outer radius when the collar inner radius is fixed as 27 mm. **b** Effect of collar inner radius when the collar outer radius is fixed as 90 mm. The legends for different lines are listed on the plots. Collar C32 in Table 5.2 is used

becomes weak (see the interference at 10 kHz in Fig. 5.55a). Similar to the monopole case, collar density hardly changes the dispersion curves in the dipole wavefields.

Figure 5.56 shows the dispersion curves for a collar with various material velocities. The S-wave velocity of the collar material is listed above each plot. The formation flexural, leaky P-waves and the first-, second- and third- order collar flexural waves are shown in the figures. The collar F1 interferes with both the leaky P and formation flexural waves in the low frequency range (approximately 6 kHz for the leaky P-wave, and 2 kHzfor the formation flexural wave in Fig. 5.56a), making the P- and S-wave velocity determinations difficult.

When the S-wave velocity of the collar material increases (Fig. 5.56b), the cutoff frequencies of the collar F2 and F3 modes increase and the gap between those two modes becomes larger. The interference between the first order collar flexural wave

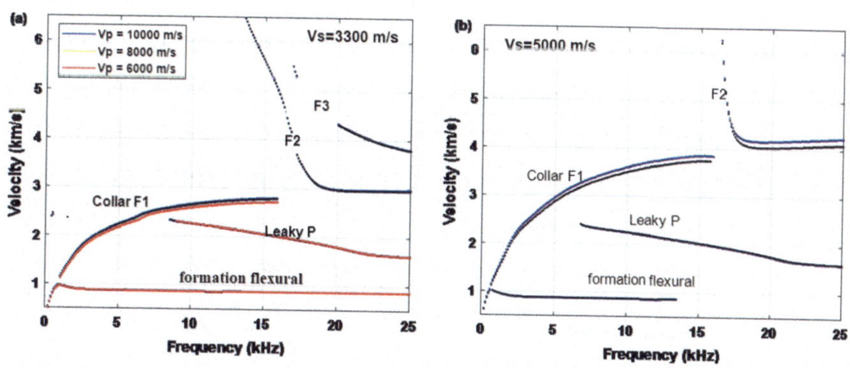

Fig. 5.56 Dispersion curves for the dipole ALWD wavefields for various collar material P-wave velocities. Collar material S-wave velocity is listed above each plot. Lines are color coded to indicate material P-wave velocity. Formation is S1 (see Table 5.2)

and formation modes becomes weaker. It is possible to determine the formation P-wave velocity (2300 m/s) at the cutoff frequency of the leaky P-wave. Formation S-wave velocity (1000 m/s) can be determined at the cutoff frequency of the formation flexural wave in Fig. 5.56b.

Figure 5.57 shows mode dispersion curves for various collar material S-wave velocities for two P-wave velocities. The P-wave velocity of the collar material is shown above each plot. The phase velocity of the collar flexural wave increases with the increasing S-wave velocity of the collar material, when the collar material P-wave velocity is fixed. From the dispersion characteristics of flexural collar waves in Figs. 5.56 and 5.57, it can be seen that even through both the P- and S-wave velocities of the collar material affect the velocities of the collar flexural waves, the S-wave velocity is the dominant factor. The increased phase velocity of the collar F1, which results from the increasing collar material S-wave velocity, weakens the interference from the collar F1 mode on the formation waves.

Figure 5.58a shows the array waveforms for the three collars, C32, C22, and C12 in slow formation S1. Although the collar flexural waves always disturb the formation flexural waves to some extent, the interference becomes weaker, and the arrival of the formation flexural wave can be determined when the P and S velocities of the collar material are high as for C12 and C22. The collar flexural wave with velocity ranging from 0.8 to 2.5 km/s (Fig. 5.58b) is very dispersive. Although the determined formation flexural-wave velocity is under 1 km/s, it approaches the formation S-wave velocity when the collar velocities increase. In this situation, it is possible to determine the formation S-wave velocity from the dispersion of flexural waves at the cutoff frequency.

A collar with large velocities (especially one with an S-wave velocity of more than 4 km/s) and large density enables the dipole tool to measure the formation P and S velocities in slow formations. A high impedance collar is better in very slow formations.

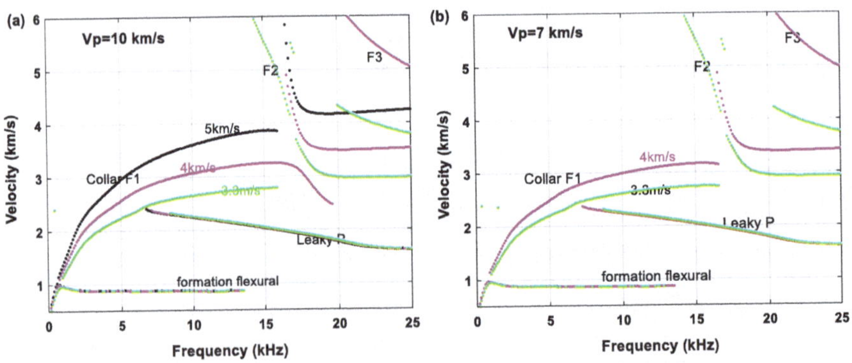

Fig. 5.57 Dispersion curves in the dipole ALWD wavefield with various collar material S-wave velocities. P-wave velocity of the collar material is listed in each plot. The black, magenta, and green dash lines are for collar S-wave velocities of 5000, 4000, and 3300 m/s. Formation is S1 (see Table 5.2)

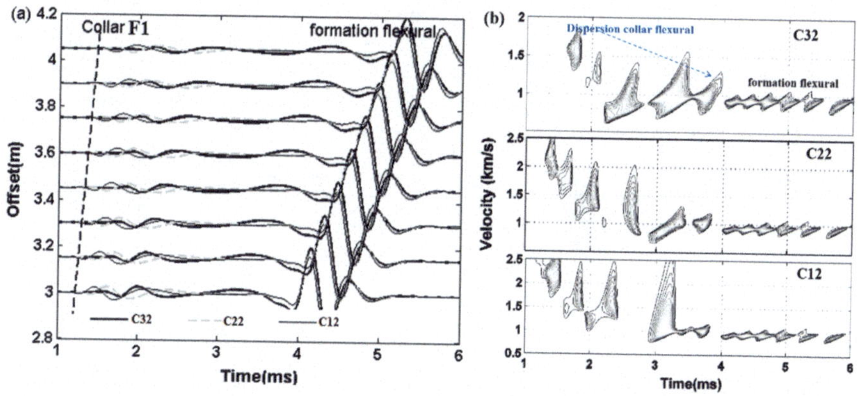

Fig. 5.58 The logging response of the dipole ALWD tools in a slow formation S1. **a** Array waveforms. Different curves indicate collars with different velocities. **b** Velocity-time semblance contour plots for array waveforms for different collars

5.2.2.2 S-Wave Velocity Measurement from the Low Frequency Quadrupole System

As mentioned in Sect. 5.2.1.2, only the outer and inner radii of the collar affect the dispersion curves of the collar screw modes in a slow formation. Examples of the effect of the inner and outer radii of collar C32 (listed in Table 5.2) on the dispersion curves of different modes are shown in Fig. 5.59.

The formation screw mode is affected only by the outer radius. The formation "screw" (or quadrupole) mode has a lower cutoff frequency and is more dispersive when the collar outer radius is large. Either decreased outer radius or increased inner radius moves the collar screw towards a lower frequency range. This indicates that

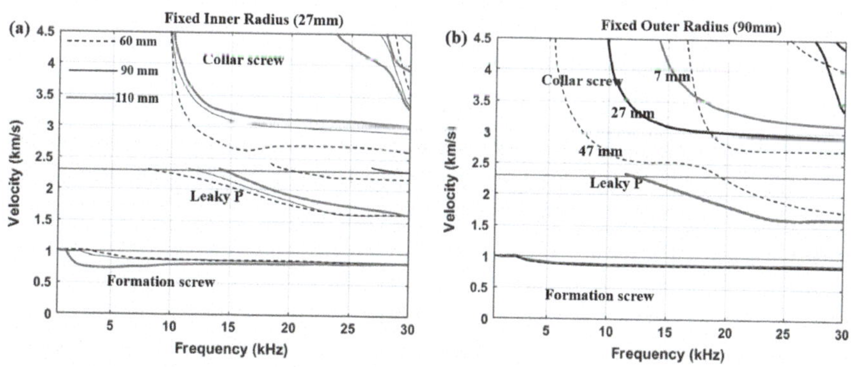

Fig. 5.59 The effect of collar radii on dispersion in quadrupole measurements in a very slow formation. **a** Different collar outer radii when the collar inner radius is fixed as 27 mm. **b** Different collar inner radii when the collar outer radius is fixed as 90 mm. The legend for different lines is listed on the plot. Collar C32 and formation S1 in Table 5.2 are used

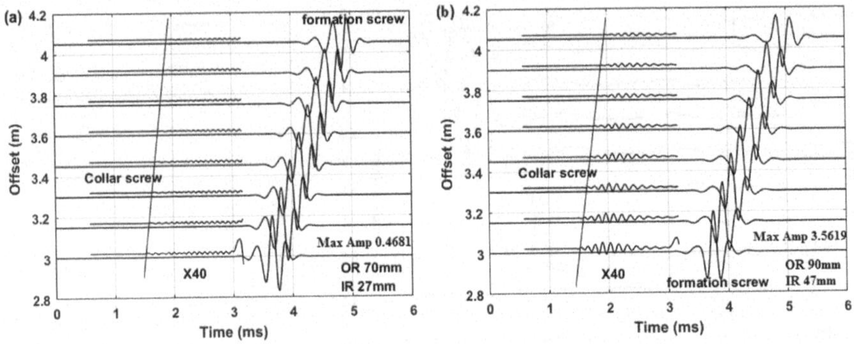

Fig. 5.60 The effect of various collar radii on quadrupole waveforms (a 4-kHz source frequency). Collar radii are listed on each plot. Formation S1 in Table 2.1 and collar C32 in Table 5.2 are used here. The borehole radius is 117 mm. The maximum amplitude of each plot is listed

a thinner collar lowers the cutoff frequency of the collar screw mode. The increased inner radius moves the cutoff frequency much more than that from the decreased outer radius. The lower cutoff frequency of the collar screw mode makes the collar screw wave the first arrival, which could be mistaken for the formation screw wave.

Two examples of quadrupole waveforms (a 4-kHz source frequency) are shown in Fig. 5.60 where the collar radii are listed on each plot. Collar C32 and formation S1 in Table 5.2 are used for the calculation. Borehole radius is 117 mm. The collar screw mode appears in both simulations. A larger radius (either outer or inner radii) collar generates larger amplitude formation waves (shown in Table 5.4). The effect on the amplitude of the outer radius is much greater than the change of the inner radius.

To avoid the influence of the collar screw modes on the formation S-wave velocity measurement, a low source frequency, approaching the cutoff frequency of the formation screw mode would be helpful. However, the frequency range of the drilling noise, from 1 to 3 kHz (see Figs. 5.2 and 5.3) could reduce signal-to-noise ratio and affect the low frequency quadrupole measurements.

Table 5.4 Maximum amplitude (**MA**) of the formation screw wave (4 kHz quadrupole) with various collar radii. OR: outer radii, IR: inner radii. Maximum amplitude of the source is set as 10,000

NO.	1	2	3	4	5
IR (mm)	27	27	27	7	47
OR (mm)	70	90	100	90	90
MA	4.7	34.8	100.7	34.6	35.6

5.2.2.3 Formation S-Wave Velocity Determination from Low Frequency Hexapole Source

The source configuration of hexapole is similar to that of the quadrupole source shown in Fig. 5.4. For the hexapole, the ring source is divided into six alternating positive and negative phases. The cutoff frequencies of the collar hexapole modes are very high (Geerits et al. 2006; Wang et al. 2009). A low frequency hexapole source produces clear waveforms to determine the formation S-wave velocity in slow formations. An example for the 4 kHz ALWD hexapole measurement in slow formation S1with the collar C32 (Table 5.2) is shown in Fig. 5.61. There is only one event in the array waveforms in Fig. 5.61a, corresponding to the formation hexapole wave. The formation hexapole wave propagates with the formation S-wave velocity. The S-wave velocity is determined from the velocity-time semblance analysis as shown in Fig. 5.61b.

The influence of the collar radii on the dispersion curves for a hexapole source are also investigated as shown in Fig. 5.62. The collar inner radius has a similar effect on the cutoff frequencies of collar hexapole modes as that on the collar screw modes. The influence of the outer radius exhibits a different behavior from that of the quadrupole measurements. Larger outer radius lowers the cutoff frequencies of collar hexapole modes.

One problem for the hexapole measurement is its low amplitude. The maximum amplitudes of hexapole waveforms (4 kHz) for different collar radii are listed in Table 5.5. Increasing the outer radius increases the amplitude of the formation hexapole wave. The amplitudes listed in Table 5.5 are 3 to 8 times lower than those of quadrupole case (Table 5.4) and much lower than those of the monopole case (Table 5.3).

In summary, a low frequency hexapole source provides an effective means to determine the formation S-wave velocity in slow and very slow formations.

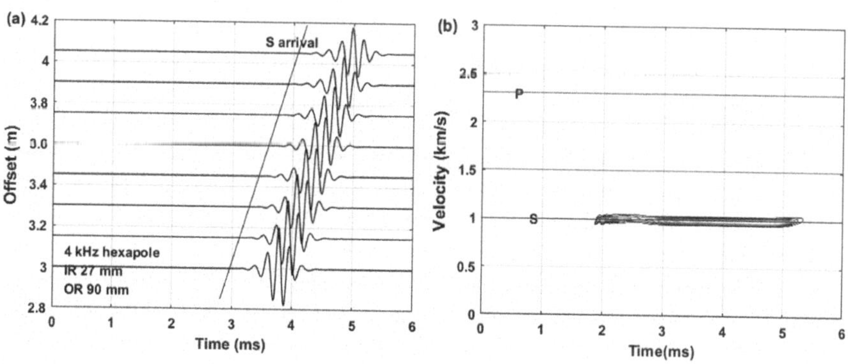

Fig. 5.61 ALWD hexapole wavefield (4 kHz center frequency) in slow formation S1 (listed in Table 5.2). **a** Array waveforms. **b** Velocity-time semblance contour plot for the array waveforms in **a**

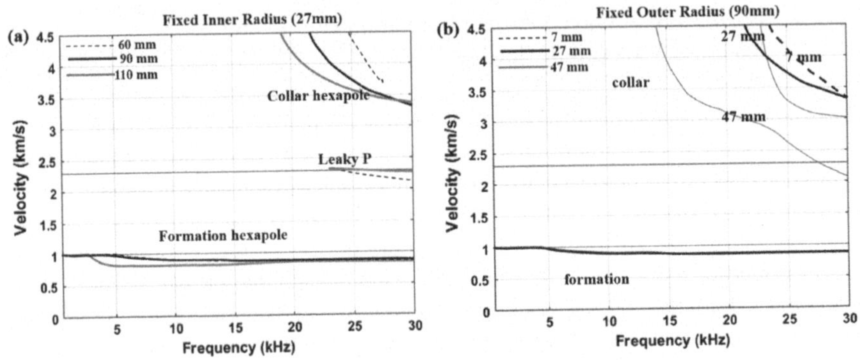

Fig. 5.62 The effect of collar radii on dispersion curves of different modes in the hexapole wavefield. **a** Effect of collar outer radius (see legend) when the collar inner radius is fixed as 27 mm. **b** Effect of collar inner radius (line labeled) when the collar outer radius is fixed as 90 mm

Table 5.5 Maximum amplitude (**MA**) of formation wave (4 kHz hexapole) with various collar radii. OR: outer radius, IR: inner radius. Maximum amplitude of the source is set as 10,000

NO.	1	2	3	4	5
IR (mm)	27	27	27	7	47
OR (mm)	70	90	100	90	90
MA	0.583	6.881	30.67	6.876	6.929

5.3 Field Examples of ALWD Measurements

Two examples of ALWD data are shown in Figs. 5.63, 5.64 and 5.65 (Qleibo 2012). The tool has a monopole and a quadrupole source. The source frequency ranges from 1 to 20 kHz. There are 48 receivers on four axial arrays and each with 4 inch intervals. Receiver azimuths are 0°, 90°, 180°, and 270° (Mohammed et al. 2011). Figure 5.63 shows the data collected by the high-frequency monopole tool. The formations above XX25 m are relatively clean sands with low Gamma Ray (GR) counts. P and Stoneley waves are easy to identify on the full waveform recording in Track 3. Below XX25, the formations with high GR counts are shales with lower velocity. The S-wave cannot be seen in the shale formation below the depth of XX30 m. The waveforms at two depths are shown in Figs. 5.64 (XX10 m) and 5.65 (XX30 m). The collar waves were eliminated by an isolator.

In the slow formation, the quadrupole tool is used to obtain the formation S-wave velocity. The S-wave velocity is determined at the cut off frequency of the formation quadrupole wave (screw mode). Figure 5.66 shows a field data example collected using a quadrupole ALWD tool (Modified from presentation document of Qleibo [2012]). The arrival times of formation quadrupole waves are marked by a red dashed line. The leaky P-waves are dispersive, and are the first arrivals. The P and S velocities can be simultaneously determined, as shown in Tracks 2 and 3. Figure 5.67 shows

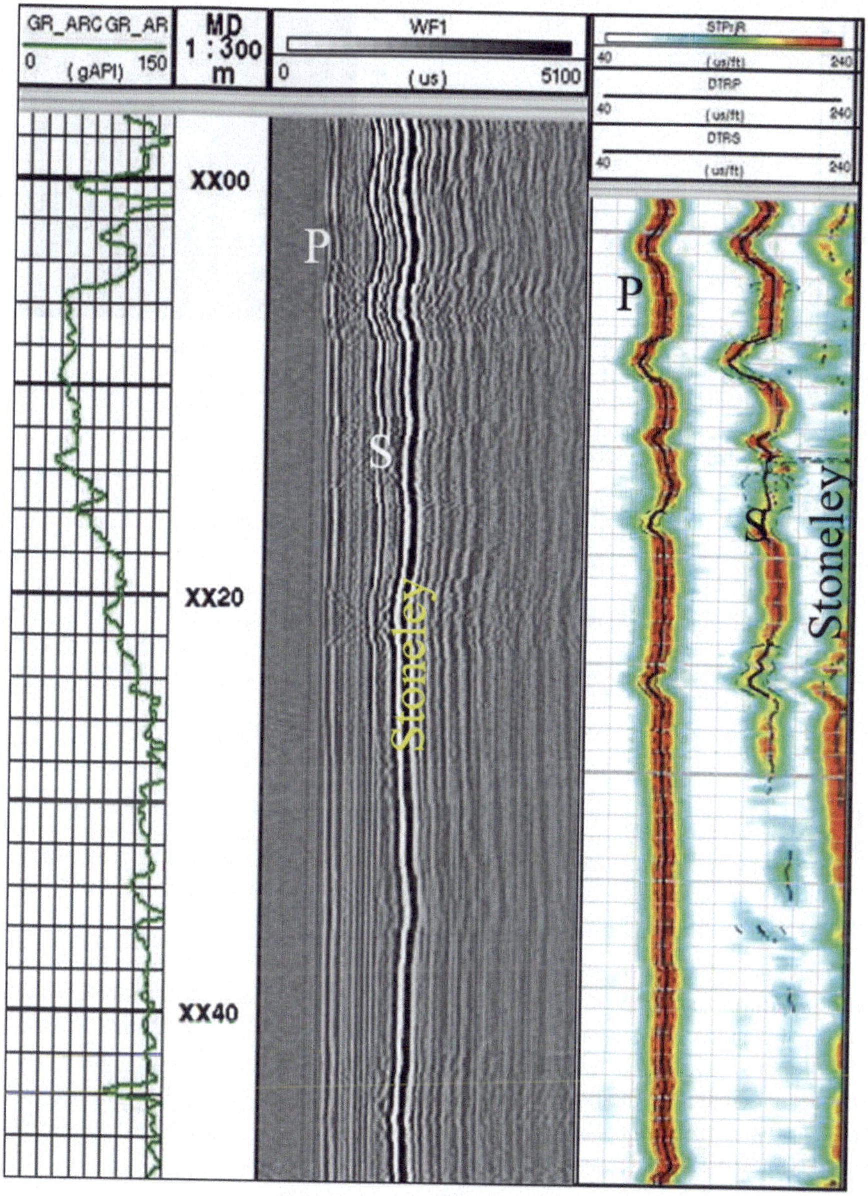

Fig. 5.63 Field data collected by a monopole ALWD tool. Track 1 is Gamma Ray (GR) log. Track 2 is depth. Track 3 shows recorded waveforms. Track 4 is extracted slowness (reciprocal of velocity) of P, S, and ST waves at different depths (Modified from presentation document of Qleibo [2012]. https://www.fesaus.org/webcast/2012/09/NTF/2_SonicScope_Schlumberger.pdf. Access on April 25, 2018)

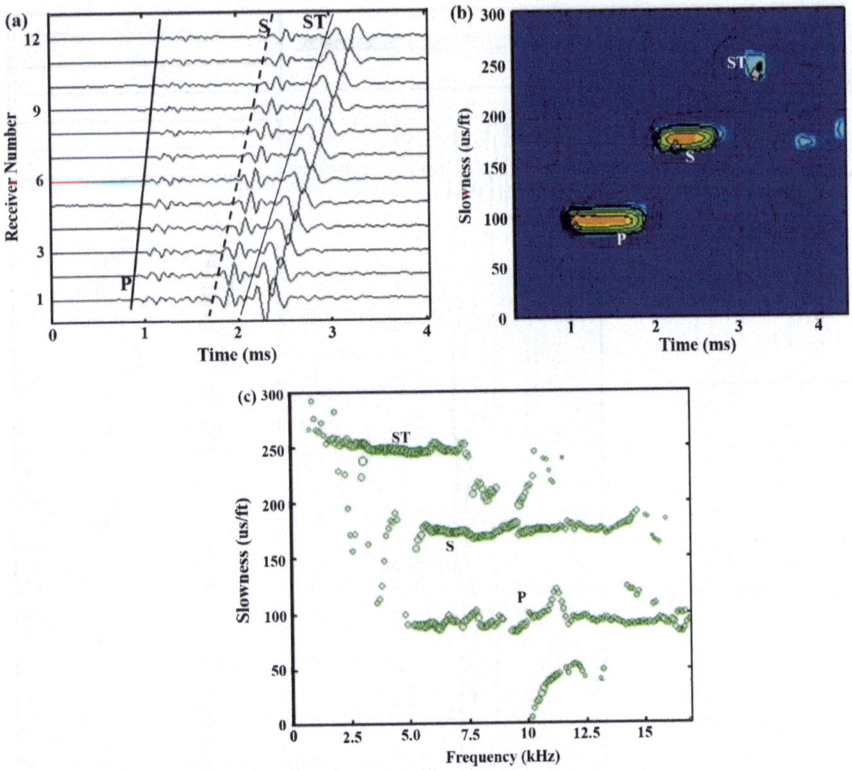

Fig. 5.64 Monopole waveforms and velocity determinations at depth of XX10 m in Fig. 5.63. **a** Array waveforms. **b** Slowness-time semblance result. **c** Dispersion analysis by the Matrix-Pencil method (see Sect. 3.2.2). The velocities for P, S, and ST waves are 3430, 1740, and 1230 m/s (Modified from presentation document of Qleibo [2012]: https://www.fesaus.org/webcast/2012/09/NTF/2_SonicScope_Schlumberger.pdf. Access on April 25, 2018)

the waveforms at depth of XX60 m. The leaky P and formation quadrupole waves are visible in the waveforms. The leaky P-wave has a high frequency and formation quadrupole wave has a low frequency. The P and S velocities, determined from the Matrix-Pencil method, are marked by dashed lines in Fig. 5.67b. The velocities are 2430 and 1095 m/s. The cutoff frequency of the formation quadrupole wave is approximately 2 kHz.

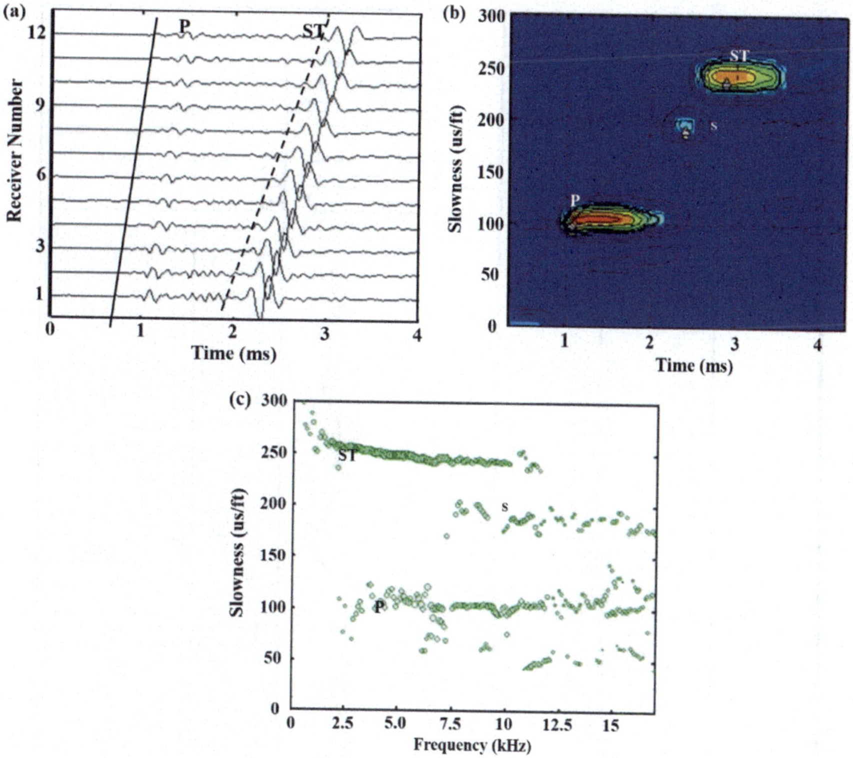

Fig. 5.65 Monopole waveform and velocity determinations at depth of XX30 m. **a** Array waveforms. **b** Slowness-time semblance result. **c** Dispersion analysis (see Matrix-Pencil method in Sect. 3.2.2). The velocities for P and ST waves are 2885 and 1250 m/s. The S-wave velocity is approximately 1570 m/s (Modified from presentation document of Qleibo, 2012: https://www.fesaus.org/webcast/2012/09/NTF/2_SonicScope_Schlumberger.pdf. Access on April 25, 2018)

5.4 Summary

In this chapter, we discussed borehole acoustic logging-while-drilling. The factors affecting the wavefields and performance of ALWD in fast and slow formations were investigated. Based on the wavefields' characteristics, the challenges and potential solutions for accurate velocity measurements were investigated.

Monopole ALWD measurements:

(1) Formation properties do not affect the velocities of collar waves, but they affect the amplitudes of collar waves.
(2) In fast formations, the collar waves could interfere with formation P-wave. This is especially true for the formations with S-wave velocity above 2800 m/s.
(3) Properties of the collar, such as radii, thickness, density, and elastic moduli, affect collar waves. The slotted collar and attenuators reduce the amplitudes of collar waves. Collar material properties affect the velocity of the collar waves.

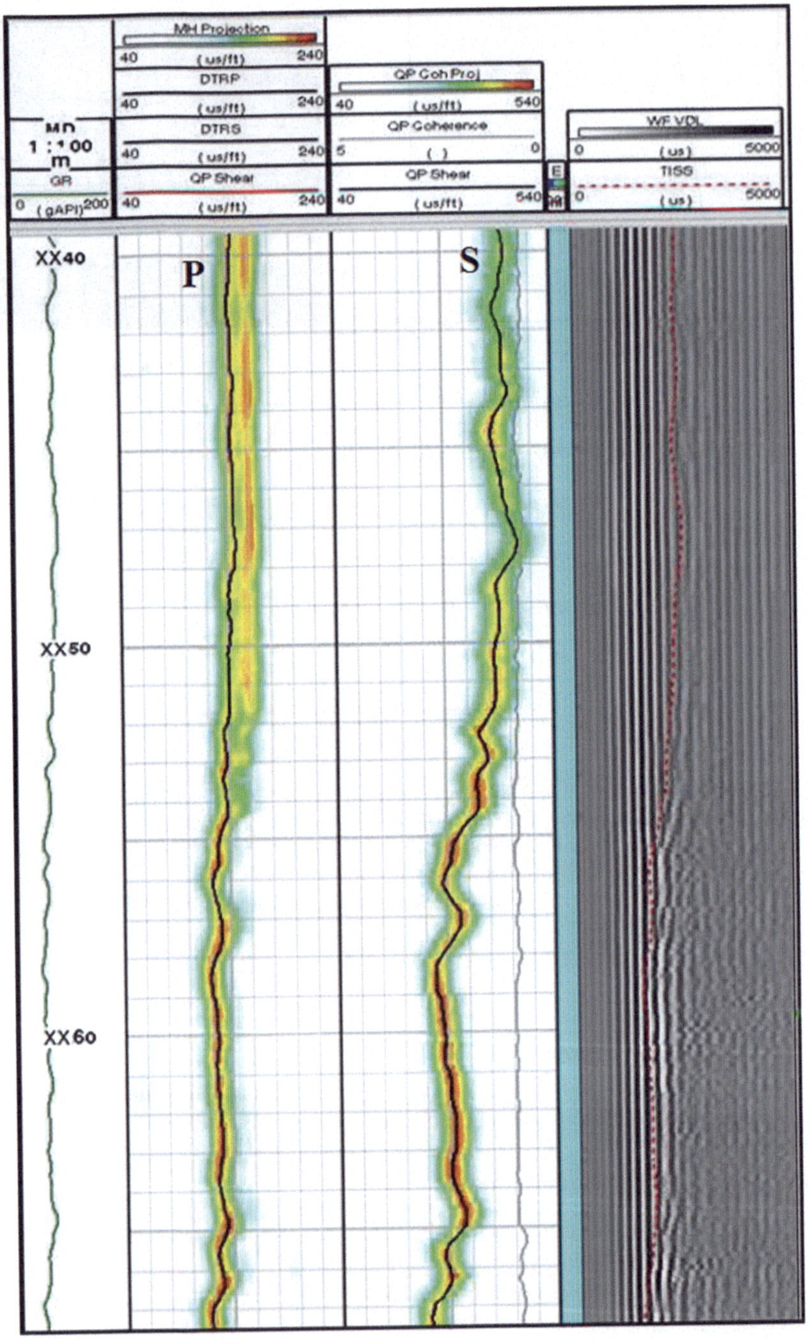

Fig. 5.66 Field data collected by a quadrupole ALWD tool. Track 1 is Gamma Ray (GR) log. Track 2 obtained P-wave slowness. Track 3 is obtained S-wave slowness. Track 4 shows the waveforms (Modified from presentation document of Qleibo [2012]: https://www.fesaus.org/webcast/2012/09/NTF/2_SonicScope_Schlumberger.pdf. Access on April 25, 2018)

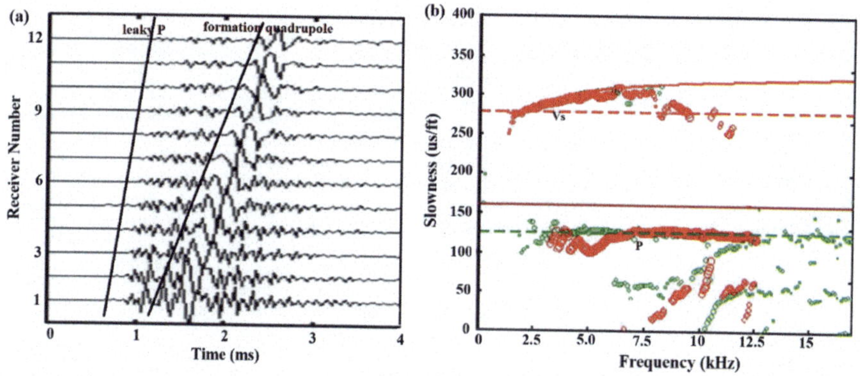

Fig. 5.67 Quadrupole measurement at XX60 m. **a** Array waveforms. **b** Dispersion analysis by the Matrix-Pencil method (see Sect. 3.2.2). The obtained velocities for P- and S-waves are 2430 m/s and 1096 m/s. The cutoff frequency of the formation quadrupole is approximately 2 kHz. The low slowness (high velocity) region above approximately 8 kHz may be higher modes (Modified from presentation document of Qleibo, 2012: https://www.fesaus.org/webcast/2012/09/NTF/2_SonicScope_Schlumberger.pdf. Access on April 25, 2018)

ALWD dipole and quadrupole measurements:

(1) Leaky P-waves in slow formations enable the formation P-wave velocity determination.
(2) The collar flexural wave could interfere with formation flexural waves in slow formations in dipole measurements, affecting the S-wave velocity determination. The interference becomes weaker when the formation velocity becomes lower. A low frequency dipole tool is effective to obtain S-wave velocity in very slow formations.
(3) The collar quadrupole modes have high cutoff frequencies. This benefits the S-wave velocity determination in slow formation from formation quadrupole mode.
(4) At very low frequencies, drilling noise becomes an impediment for velocity determination.

Table 5.6 gives a detailed summary about formation velocity determination capabilities of different ALWD sources in different formation.

Table 5.6 Summary of different ALWD sources for formation velocity measurements

	Monopole	Dipole	Quadrupole	Hexapole
Very slow formation	Vp can be determined from leaky P at the cutoff frequency. Vs cannot be determined directly.	Due to the weak interference of the collar wave on the formation waves. Vp and Vs can be determined at the same time from leaky P and formation flexural wave at the cutoff frequencies, respectively.	Vp cannot be determined because of the very weak leaky P wave amplitude. Vs can be determined directly when the source frequency is lower than the cutoff frequency of collar wave. Collar wave generated by a thin collar may have a low cutoff frequency which may affect the Vs determination.	Vp cannot be determined. Vs can be determined directly when the source frequency is lower than the cutoff frequency of the collar wave. The challenge of Vs determination is the low amplitude of the formation hexapole wave
Slow formation	Same as above	Vp can be determined from leaky P at the cutoff frequency. Vs determination cannot be made because of the interference of the collar wave on the formation flexural wave, but it is possible when collar material has large impedance where the interference becomes weak.	Same as above	Same as above
Slow-fast formation	Vp determination is a challenge because of the interference of the collar wave. Acoustic isolation and collar material with large impedance may help. Vs can be determined directly.	Formation flexural and collar flexural affect each other which means the Vs cannot be determined directly.	Vs can be determined directly when the source frequency is lower than the cutoff frequency of the collar wave	Not discussed
Fast-fast formation	Same as above	Formation flexural is faster than the collar flexural which means the Vs can be determined directly.	Same as above	Not discussed

Chapter 6
Effects of Tool Eccentricity on Acoustic Logs

In previous chapters, we presented the wavefields acquired with centralized acoustic logging tools, where the tool was centered within a cylindrical borehole, as shown in Fig. 6.1a. Even with centralizers, the tool axis may not perfectly align with the borehole axis. The simplest case of eccentering is when the tool and borehole axes are parallel but do not coincide (Fig. 6.1b). In this case, a single vector defines the azimuth (θ) and magnitude (d) of the eccentering (in Fig. 6.2). In another case, the tool might be tilted inside the borehole (Fig. 6.1c). This is more likely to occur in deviated boreholes.

The acoustic logging-while-drilling (ALWD) tool is often off center due to drill string movements. In this chapter, the wavefields of an eccentered tool in both wireline and ALWD cases are discussed. Methods to estimate the tool position in the borehole are covered in Sect. 6.3.

6.1 Wavefields of an Eccentered Multipole Wireline Tool

The most commonly used tools have a single array of receivers. If a wireline tool in a fluid-filled borehole is not centralized, the waveforms differ from those in the centralized case. Figures 6.3 and 6.4 show waveforms for the centralized case, where monopole source A and receiver are at the borehole center, and eccentered case, where the source and receiver are offset 46 mm from the center of the fluid-filled borehole. The elastic parameters and geometry of the borehole are listed in Table 2.1.

In a fast formation, the tool offset mostly affects the waveforms after 2.5 ms. The pR waves become complex. High-order modes appear and become mixed with the fundamental modes, making the waveforms more complex.

In a slow formation, the effect of tool offset on the waveforms (Fig. 6.4) becomes more serious than in the fast formation case. Both the leaky P and ST waves are affected by tool eccentering. The leaky P-waves in the eccentered case are no longer strong as in the centered case. A high-frequency wave appears within the ST wave

H. Wang et al., *Borehole Acoustic Logging—Theory and Methods*,
Petroleum Engineering, https://doi.org/10.1007/978-3-030-51423-5_6

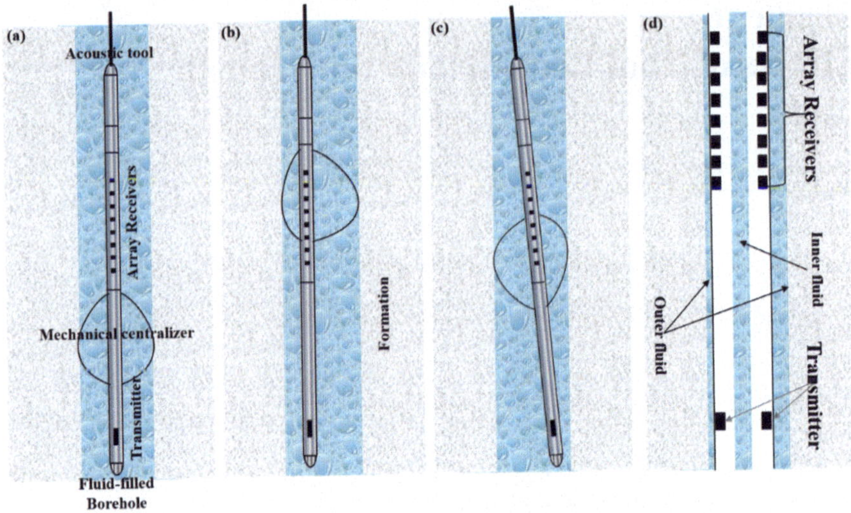

Fig. 6.1 Tool positions in the borehole. **a** Ideal case, centralized tool without tilt. **b** Eccentered tool without tilt. **c** Tilted tool. **d** Eccentered ALWD tool. The tilted tool case is not discussed in this chapter

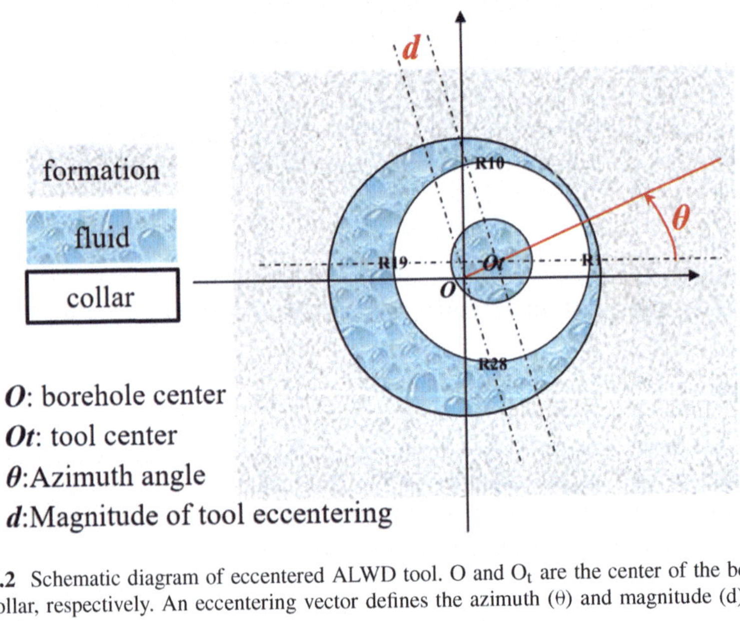

Fig. 6.2 Schematic diagram of eccentered ALWD tool. O and O_t are the center of the borehole and collar, respectively. An eccentering vector defines the azimuth (θ) and magnitude (d) of the eccentering

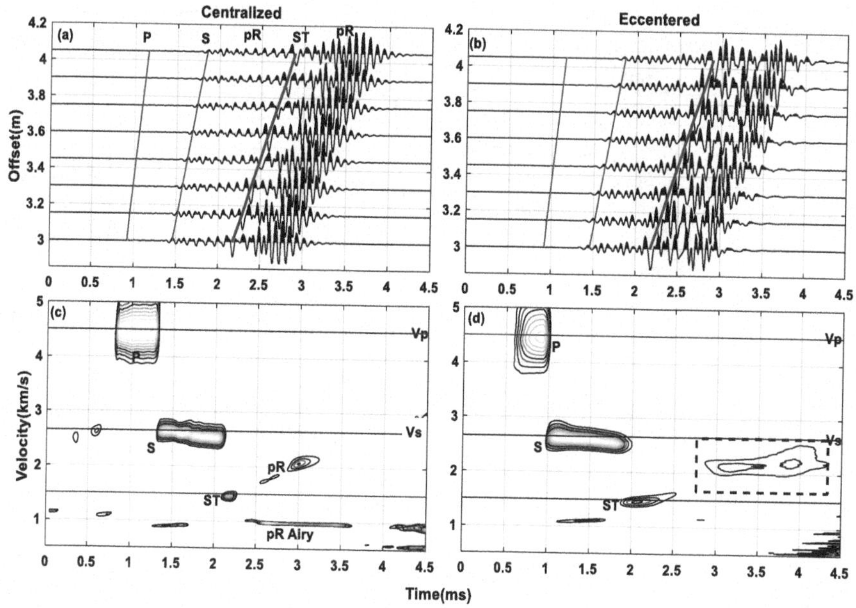

Fig. 6.3 Waveform comparison between the centralized and eccentered 10 kHz point monopole source in formation F1 (see Table 2.1). Borehole is fluid filled. **a** Waveforms in a centralized case. **b** Waveforms in an eccentered case. **c** Velocity-time semblance for array waveforms in **a**. **d** Velocity-time semblance for array waveforms in **b**. Arrival times of P, S, and ST waves are marked with blue lines

train when the tool is eccentered. The velocity of the "Stoneley" wave after 3 ms becomes slower and more dispersive than the ST wave in the centered case (Fig. 6.4b).

When the source A is offset from the borehole center, the waveforms at different azimuths become different. Figure 6.6 shows the configuration of the sources and receivers. Figure 6.5 shows the waveforms at azimuthal receivers A, B, C, and D when source is at position A. The waveforms in blue in Fig. 6.5 are at azimuthal receivers A, B, C, and D in subfigures a, b, c, and d, respectively. The black lines in the subfigures show the waveforms (listed in each plot) of the monopole, dipole and quadrupole modes. Note that the dipole modes at receivers B and D are zero by symmetry.

Modern acoustic logging tools contain azimuthally distributed sources and azimuthally distributed receivers (Close et al. 2009). The wavefield of an eccentered multipole tool can be modeled as the linear combination of a number of source and receiver locations. The waveforms for each receiver can be decomposed into their modes using Eq. 2.19. Figure 6.6 shows an eccentered wireline tool in a fluid-filled borehole. Letters A, B, C, and D in Fig. 6.6b, c depict the positions of the sources and receiver arrays. According to the vector defined in Fig. 6.2, the radii of the point sources at A, B, C, and D with respect to the borehole axis are,

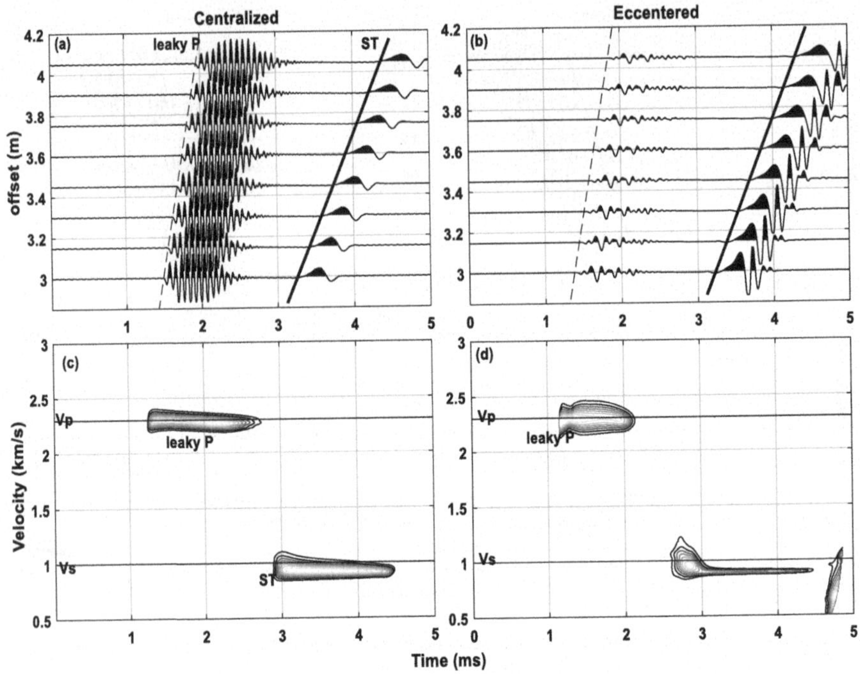

Fig. 6.4 Waveform comparison between the centralized and eccentered 10 kHz point monopole source in slow formation S1 (see Table 2.1). Borehole is fluid filled. **a** Waveforms for a centralized case. **b** Waveforms for an eccentered case. **c** Velocity-time semblance for array waveforms in **a**. **d** Velocity-time semblance for array waveforms in **b**. Arrival times of P and ST waves are marked with lines

$$r_A = \sqrt{r_t^2 + d^2 + 2dr_t \cos\theta}, \tag{6.1}$$

$$r_B = \sqrt{r_t^2 + d^2 + 2dr_t \sin\theta}, \tag{6.2}$$

$$r_C = \sqrt{r_t^2 + d^2 - 2dr_t \cos\theta}, \tag{6.3}$$

$$r_D = \sqrt{r_t^2 + d^2 - 2dr_t \sin\theta}, \tag{6.4}$$

where r_t is the radius of the tool and d is the eccentering magnitude.

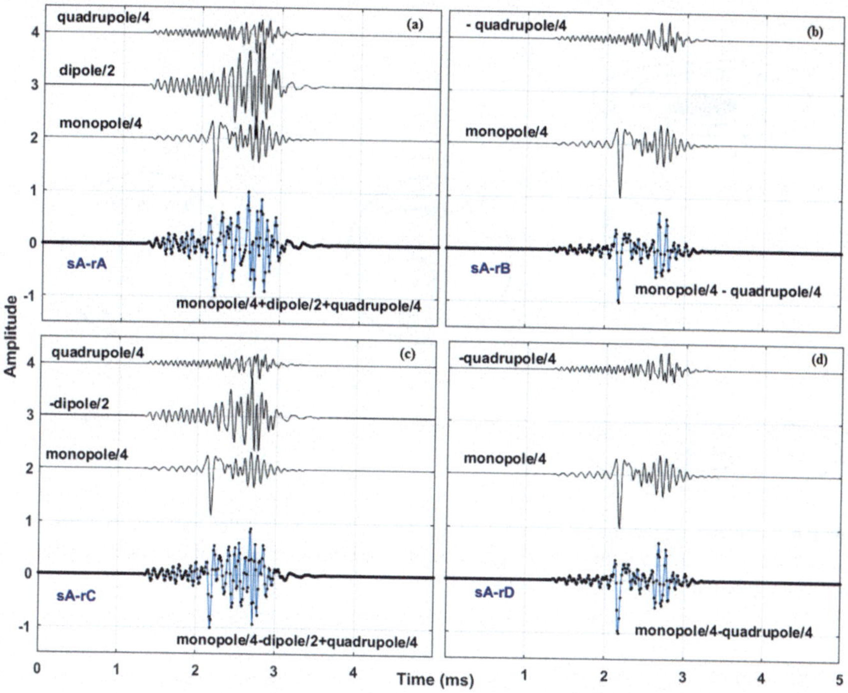

Fig. 6.5 The waveforms at the different azimuthal receivers for a centralized tool when there is a single point source located at A. Borehole radius is 100 mm and the tool radius is 46 mm. Formation is F1 (Table 2.1). Waveform and its decomposition at receiver A (**a**), at receiver B (**b**), at receiver C (**c**), and at receiver D (**d**). Letters "s" and "r" designate the source and receivers. The source-receiver distance along the z direction is 3 m. The waveforms are calculated using discrete wavenumber integration (Eqs. 2.16 and 2.18). The modes are decomposed using all azimuthal waveforms according to Eq. 2.19

6.1.1 Monopole Wavefields

6.1.1.1 Fast Formation

If a monopole tool is eccentered, cylindrical symmetry no longer holds. The wavefield of an eccentered point source can be calculated using Eq. 2.16. By combining the wavefields of four eccentered point sources, the waveforms at one receiver for an array of 4 sources are obtained. For example, if a monopole tool is eccentered 23 mm and 0° from the borehole axis, the waveform at receiver A is the summation of contributions from four eccentered sources A, B, C, and D. Figure 6.7 shows the waveforms calculated for different receiver positions. Four sources are used, located at A, B, C, and D. The source-receiver distance is 3 m and the formation is F1 (Table 2.1). The arrival times of the P- and S-waves are marked with dashed and solid lines, respectively. The P-wave is magnified 20 times. The figure shows that

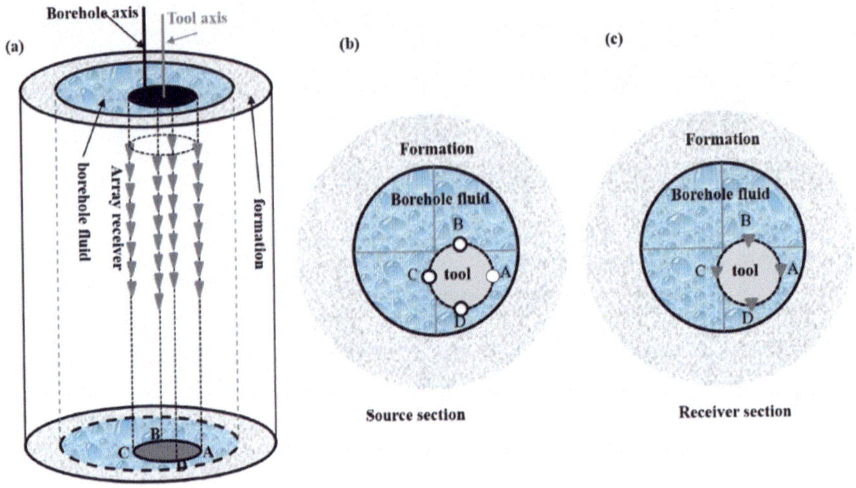

Fig. 6.6 Schematic diagram of an eccentered wireline tool in a fluid-filled borehole. **a** Perspective view of the eccentered wireline tool. **b** and **c** are the top-down views of the source and receiver sections, where the eccentering azimuth is 315°

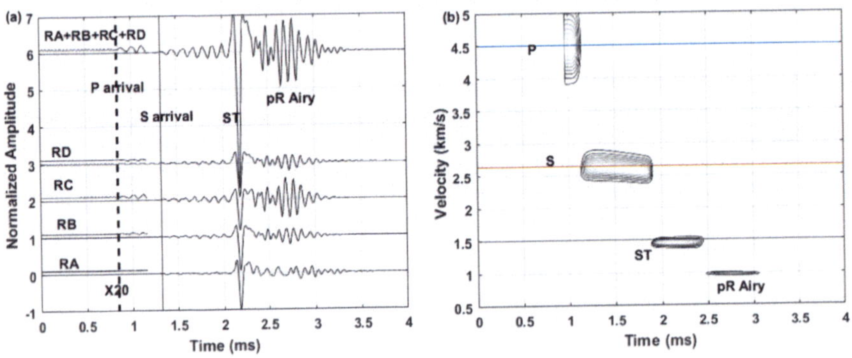

Fig. 6.7 The wavefield at 3 m offset of a 10 kHz eccentered monopole source in a fluid-filled borehole surrounded by formation F1 (see Table 2.1). Eccentering parameters d and θ are 23 mm and 0°. The P-wave is magnified 20 times for display. **a** Waveforms at each azimuthal receiver and the sum of waveforms from four azimuthal sources. **b** The velocity-time semblance contour determined from the sum of the four waveforms at each offset. Offset ranges from 3 to 4.05 m and the interval between two adjacent receivers is 0.15 m

the waveforms are different at different azimuths. The waveforms at receivers B and D are identical. This is a result of the symmetry of the two receivers relative to the borehole center. The P-wave at receiver C is largest among all the receivers because receiver C is closest to the borehole center. The P-wave at receiver A is the smallest. The same phenomenon can be seen in the amplitudes of the S-waves, where the maximum and minimum amplitudes of the S-wave are at receivers C and

A, respectively. For the ST wave, the amplitude differences are very small. The most obvious difference in the waveforms at these four azimuthal receivers is the Airy phase of the pR wave. The Airy phase is small at receivers A, B, and D. This is because of the reduced distance between the borehole wall and receivers at these azimuths. This moves the pR wave to a higher frequency range and leads to weaker excitation at the 10-kHz source frequency at these azimuths.

By adding all waveforms from azimuthal receivers A, B, C, and D, an average monopole trace, shown in Fig. 6.7a, is obtained. The velocities of different modes can be obtained from the velocity-time semblance analysis. Figure 6.7b is obtained from an array of receivers spaced from 3 to 4.05 m. The distance between two adjacent receivers is 0.15 m.

Modal composition of the waveforms at different azimuthal receivers is analyzed based on Eq. 2.18. Figure 6.8 shows the decomposed modes at different azimuthal receivers. The asymmetric modes, such as flexural ($n = 1$), screw ($n = 2$), and hexapole ($n = 3$) modes, appear when the radial symmetry of the sources is broken. The symmetric modes ST and pR are for $n = 0$. The strongest asymmetric mode at receiver A (Fig. 6.8b) is the flexural mode. The second significant asymmetric mode appears at receiver C (Fig. 6.8d). Although there are asymmetric modes (screw mode)

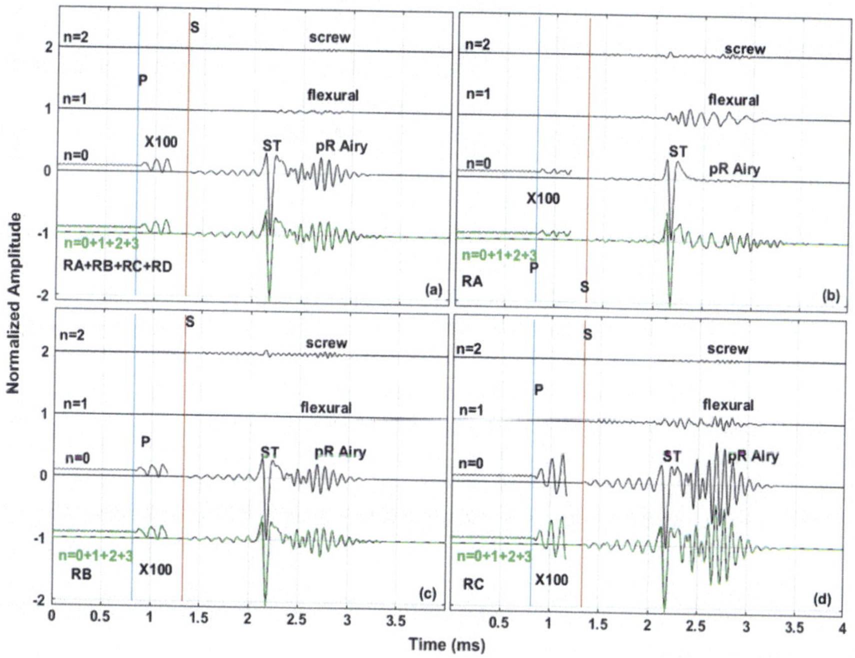

Fig. 6.8 Mode decomposition for the waves at different azimuthal receivers for an eccentered tool with parameters d and θ of 23 mm and 0° in formation F1. **a** Sum of waveforms at four azimuthal receivers. **b** Receiver A. **c** Receiver B. **d** Receiver C. The P-wave is magnified 100 times for display. The green curves are the sum of the waveforms of the first 4 orders

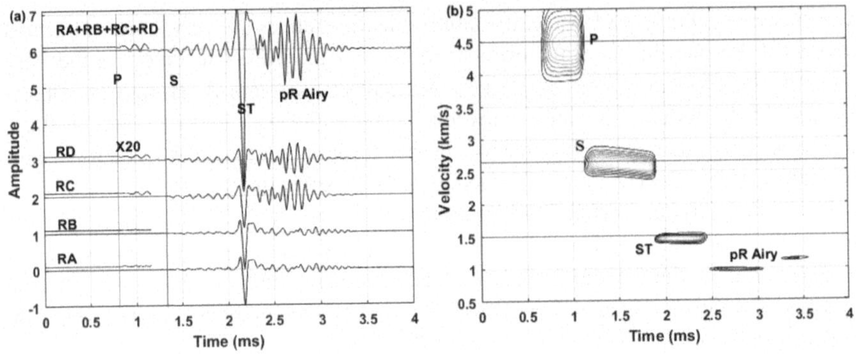

Fig. 6.9 The wavefield of an eccentered monopole tool at 10 kHz in a fluid-filled borehole surrounded by formation F1 (see Table 2.1). Eccentering parameters d and θ are 23 mm and 45°. **a** Waveforms at different azimuthal receivers and the sum of all four receivers. The P-wave is magnified 20 times for display. **b** The velocity-time semblance contour plot for the sum of all four receivers. Waveforms at offsets (source-receiver distance) between 3 and 4.05 m with 0.15 m interval are used. Waveforms at azimuthal receivers are different from those with 0° eccentering (Fig. 6.7). The summation waveforms are the same for both 0 and 45° cases

at receivers B (Fig. 6.8c) and D, they are small. The P-wave is not affected by the asymmetric modes because of the large velocity difference between the P- and S-waves. The asymmetric modes, at receivers A and C, are clear. These modes mix with the pR wave, resulting in a dispersive feature of the S-wave. The pR Airy phase is also mixed with asymmetric modes at receivers A and C.

Figure 6.9 shows the waveforms at different azimuthal receivers, the sum of the waveforms and their velocity analysis when the tool eccentering parameters d and θ are 23 mm and 45°. The waveforms at receivers A and B are the same. The waveforms at receivers C and D are also the same. The P-wave at receivers C and D is larger than at the two other receivers because of their shorter distance to the borehole center. The contribution of the P-wave in the sum of the waveforms is mainly from receivers C and D. This is also the same for the pR Airy phase. The velocity determination using the sum of the waveforms is similar to that in Fig. 6.7b. The P-wave velocity is accurately determined, but the S-wave velocity is underestimated, due to the dispersive asymmetric modes induced by the eccentered tool. Additional details are shown in Fig. 6.10, where the modal decomposition results are shown. The amplitude of the screw mode is very small at all receivers. This is different from the waveforms in Fig. 6.8. The induced asymmetric modes depend on tool eccentering parameters d and θ.

The eccentering parameters d and θ are the key factors that affect the amplitudes of the different modes.

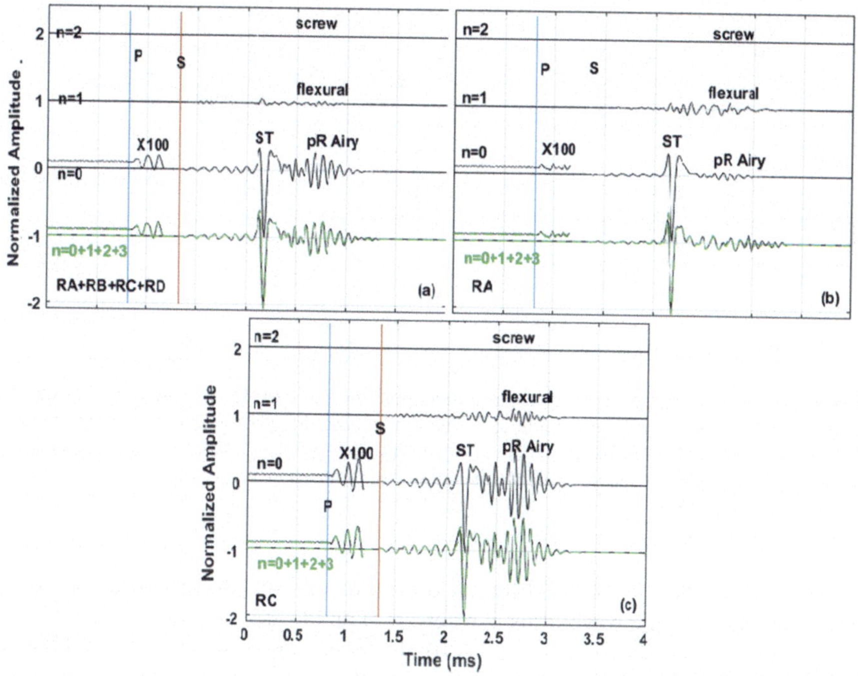

Fig. 6.10 Mode decomposition for the waves at different azimuthal receivers for the case shown in Fig. 6.9. **a** Sum of four receivers. **b** Receiver A. **c** Receiver C. The P-wave is magnified 100 times for display. The green curves are the summation waveforms of the first 4 modes (n = 0, 1, 2, 3)

6.1.1.2 Eccentered Tool in a Slow Formation

In Chap. 2, we discussed the generation of leaky P and ST waves in slow formations with a centralized monopole tool. We now investigate the effects of tool eccentering.

Figure 6.11 shows synthetic waveforms at four azimuthal receivers when the monopole tool has an eccentering angle of 0° and a displacement of 23 mm in formation S1. The summation of the waveforms (RA + RB + RC + RD) from the four azimuthal receivers is shown in Fig. 6.11a. Figure 6.11b shows the velocity-time semblance contour plots of the waveforms. Figure 6.11a shows that the leaky P modes have different characteristics at different azimuthal receivers. The amplitude of the leaky P modes at receiver C is larger than that at others. This is due to the larger annulus between the borehole wall and the receiver at this position. The larger annulus moves the high-order leaky P modes to a lower frequency. The velocity obtained from the semblance plot is slightly lower than the true P-wave velocity due to dispersion of the leaky P modes. The ST wave is barely affected by the tool eccentering. However, additional modes appear after the ST mode (beyond 3.7 ms). These correspond to the higher-order modes, such as flexural, screw, and hexapole.

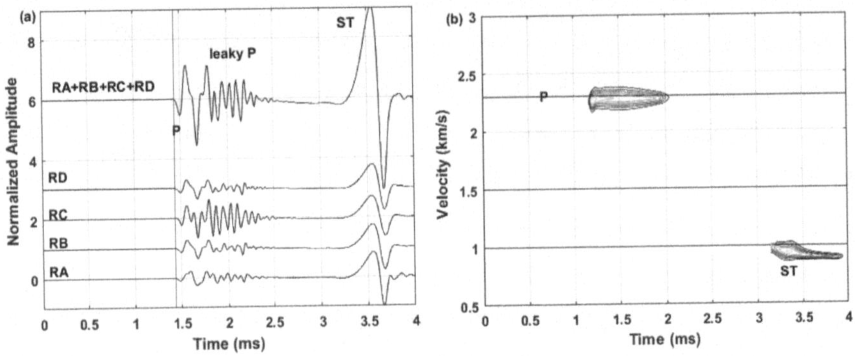

Fig. 6.11 The wavefields of an eccentered monopole tool at 10 kHz in a fluid-filled borehole surrounded by formation S1 (see Table 2.1). The tool eccentering parameters d and θ are 23 mm and 0°. **a** Waveforms at different azimuthal receivers and the sum of the four receivers. **b** The velocity-time semblance contour plot for the sum of the four receivers. Semblance calculated as described in Fig. 6.7b

The decomposition of the different modes at different receivers is shown in Fig. 6.12. Because the waveforms at receivers B and D are the same, the waveform at receiver D is not shown. Like Fig. 6.10, the green curves are the sum of the first 4 modes (n from 0 to 3). Note that there is no flexural mode at receivers B and D due to the symmetry of these two receivers relative to the borehole center. Leaky P modes at receiver C, for the dipole (n = 1) and quadrupole (n = 2), are larger than those at other receivers because the eccentered tool places receiver C far from the borehole wall. This is equivalent to enlarging the borehole radius, which moves the leaky P modes to a lower frequency. The higher-order modes following the ST mode (beyond 3.7 ms) can be seen at different azimuthal receivers. Receiver A shows the clearest flexural, screw, and hexapole modes.

The waveforms at azimuthal receivers are also affected by the eccentering angle. Similar to Fig. 6.9, the synthetic waveforms are shown in Fig. 6.13 when the tool eccentering parameters d and θ are 23 mm and 45°. The waveforms at receivers A and B are the same and the waveforms at receivers C and D are the same. The leaky P-waves, at receivers C and D, exhibit more high-order modes and larger amplitudes than those at receivers A and B.

The mode decomposition in Fig. 6.14 shows the modal distribution for eccentering angle 45°. They are different from the 0° case. Also note the contribution from higher-order modes following the ST wave. They are small and unlikely to affect the dispersion characteristics of the ST wave.

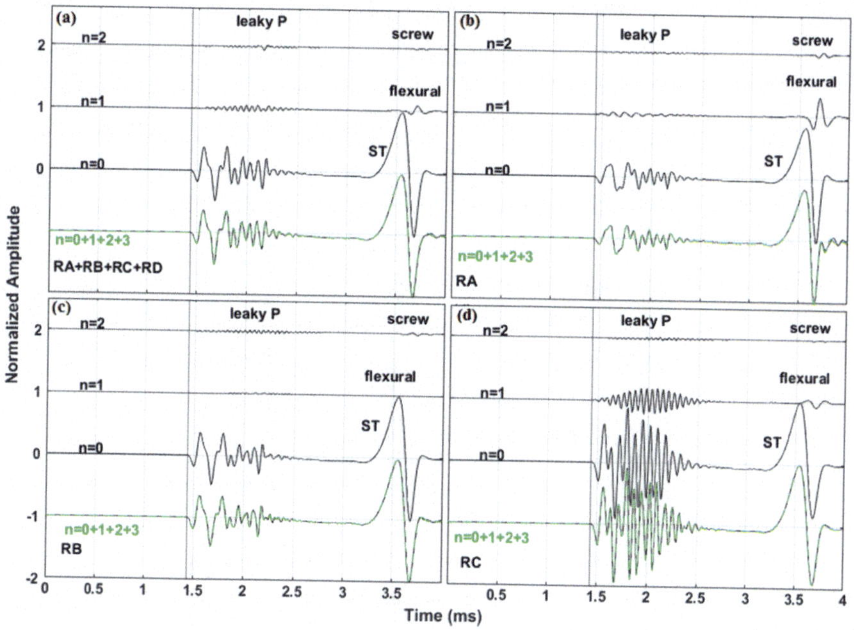

Fig. 6.12 Modal decomposition for the waves at different azimuthal receivers for eccentered monopole tool as in Fig. 6.11 in slow formation S1. **a** Sum of the four receivers. **b** Receiver A. **c** Receiver B. **d** Receiver C. Green curves are the sum of the first 4 order modes

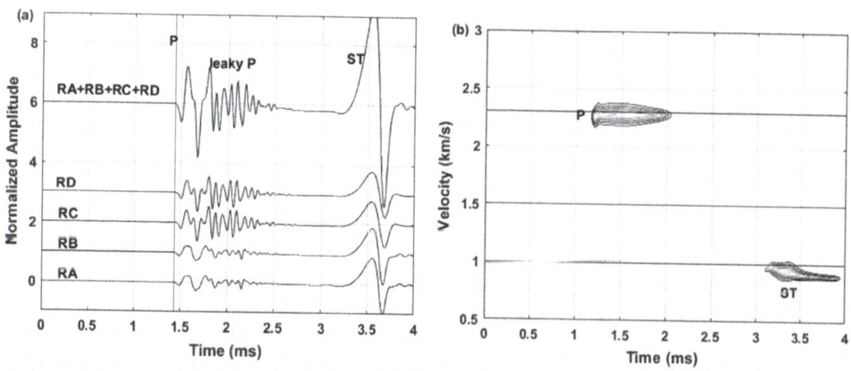

Fig. 6.13 The wavefield of an eccentered monopole tool at 10 kHz in a fluid-filled borehole surrounded by formation S1 (see Table 2.1). Tool eccentering parameters d and θ are 23 mm and 45°. **a** Waveforms at different azimuthal receivers and the sum of the four receivers. **b** The velocity-time semblance plot for the sum of waveforms at the four receivers. Semblance calculated as described in Fig. 6.7b

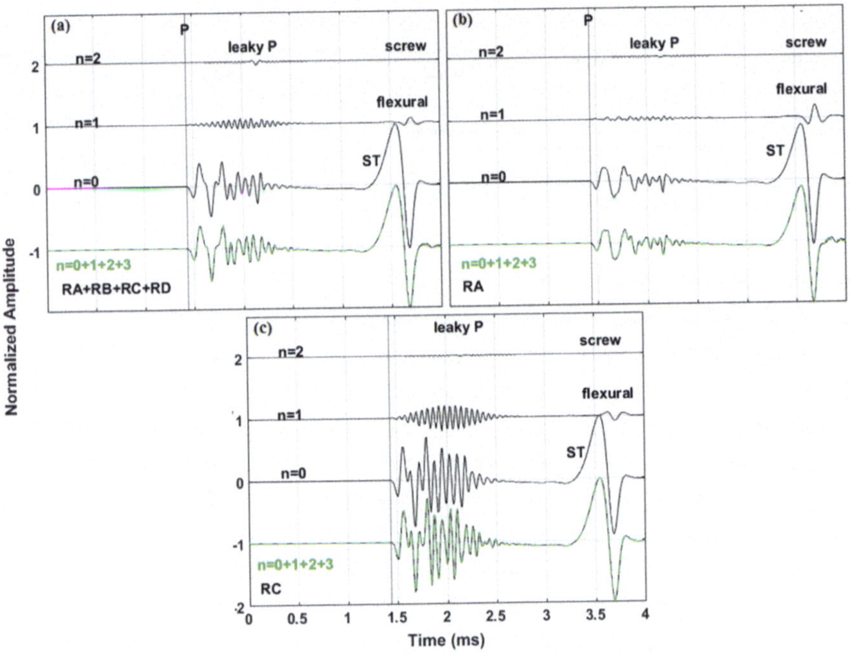

Fig. 6.14 Mode decomposition for the waves at different azimuthal receivers for eccentered monopole tool as in Fig. 6.13 in slow formation S1. **a** Sum of the four receivers. **b** Receiver A. **c** Receiver C. Green curves are the sum of the first 4 order modes (n = 0, 1, 2, 3)

6.1.2 *Dipole Wavefields in Slow Formations*

6.1.2.1 Influence of Eccentering Amount (*d*)

In Sect. 2.3.2, we discussed the wavefields of the centralized wireline dipole tool in slow formations. Dipole measurements contain the leaky P as well as flexural modes. The formation S-wave velocity can be determined from the flexural-wave velocity at the cutoff frequency (low frequency limit).

In this section, we use wavefields from a wireline dipole tool with a low source frequency (2 kHz) in slow formation S1 to investigate the effects of the tool eccentricity and azimuth of tool eccentering. The dipole source polarization is along C to A (Fig. 6.6). The output waveforms of the dipole tool are generated by subtracting the waveforms of the inline receivers, RA-RC.

Figure 6.15 shows the waveforms at different azimuthal receivers. The eccentering is 23 mm and the angle is 0°. Figure 6.15a shows the full waveforms and Fig. 6.15b shows the magnified leaky P-waves. The dashed and solid curves represent the waveforms for the centralized and eccentered cases, respectively. The figure clearly shows that the leaky P-waves for the eccentered case have larger amplitudes than those in the centralized case. The effect of the tool eccentering amount *d* can be removed by

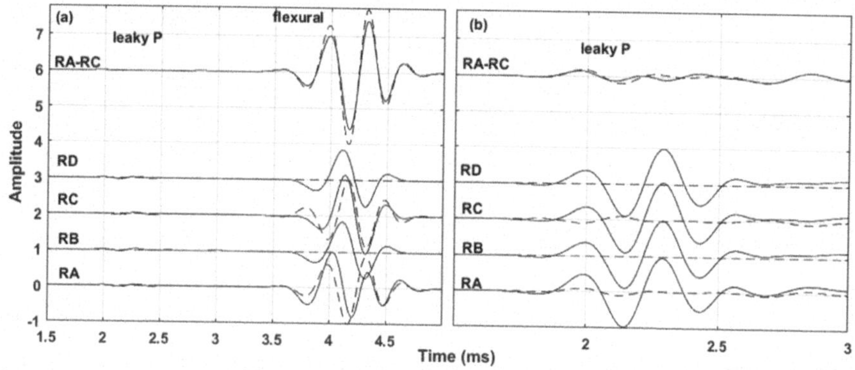

Fig. 6.15 The wavefield of an eccentered 2 kHz dipole tool in a fluid-filled borehole surrounded by formation S1 (see Table 2.1). The tool eccentering parameters d and θ are 23 mm and 0°. The dashed curves are the waveforms from a centralized tool. **a** Waveforms at different azimuthal receivers and the output waveforms obtained by subtracting waveforms at inline receivers. **b** The amplified waveforms of the leaky P-wave

subtracting the waveforms of the inline receivers. For the flexural wave, the waveforms at inline receivers (RA and RC) have small difference in both the amplitude and phase from those of the centralized tool case. The dipole characteristic remains intact. The waveforms at the crossline receivers (B and D) are different from the centralized case. They are no longer zero. The amplitudes of these waves are of the same order as those of the inline receivers. The crossline receivers are symmetric relative to the borehole center when the tool eccentering angle is zero. The strongest ST wave, after 3.5 ms, appears at the crossline receivers.

Figure 6.16 shows the decomposition of the waveforms into different modes. The ST mode (n = 0) is prominent at all azimuthal receivers. This mode may be reduced in amplitude by the taking the differences in the waveforms at the inline receivers (RA-RC). The dipole mode (n = 1) is large. There is a weak screw mode.

The tool eccentering amount d does not affect the formation S-wave velocity measurement when the tool eccentering angle is zero.

6.1.2.2 Influence of Tool Eccentering Angle

Unlike the monopole case, the dipole and quadrupole wavefields are affected by the tool eccentering angle because the source is directional.

Figure 6.17 shows the waveforms at different azimuthal receivers when the tool eccentering angle is 45°. Different from the monopole case, no receiver pair has identical waveforms. The amplitudes of leaky P-waves vary at different azimuthal receivers. After 3.5 ms, the influence of eccentering on the waveforms at the inline receivers is small and the waveforms show good dipole characteristics. The formation S-wave velocity determination is not affected in a significant way. The waveforms after 3.5 ms in the crossline receivers show the same characteristics as those with

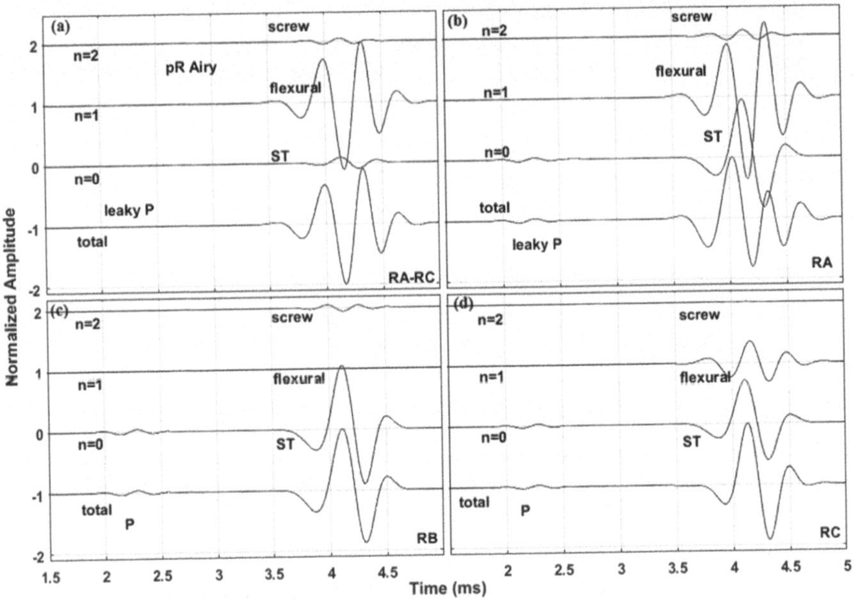

Fig. 6.16 Mode decomposition of the waves at different azimuthal receivers for eccentered dipole tool in Fig. 6.15 in slow formation S1. **a** Differences in inline receivers (RA-RC). **b** Receiver A. **c** Receiver B. **d** Receiver C. Waveforms at receiver D are the same as those at receiver B

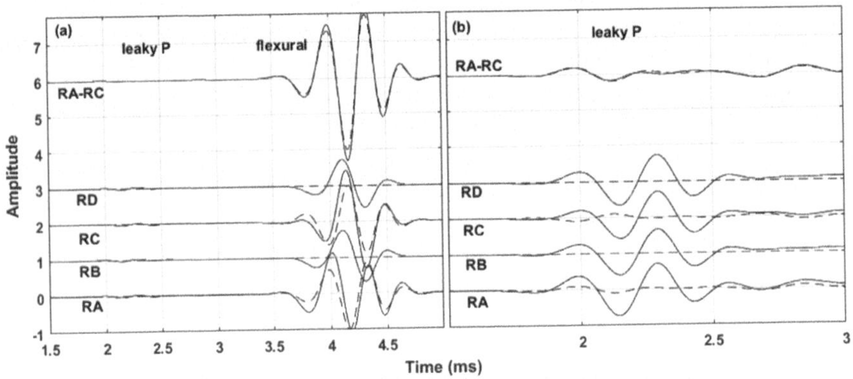

Fig. 6.17 The wavefields of an eccentered 2 kHz dipole tool in a fluid-filled borehole surrounded by formation S1 (see Table 2.1). The tool eccentering parameters d and θ are 23 mm and 45°. The dashed curves are the waveforms for a centralized tool. **a** Waveforms at different azimuthal receivers and the output waveforms after subtracting the waveforms of inline receivers. **b** Amplified waveforms for early times to highlight the leaky P-waves

a tool eccentering angle of 0°. They are of the same magnitude as the waves at the inline receivers.

Figure 6.18 shows the modal decomposition of the waveforms at different azimuthal receivers. Waveforms after 3.5 ms include additional modes compared to the centralized tool case, the distribution of which depends on the azimuth of the receiver. The strong ST wave appears at all azimuthal receivers. The flexural wave is not excited at receivers B and D. The screw mode is very weak at azimuthal receivers C and D. The waveforms at receivers B and D are nearly the same.

When the tool is eccentered along the crossline direction (θ = 90°), the wavefield changes very little from that in the centralized case, where there are no waves at crossline receivers and the waveforms at the inline receivers keep the dipole features.

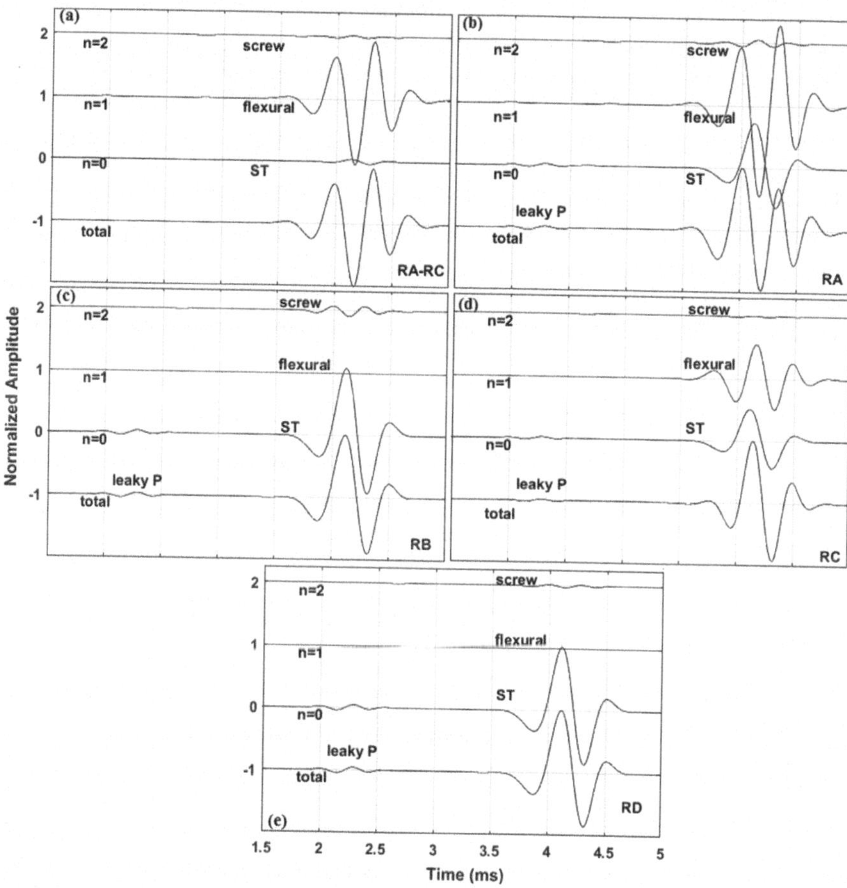

Fig. 6.18 Mode decomposition for the waves at different azimuthal receivers for the dipole tool case shown in Fig. 6.17 in slow formation S1. **a** Difference of the two inline receivers. **b** Receiver A. **c** Receiver B. **d** Receiver C. **e** Receiver D

Generally, formation velocity measurement is barely affected by dipole tool eccentering.

6.1.3 Quadrupole Wavefields in Slow Formations

An eccentered quadrupole tool generates leaky P-waves in addition to formation screw waves. The leaky P-wave amplitude increases with increasing source frequency and the amplitude of the formation screw wave decreases with increasing source frequency. A low frequency source (such as 2 kHz) efficiently excites the screw wave. Because two azimuthal sources, with alternate phases, are used in quadrupole measurements, the waveforms at the azimuthal receivers also exhibit alternating phases, as shown by dashed curves in Fig. 6.19.

When the tool is eccentered along the axis containing sources C and A (eccentering angle of 0), the waveforms (solid curves) in Fig. 6.19 at azimuthal receivers B and D are the same. The waveforms are slightly changed from the waveforms of the centralized tool. The leaky P-wave is enhanced when the tool is eccentered. Waveforms at receivers A and C are strongly affected by tool eccentering because of broken symmetry. The output trace (solid curve) of the quadrupole tool (RA-RB + RC-RD) is different from the centralized tool (dashed curve). The S-wave velocity obtained from the velocity-time semblance (Fig. 6.20) is accurate. The P velocity can also be obtained even though the semblance contour has a large spread. Dispersion correction improves the P-wave velocity determination.

The mode decomposition shown in Fig. 6.21 shows why the S-wave velocity

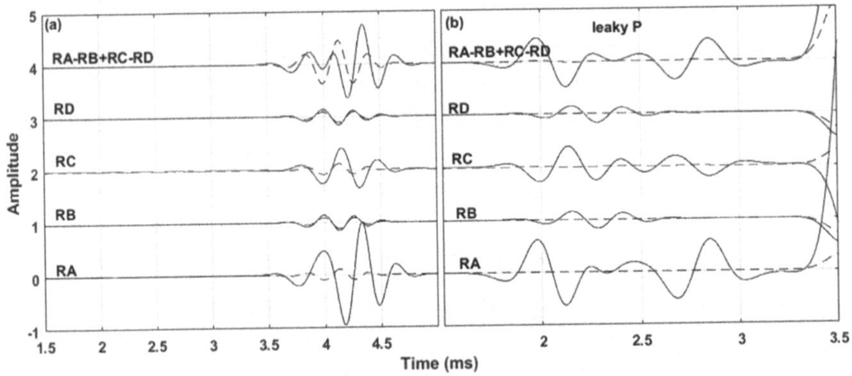

Fig. 6.19 The wavefield of an eccentered 2 kHz quadrupole tool in a fluid-filled borehole surrounded by formation S1 (see Table 2.1) when the tool eccentering parameters d and θ are 23 mm and 0°. **a** Waveforms at different azimuthal receivers and the output waveforms obtained by subtracting waveforms at crossline receivers from those at inline receivers. **b** The amplified waveforms before 3.5 ms in **a**. The dashed curves are the waveforms for a centralized tool. Note that the amplitude of the leaky P-wave for the centralized tool case is very low but is not zero

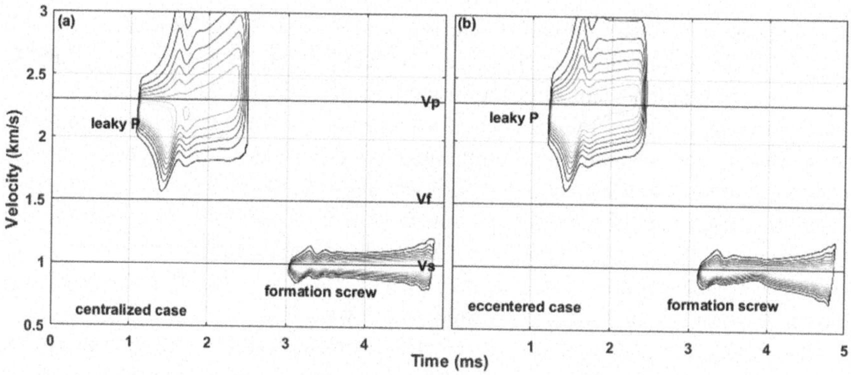

Fig. 6.20 Velocity-time semblance plots for synthetic array waveforms from a 2 kHz quadrupole tool in slow formation S1. **a** Centralized tool. **b** Eccentered tool. Tool eccentering parameters d and θ are 23 mm and 0°

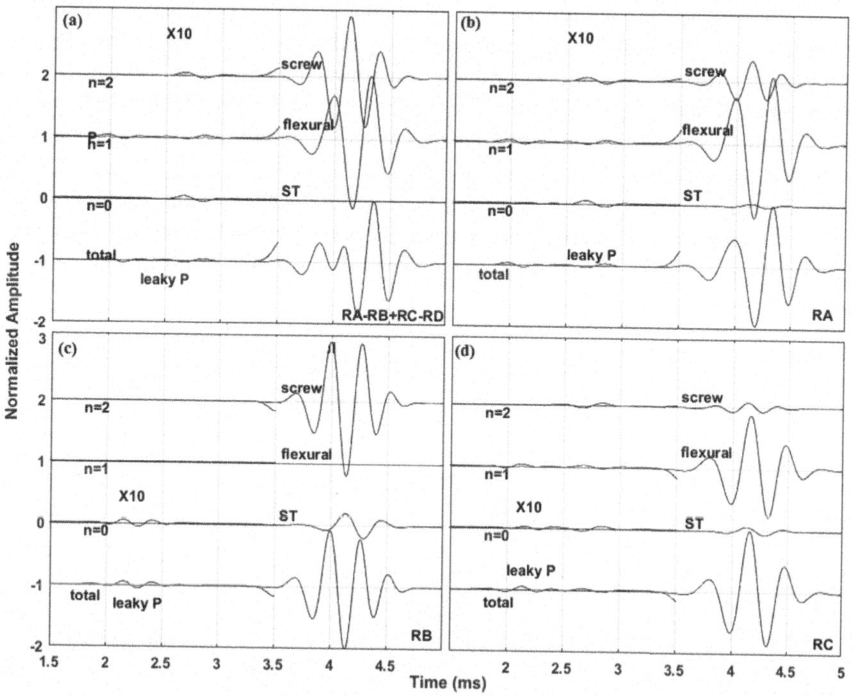

Fig. 6.21 Mode decomposition for the waves at different azimuthal receivers for the case shown in Fig. 6.19. **a** Output trace of the quadrupole tool (RA-RB + RC-RD. **b** Receiver A. **c** Receiver B. **d** Receiver C. The waveforms at azimuthal receivers B and D are exactly the same. The waveforms before 3.5 ms are magnified 10 times

determination is not affected by eccentering. The leaky P-waves at receivers B and D are from the monopole (n = 0) mode. The leaky P modes at receivers A and C have contributions from not only the monopole (n = 0), but also the dipole (n = 1) and quadrupole (n = 2) modes. The formation screw wave is no longer the pure screw mode after 3 ms and it is contaminated by other modes. Among these modes, the flexural mode has the greatest amplitude at receivers A and C. Tool eccentering affects the radiation pattern of the quadrupole source. Waveforms at receivers B and D have no flexural waves and the formation screw mode dominates with a weak ST wave. The waveform after 3 ms, in the trace of RA-RB + RC-RD in Fig. 6.21a, consists of strong flexural and screw modes and weak ST.

In summary, even though the waveforms from an eccentered quadrupole tool are more dispersive than those in the centralized case, the S-wave velocity can be obtained.

Figure 6.22 shows the waveforms at azimuthally distributed receivers when the tool eccentering angle is 45°. Waveforms (solid curves) at all azimuths dramatically change compare to the centered tool case (dashed curves). All eccentered waveforms are larger than the centralized waveforms due to the strong flexural waves. In the centered quadrupole case, the waveforms at azimuthal receivers A and C (also the receiver pair B and D) are the same. The waveforms at receivers A and C are out of phase for this eccentering case, similar to that observed in the case of tool eccentering angle of 0°. Waveforms at receivers B and D are also out of phase for 45° eccentering angle. The dipole (n = 1) signatures can be found for receiver pairs A-B and C-D. The anti-symmetry characteristic is discussed in more detail in Sect. 6.3.3. The mode decomposition in Fig. 6.23 shows that the flexural mode dominates the waveforms even for a quadrupole source.

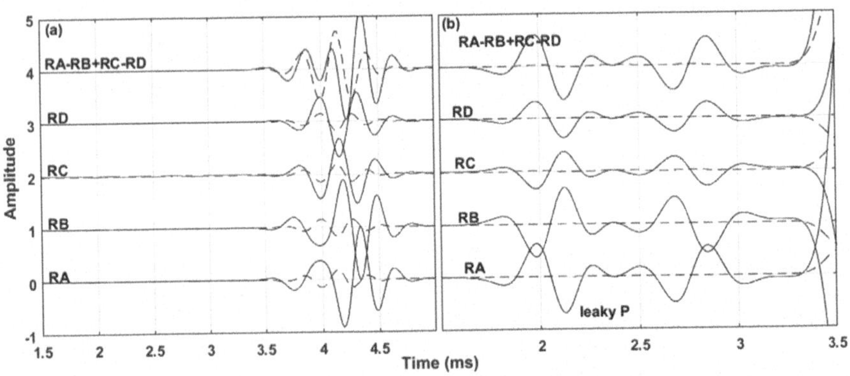

Fig. 6.22 The wavefield (solid curves) of an eccentered 2 kHz quadrupole tool in a fluid-filled borehole surrounded by formation S1 (see Table 2.1) when the tool eccentering parameters d and θ are 23 mm and 45°. The dashed curves are the waveforms from the centralized tool. **a** Waveforms at different azimuthal receivers and the output waveforms by subtracting waveforms at crossline receivers from those at the inline receivers. **b** The amplified waveforms

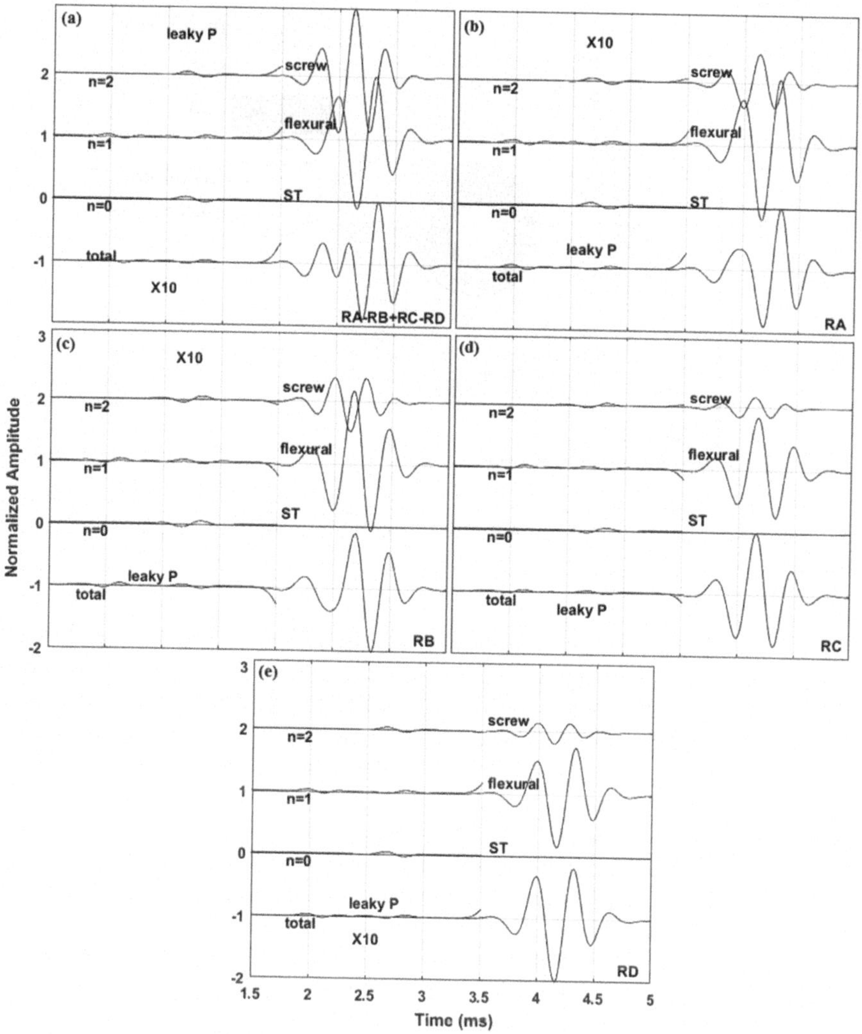

Fig. 6.23 Mode decomposition for the waves at different azimuthal receivers for the quadrupole case in Fig. 6.22. The quadrupole tool eccentering parameters d and θ are 23 mm and 45°. **a** Output trace of the quadrupole tool (RA-RB + RC-RD). **b** Receiver A. **c** Receiver B. **d** Receiver C. **e** Receiver D. The waveforms before 3.5 ms are magnified 10 times

6.2 Eccentered Monopole ALWD Tool in Fast Formations

During drilling, the drill collar is most likely to be off-centered. Figure 6.24 shows a schematic diagram of an eccentered monopole ALWD tool in a fluid-filled borehole.

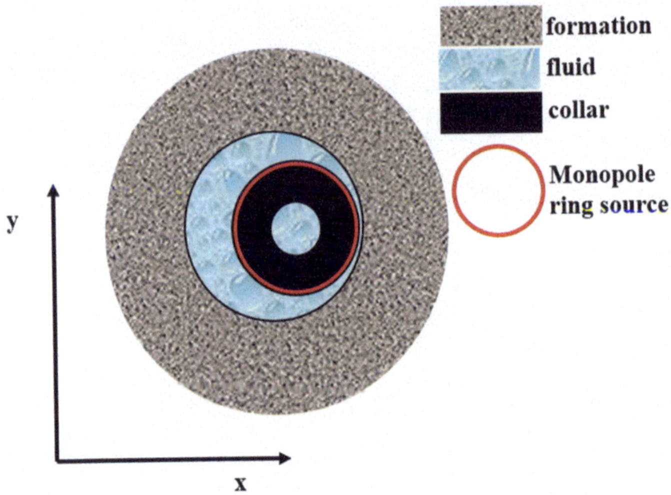

Fig. 6.24 Schematic figure of an eccentered monopole ALWD tool. Tool eccentering angle is 0° along x direction

Figure 6.25 shows wavefield snapshots for formations F1 and F2 where the drill collar is shifted 24 mm along the x direction. The smallest thickness of fluid between collar (receiver 1) and borehole wall is 3 mm and the maximum (receiver 19, see receiver index in Fig. 5.7a) is 51 mm. Collar longitudinal (distance range marked with black bars in Fig. 6.25a, b) P, S, pR, and ST modes are clearly visible and marked in the figure. There is an additional mode visible in the x-z snapshot (distance range marked with gray bars, from 1.1 to 2.1 m (Fig. 6.25a) and 1.3–2.4 m (Fig. 6.25b)). This mode is the induced collar flexural wave. The S and pR waves are contaminated by the collar flexural mode. The P-wave in the F1 model (Fig. 6.25a) and S-wave in the F2 (Fig. 6.25b) model are not affected by tool eccentering.

The snapshots for the top-down view, x-y profile at z = 1.53 m, are shown in Fig. 6.25c (F1) and 6.25d (F2). The two small white circles in the figures denote the inner and outer boundaries of the collar and the largest white circle is the borehole boundary. The azimuthal variation of color in the snapshots corresponds to the flexural mode.

The collar flexural wave interferes with the P-wave in the slower formation (F2) and with the S-wave in the faster formation (F1) when the monopole tool is eccentered.

6.2.1 *Fast Formation (F1)*

This section focuses on the effect of tool eccentering on the ALWD measurement in formation F1 (Table 5.2). Figure 6.26 shows the waveforms at various azimuths

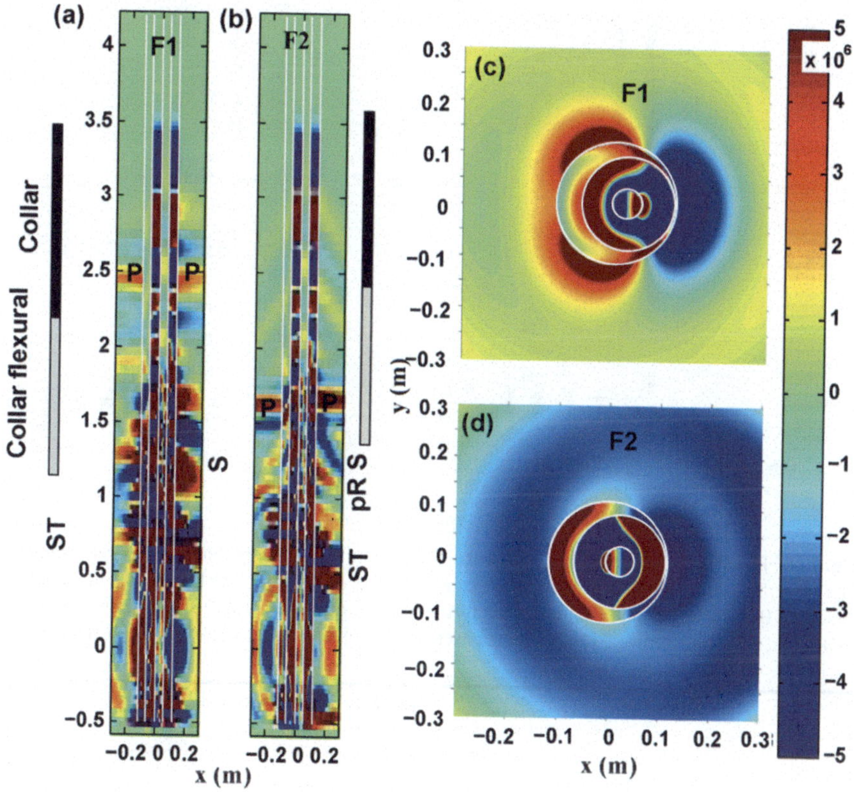

Fig. 6.25 Pressure snapshots at 0.7 ms for different profiles when the eccentering vector is 24 mm in formations F1 and F2. Collar C32 in Table 5.2 is used. **a** and **b** are x-z profiles, and **c** and **d** are x-y profiles at z = 1.53 m. Two small white circles in the figures **c** and **d** denote the inner and outer boundaries of the collar and the largest white circle is the boundary of the borehole. The color bar indicates the relative amplitudes. Model geometry is shown in Fig. 6.24 (Fig. 2 in Wang et al. [2017])

when the tool offset varies from 0 to 24 mm. The receivers are located 3 m from the source. The waveforms between 1.1 and 1.5 ms contain the S and pR waves when the tool is centralized (Fig. 6.26a). The tool eccentering makes these modes dispersive and they lose coherence as shown in Fig. 6.26c–f. It becomes difficult to identify S-wave arrival times due to the overlap between the collar flexural and S waves. In the azimuths from 90 to 270°, coda waves appear in the latter part of the traces (after 1.5 ms) when the tool is offset from the borehole center. The S-wave velocity can be determined when the tool offset is less than 6 mm but with some difficulty. As the offset increases, the collar flexural overwhelms the S-wave.

Figure 6.27 shows velocity-time semblance plots for the portions of the waveforms between 1.1 and 1.9 ms at different tool offsets and different receiver azimuths (0, 90 and 180°). The coherence of the array waveforms at azimuth 0° (Fig. 6.27a) decreases when the tool offset exceeds 2 mm. There are also additional coherent areas, ahead of or behind the S-wave arrival, due to the dispersive collar flexural waves.

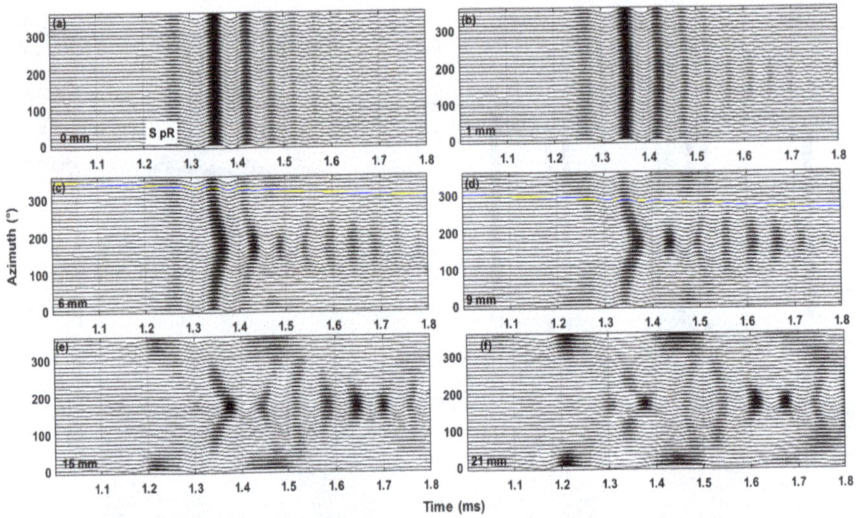

Fig. 6.26 Effect of tool eccentricity on S and pR wave at different azimuthal receivers in formation F1 (Table 5.2). Spacing between the receiver and source is 3 m. **a–f** are the waveforms for tool offsets of 0, 1, 6, 9, 15, and 21 mm. Waveforms in each plot are normalized by their maximum amplitude. Each plot shows 36 waveforms covering azimuths from 0 to 360°. Tool is offset at 0°

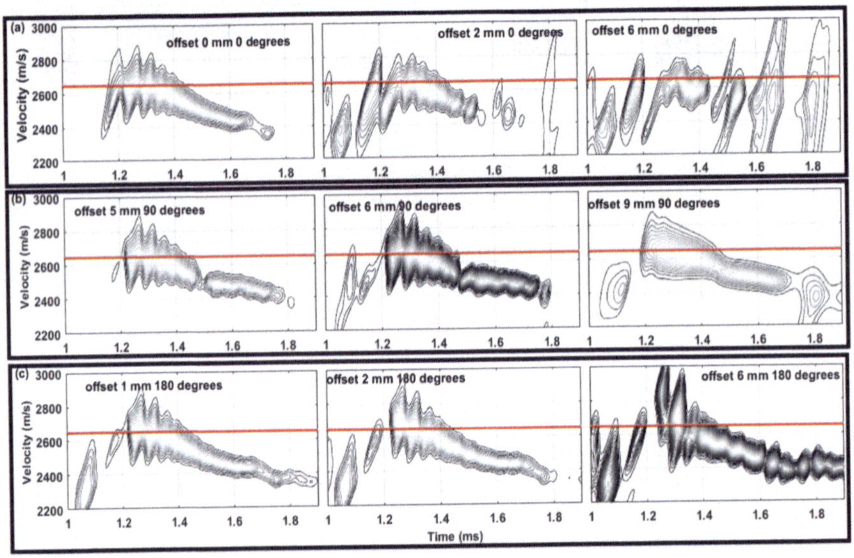

Fig. 6.27 Velocity-time semblance for the S-wave velocity at array receivers with different azimuths for different tool offsets in formation F1. The S-wave velocity of the model is marked with red line. Receiver array used is the same as in Fig. 6.7b

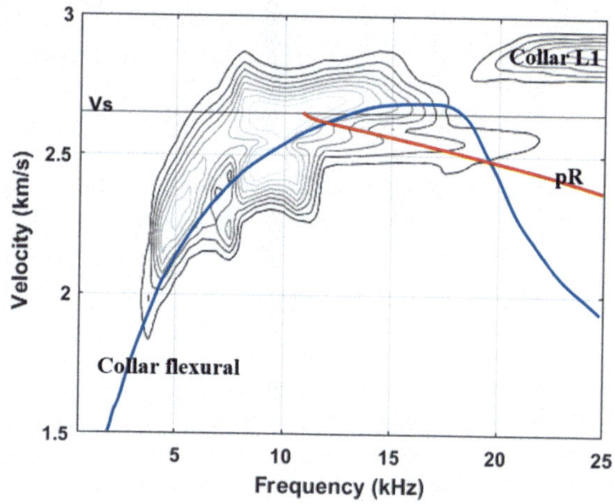

Fig. 6.28 Dispersion analysis for monopole source waveforms at 180° azimuthal receiver when tool offset is 6 mm in formation F1. Red and blue curves are the dispersion curves of pR and collar flexural for the centralized monopole and dipole tools, respectively. The source-receiver distance is from 3 to 4.05 m with 0.15 m interval

For the waveforms at 90° (Fig. 6.27b), the S-wave velocity can be determined from the maximum value of semblance when the tool offset is 5 mm or less. As can be seen in Fig. 6.27c (azimuth of 180°), the S-wave velocity measurement becomes difficult even for a very small tool eccentering.

Dispersion analysis demonstrates that the S and pR waves are overriden by the collar flexural wave. Figure 6.28 shows an example of the dispersion analysis for the waveforms at 180° azimuthal receiver when the tool offset is 6 mm. The contour plot is the extracted dispersion from the waveforms between 1.1 and 1.8 ms. Red and blue curves are the analytical dispersion curves of pR (Fig. 5.5) and collar flexural (Fig. 5.12) for the centered monopole and dipole tools, respectively. The wave for the eccentered monopole tool is a mixture of collar flexural and pR waves.

6.2.2 *Fast Formation (F2)*

6.2.2.1 Tool Eccentering Effects on P-Wave

In formation F2 (Table 5.2), the effect of tool eccentering is more serious than in the case of the higher velocity formation F1. Figure 6.29 shows the simulated waveforms at different azimuthal receivers with different tool eccenterings (or offsets) for formation F2.

Receiver 1 is at 0°, in the direction of tool offset. The variation of waveforms can be seen by comparing different panels in Fig. 6.29. Waveforms in each plot are normalized by their maximum amplitudes. The arrival time of the collar longitudinal (L) wave is at approximately 0.6 ms. The P-wave arrival times are at approximately 1.1 ms. The eccentering does not change the arrival times of the collar wave. Figure 6.30 shows the collar L and P waves at receiver 19 (azimuth of 180°) for

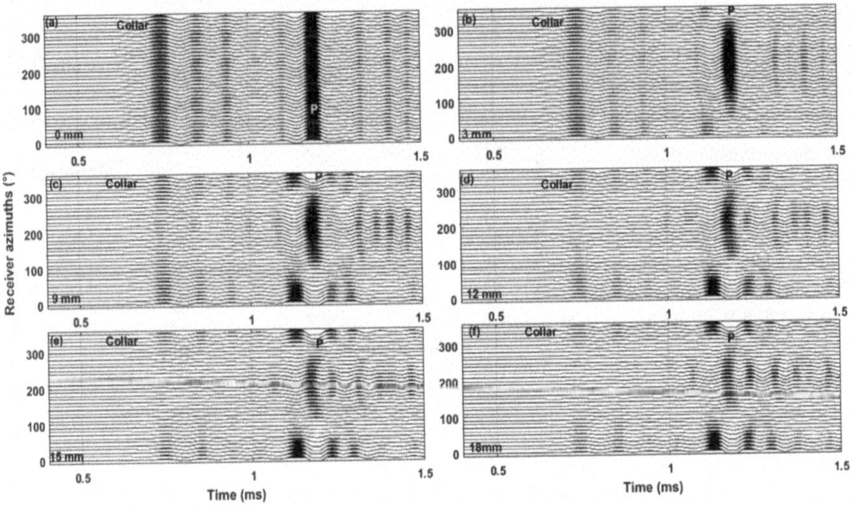

Fig. 6.29 Collar wave and P-wave at different azimuthal receivers with different tool eccenterings (or offsets) in formation F2 (Table 5.2). Collar C32 in Table 5.2 is used. Spacing between the receiver and source is 3 m. **a–f** are the waveforms for tool offsets of 0, 3, 9, 12, 15, and 18 mm. Waveforms in each plot are normalized by their maximum amplitude

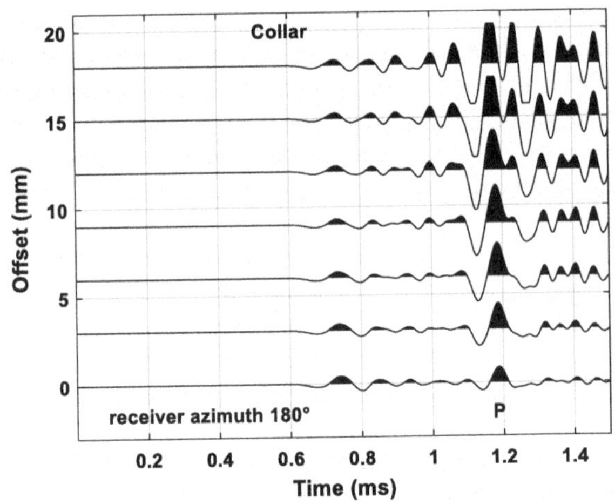

Fig. 6.30 Collar and P waves at receiver 19 (azimuth of 180°) for different tool eccenterings

different tool offsets. The amplitude of the collar wave is barely sensitive to tool offset on receiver 19. P-wave arrival time is barely affected by the tool eccentering. The P-wave amplitude increases with the increasing eccentering (see Fig. 6.30). For azimuths where the fluid annulus is small (near receiver 1 at 0°), the amplitudes become larger as tool eccentering increases. The amplitudes slightly change in the orthogonal direction to the tool offset (near receivers 10 at 90° and 28 at 270°). The interference between the collar and P waves become stronger as eccentering

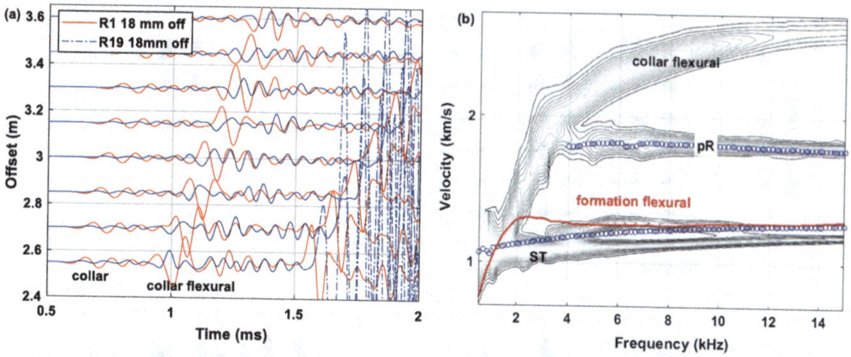

Fig. 6.31 **a** Monopole waveform at azimuthal receivers R1 and R19 when the tool eccentering is 18 mm in formation F2 (Table 5.2). **b** Dispersion analysis for waves at R1 using receivers at eccentering of 18 mm. Only the modes with velocity below 3000 m/s are shown. The circles and red curve indicate the expected dispersion for centralized monopole and dipole tools, respectively

increases, especially for receivers near the offset direction. This interference results from the appearance of a collar flexural mode between the collar longitudinal and formation P-wave. It makes P-wave velocity determination more difficult. There is little interference in the orthogonal direction to tool offset. This can be seen in Fig. 6.29 from the difference between the waveforms at receivers 1 (0°) and 19 (180°), as well as the difference between the waveforms at receivers 10 (90°) and 28 (270°).

Figure 6.31a shows the first 2 ms of waveforms at two azimuths, 0 and 180° in formation F2. The opposite phases of the flexural collar waves at the two azimuths are clear. Figure 6.31b shows the dispersion analysis for waves at receiver 1. Circles show the dispersion for a centralized monopole tool case obtained using the filtered frequency semblance method (Rao and Toksöz 2005). The red solid curve is the dispersion curve of formation flexural wave when the dipole tool is centered.

For P-wave velocity measurement, waveforms at receivers at four different azimuths can be used. Figure 6.32 shows the waveforms at three different azimuth receivers (0°, 90°, and 180°) as the tool offset increases from 0 to 24 mm. Trace amplitudes are normalized to the maximum amplitude in each plot. The arrival time and shape of the collar longitudinal wave, labeled on the left panel of Fig. 6.32a, do not change but the amplitude becomes larger when the tool eccentering increases. The P-wave, which follows the collar wave, arrives earlier and its amplitude becomes larger as tool offset increases. When offset is less than 6 mm, the P-wave arrives earlier when the tool offset increases (see Fig. 6.32a). When the tool offset is greater than 6 mm, the P-wave amplitude continues to grow with the increasing offset. Figure 6.32b shows the waveforms at receiver 10 (azimuth of 90°) with various tool offsets. When the tool offset is less than or equal to 6 mm, the observed modes are nearly the same as those in the centralized tool case (offset 0 mm). For tool offsets larger than 9 mm, the collar wave amplitude increases with the increasing tool offset. If tool offset is greater than 15 mm, the collar flexural wave interferes with the P-wave.

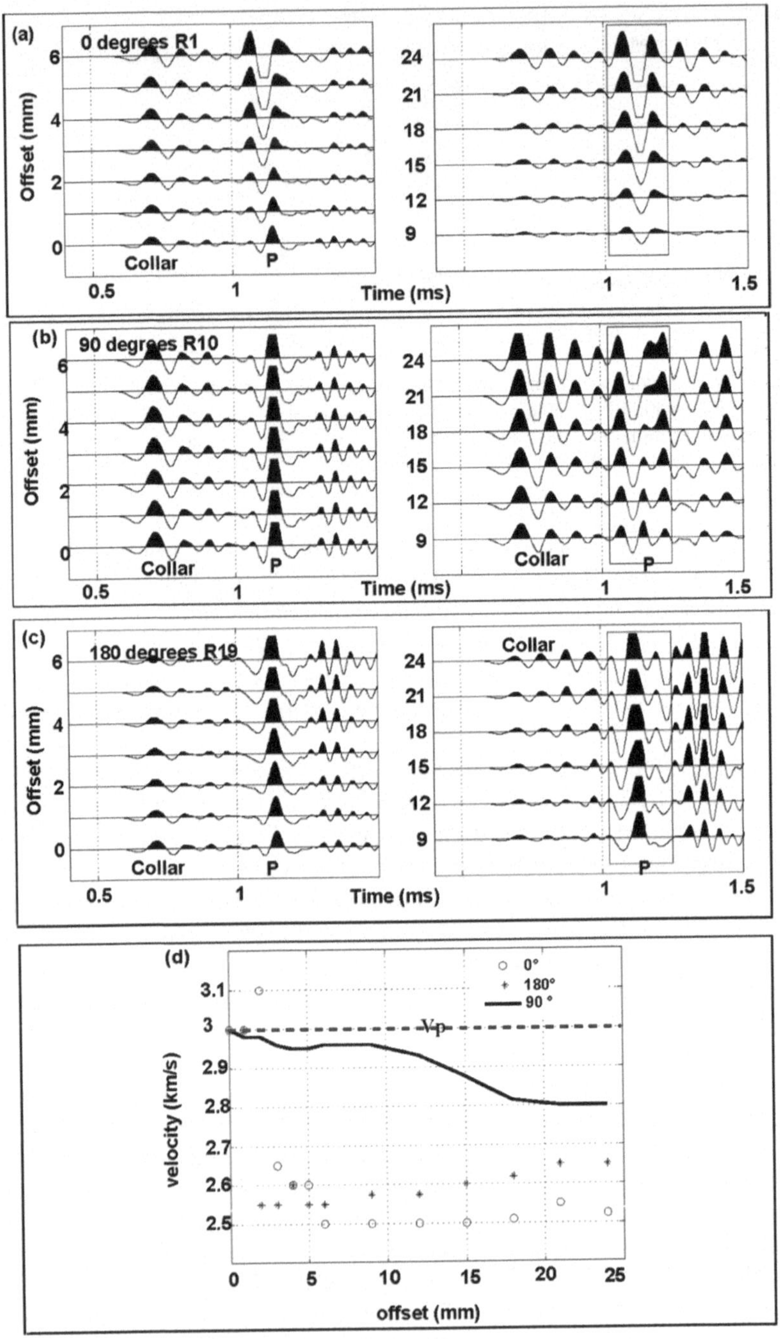

Fig. 6.32 Waveforms at receivers at three azimuths, in formation F2, for different tool eccenterings. Distance from receiver to source is 3 m. Receiver azimuths are different for each panel. Tool is offset towards 0°. **a** 0° azimuth; **b** 90° azimuth; **c** 180° azimuth; **d** P-wave velocity obtained from waveforms at 0, 90, and 180° azimuths. The velocities are obtained from the velocity-time semblance analysis (Fig. 3 in Wang et al. [2017])

We use a velocity-time semblance method (described in Sect. 3.1.2) to calculate the P-wave velocity from array waveforms at receivers with different azimuths at tool offsets from 0 to 24 mm. Figure 6.32d shows the P-wave velocities obtained for different tool offsets using waveforms at azimuth receivers (0, 90 and 180°). Except for the smallest eccentering (1–3 mm), the estimated P-wave velocity from the waveforms at azimuth receivers 0 and 180° is significantly lower than both the velocity of the collar flexural wave approximately 2800 m/s at 10 kHz and the formation P-wave velocity of 3000 m/s. This could cause misinterpretation of field measurements as it is difficult to limit the offset to below 3 mm. For the waveforms at 90° azimuth, the errors in the P-wave velocity determination are much smaller (less than 3%) for eccentering equal to or smaller than 10 mm.

In summary, for waveforms at azimuth receivers 0 and 180° in the formation F2, the P-wave is affected by the induced collar flexural wave when the monopole tool is eccentered.

6.2.2.2 Tool Eccentering Effects on S and ST Waves

Eccentering of a monopole tool affects the S and ST waves. Calculated waveforms at different azimuthal receivers are shown in Fig. 6.33. Waveforms at 180° to the direction of tool offset significantly change with offset. The S-wave amplitudes decrease with the increasing offset. The ST wave amplitude at receiver 19 (180°) becomes smaller as tool offset increases.

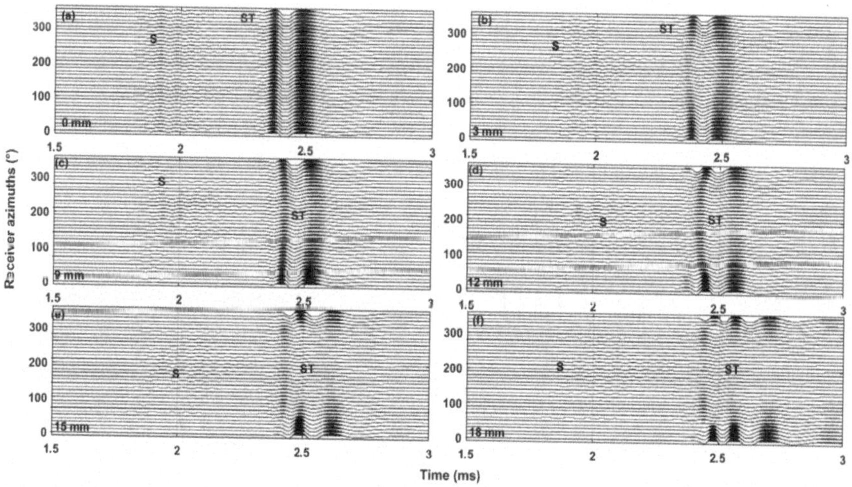

Fig. 6.33 Full waveforms for receivers in formation F2. Collar C32 in Table 5.2 is used. **a**–**f** are the waveforms for the tool offset of 0, 3, 9, 12, 15, and 18 mm. Waveforms in each plot are normalized by their maximum amplitude. Tool is offset towards 0°

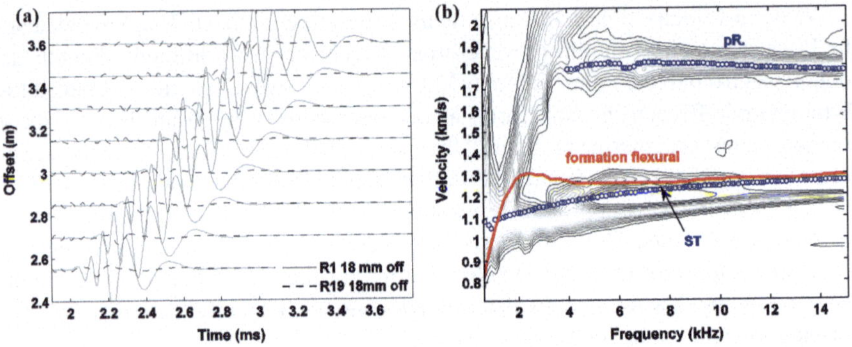

Fig. 6.34 **a** Waveform after 2 ms at azimuth receivers R1 and R19 in formation F2 when tool offset is 18 mm. **b** Dispersion analysis (contour plot) for the waveforms at receiver 1 in **a**. Blue circles are the dispersion characteristics for the centralized monopole tool. Red curve is the formation flexural dispersion curve for a centralized dipole tool. The contoured region with slow velocity (<1.4 km/s) is a mixture of formation flexural and ST modes

ST waves are prominent at receiver R1 (0°) and much smaller at receiver R19 (180°) when receiver offset is 18 mm (Fig. 6.34a). The polarities of all waves at the two receivers are opposite. There is also small amplitude formation flexural wave (Fig. 6.34a) due to multipole mode excitation because of loss of azimuthal symmetry. The details of the dispersion of the ST and flexural waves are shown in Fig. 6.34b.

The tool eccentering does not strongly affect the waveforms for the receivers in the direction orthogonal to the tool offset. Figure 6.35a shows the waveforms at receiver 10 for tool eccentering of 18 mm. The curves labeled "R10 tool centralized" and "R10 tool off 18 mm" are the waveforms at R10 (90°) in the centered tool and

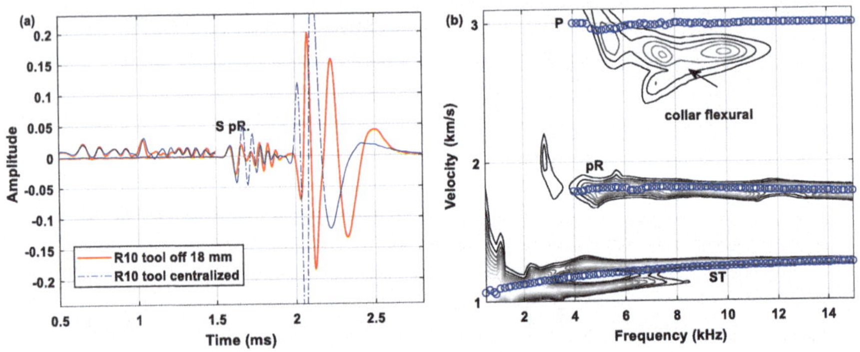

Fig. 6.35 **a** Waveforms at receiver 10 for tool eccentering of 18 mm and for the centralized monopole tool in formation F2. **b** Dispersion analysis (contour plots) for the array waveforms at the azimuthal receiver 10. Receiver spacing are the same as in Fig. 6.34a. The dispersion curves for the centralized tool case are overlaid with circles. Only the modes with velocity below 3000 m/s are shown

the 18 mm offset eccentered tool, respectively. The waveforms at receivers R10 and R28 are identical.

The determination of P-wave velocity is not affected by the collar flexural wave at receivers R10 and R28 when the tool is eccentered. There are changes in the dispersion curves for offset tools compared to centralized tools. Figure 6.35b shows the results of dispersion analysis for the waveform at receiver 10 with a tool eccentering of 18 mm. Circles denote the dispersion curves for a centralized tool as a reference. The P-wave becomes more dispersive compared to the case of a centralized tool. It is also affected by the collar flexural wave. The ST and formation flexural waves arrive about the same time, but the ST dominates the wave train (Fig. 6.35a).

For the determination of P and S velocities from an eccentered monopole ALWD tool, measurements can benefit from usage of data from multiple azimuthal sources and receivers. Collar designs that reduce the amplitude of collar longitudinal and flexural waves could significantly aid in the determination of formation P and S velocities.

If the waveforms at receivers of all azimuths are acquired separately, it may be possible to see identical waveforms at the two receivers that have an azimuth difference of 180°. These receiver positions correspond to the direction orthogonal to the tool offset. Knowing the direction of offset enables one to determine the receiver in the direction of the smallest fluid column (0°). The knowledge of the offset, combined with waveform calculation, improves the accuracy of P- and S-wave velocity determination.

6.3 Eccentered ALWD Tools in Slow Formations

In this section, the wavefields for the eccentered tool in a slow formation (Vp = 2000 m/s, Vs = 1000 m/s, density = 2000 kg/m^3) is investigated. Collar C32 in Table 5.2 is used.

6.3.1 Monopole Tool

Figure 6.36a–d display the first 1.5 ms of the waveforms at different azimuthal receivers for tool eccenterings of 0, 6, 12, and 18 mm, respectively. Waveforms in each plot are normalized by their maximum amplitudes. The tool eccentering does not affect the arrival time of the collar wave at 0.6 ms. The amplitudes at receivers located orthogonal to the directions of the tool eccentering are affected very little.

Figure 6.37a–d show the full waveforms between 1.5 and 4 ms at different azimuths for tool eccenterings of 0, 6, 12, and 18 mm. Leaky P (marked as P at 1.5 ms) and ST (at 3.5 ms) waves are visible. The amplitudes of the P and ST waves in the directions of the smallest and largest fluid columns behave different from each other.

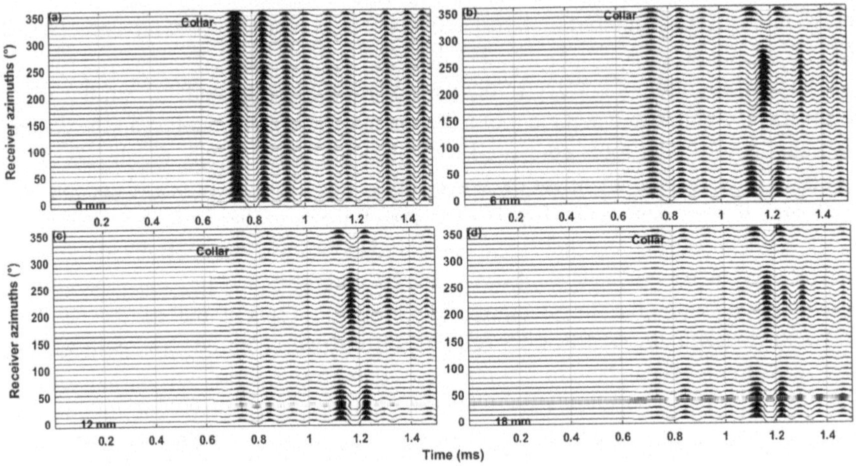

Fig. 6.36 Collar waves in a slow formation (Vp = 2000 m/s, Vs = 1000 m/s, density = 2000 kg/m^3) at azimuthal receivers with different tool eccenterings. **a**–**d** are for the tool eccenterings of 0, 6, 12, and 18 mm. Waveforms in each plot are normalized by their maximum amplitude. The source-receiver distance is 3 m

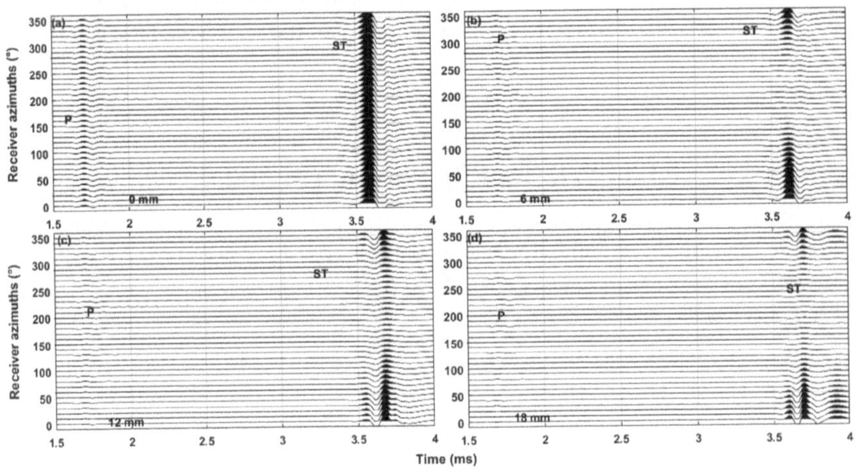

Fig. 6.37 Full waveforms in a slow formation (see Fig. 6.36) for receivers at different azimuths with different tool offsets. Time scale is from 1.5 to 4 ms. **a**–**d** are the waveforms for the tool offset of 0, 6, 12, and 18 mm. Waveforms in each plot are normalized by their maximum amplitude. Formation P and S velocities are 2000 and 1000 m/s. The source-receiver distance is 3 m

Figure 6.38 shows top view (x-y) snapshots of the wavefield at the location 2.1 m from the source position. The recording times of the snapshots are listed on each plot. The magnitude of the tool eccentering is 18 mm. The white circles in the figure denote the inner and outer boundaries of the collar and the black circle is borehole wall. The colors indicating opposite polarities about the axis along the tool offset

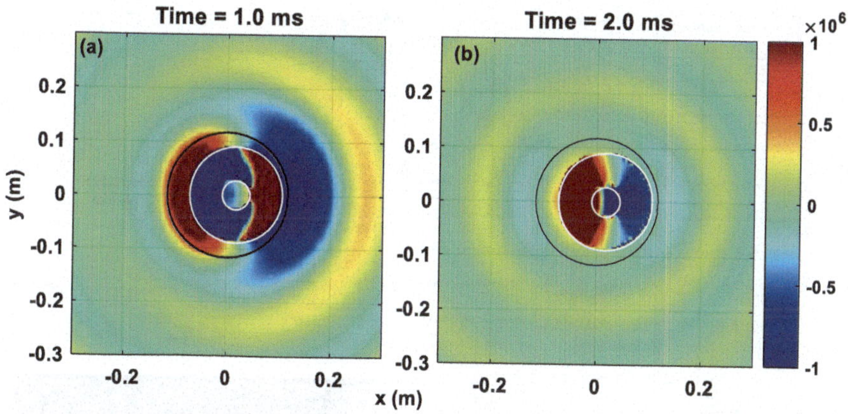

Fig. 6.38 Snapshots of x-y image of wavefield in a slow formation (Vp = 2000 m/s, Vs = 1000 m/s, density = 2000 kg/m^3) at two different times for a monopole tool eccentered 18 mm. The source-receiver distance is 2.1 m. The two small white circles denote the inner and outer boundaries of the collar and the largest black circle is the boundary of the borehole

in the collar in the snapshots at 1 and 2 ms are the collar flexural modes which are visible in Fig. 6.36.

Further evidence for the collar flexural waves generated by eccentering is shown by the waveforms. Figure 6.39 shows the array waveforms, the velocity-time semblance plot, and the dispersion analysis for waveforms at azimuthal receivers R1 and R19 for tool offset of 18 mm. The collar, and leaky P-waves can be identified in the array waveforms. The mode between the collar longitudinal and leaky P-waves is a collar flexural wave. The polarity difference between receivers 1 and 19 substantiates that it is a flexural wave. The dispersion in Fig. 6.39c further confirms the interpretation as a collar flexural wave. The collar flexural-wave velocity is approximately 2600 m/s which is determined from the velocity-time semblance plot in Fig. 6.39b. The collar flexural wave barely affects the formation P-wave velocity determination. A high source frequency (10 kHz) helps to separate the formation P and collar flexural waves. When the source frequency is below 4 kHz, the collar flexural wave interferes with the P-wave (see Fig. 6.39c). In this case, a dispersion-based filtering method may reduce the interference of the collar flexural wave and leaky P-wave.

Formation flexural and ST waves also interfere below 2 kHz (Fig. 6.39c, d). The collar flexural mode also perturbs the ST mode at very low frequencies (below 1 kHz in Fig. 6.39d). A formation quadrupole wave also appears in Fig. 6.39d.

Figure 6.40 shows the effects of tool eccentricity on the modes at receivers orthogonal to the tool offset. Waveforms at receiver 10 (90°) are shown for a centralized tool and a tool offset by 18 mm. Figure 6.40a shows the portion of the collar and leaky P-waves and Fig. 6.40b shows the ST wave. The dashed lines are the waveforms for a centered tool and the solid lines show waveforms for an eccentered tool. There is very slight difference between the collar waves for the centered and eccentered cases. The tool eccentricity does not affect the collar waves in any significant way

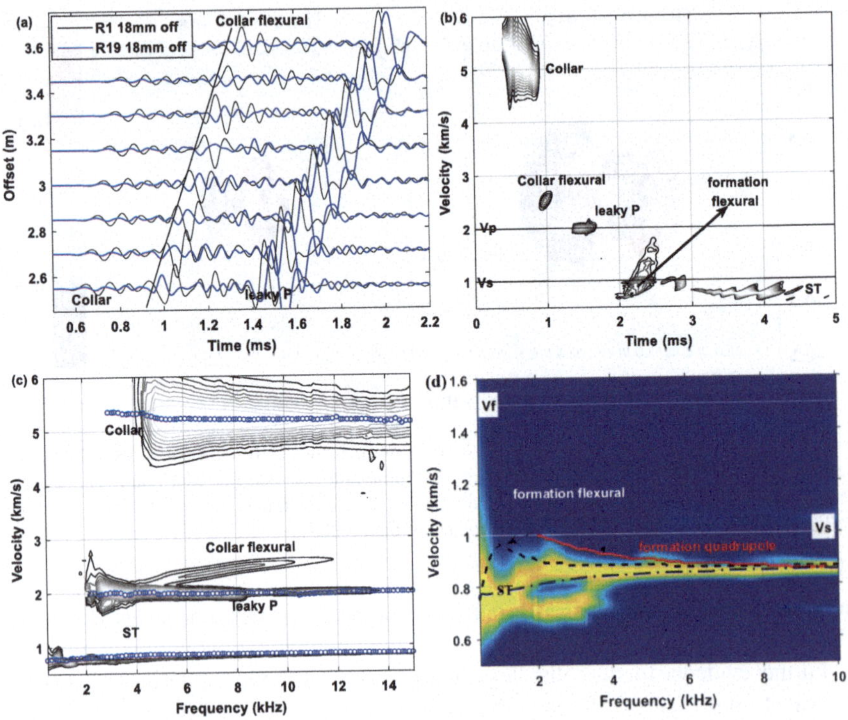

Fig. 6.39 Wavefields in a slow formation (Vp = 2000 m/s, Vs = 1000 m/s, density = 2000 kg/m^3) at azimuthal receivers 1 and 19 when the tool eccentering is 18 mm. **a** The waveforms at receivers 1 and 19 for the eccentered monopole tool. Offset in the vertical axis is the source-receiver offset. **b** and **c** Velocity-time semblance plot and the dispersion analysis of waveform at receiver R1. **d** Dispersion (color image) for the waveforms at receiver R19. Dispersion curves of ST, formation flexural, and formation quadrupole waves are shown in the expanded (velocity) plot in panel d

except for the small collar flexural wave from 1 to 1.4 ms. For the leaky P-wave, the tool offset slightly reduces the amplitude. The waveforms of ST are affected in amplitude and arrival time due to the tool offset. In addition, other modes (flexural and quadrupole) are generated and mixed with the ST wave, as shown in Fig. 6.40b and the dispersion analysis in Fig. 6.40d.

6.3.2 *Dipole Tool*

For dipole tools, both the amount and the direction of eccentering affect the waveforms.

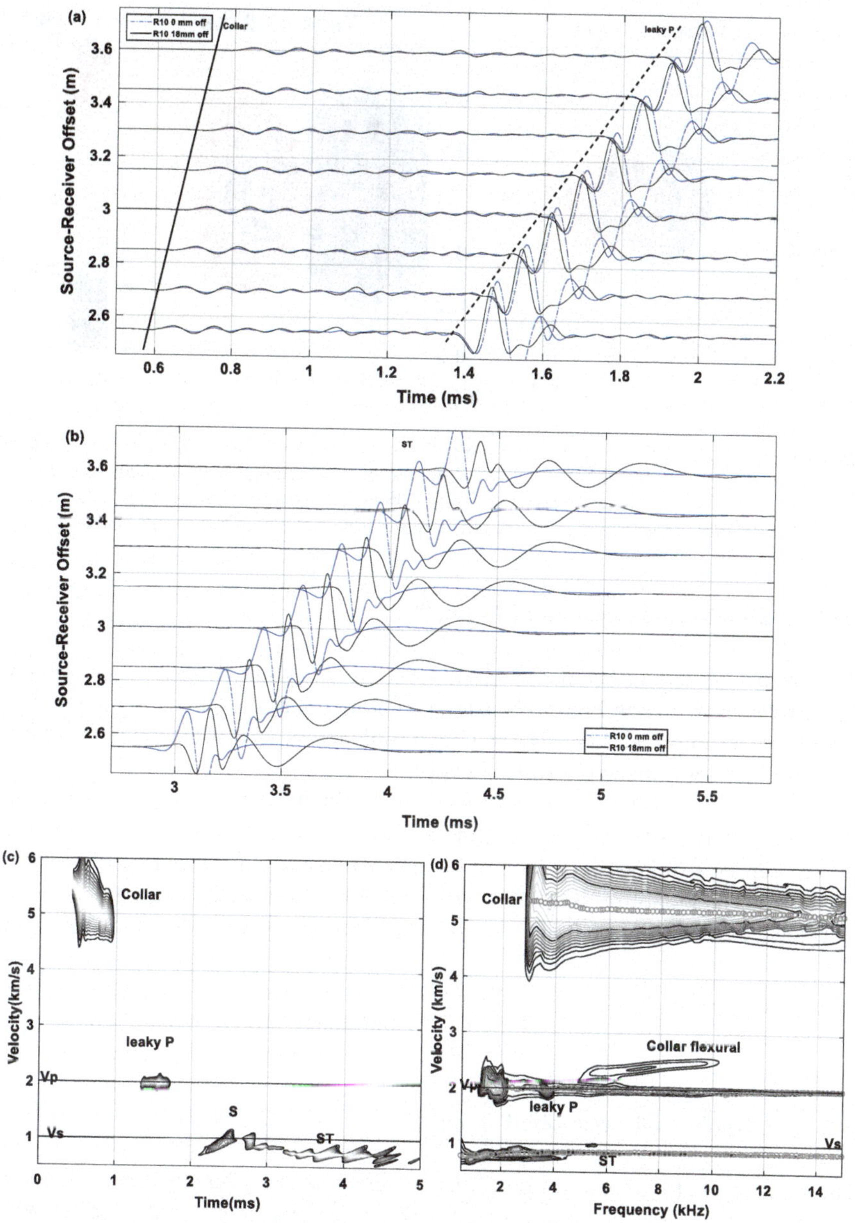

Fig. 6.40 Wavefields in a slow formation (Vp = 2000 m/s, Vs = 1000 m/s, density = 2000 kg/m^3) at azimuthal receiver 10 with different tool eccenterings. **a** A monopole source array waveform at R10 (collar wave and leaky P wave) for tool offset of 0 and 18 mm. **b** array waveform of ST wave, **c** velocity-time semblance plot, and **d** the dispersion analysis for the tool offset of 18 mm

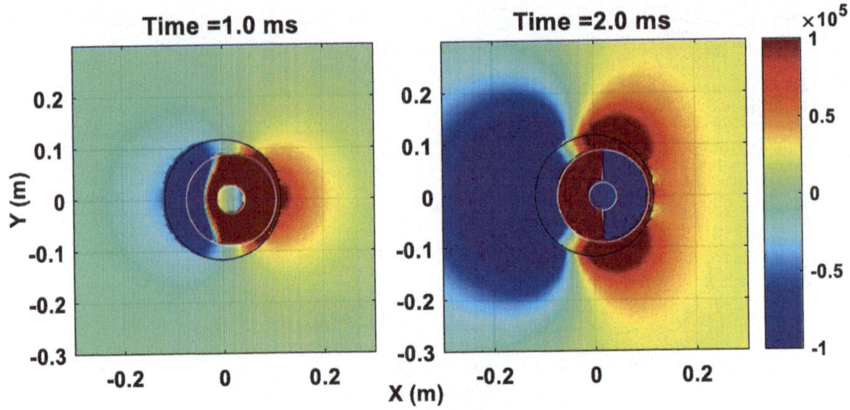

Fig. 6.41 Snapshots in the x-y plane of the wavefield in a slow formation (Vp = 2000 m/s, Vs = 1000 m/s, and density = 2000 kg/m^3) at two different times with a dipole tool eccentered 18 mm along 0° azimuth. The source-receiver distance is 1.05 m. The two small white circles denote the inner and outer boundaries of the collar and the largest circle is the boundary of the borehole

6.3.2.1 Dipole Tool Eccentered Along 0°

The radiation pattern of a centered dipole source is antisymmetric with respect to the dipole axis, as shown in Fig. 5.26. The maximum amplitudes are at azimuths of 0 and 180°. The amplitudes are zero at azimuths of 90 and 270° when the polarization of the dipole tool is along 0°–180° direction.

The symmetry of the radiation pattern breaks down when the dipole tool shifts along the direction of 0° as receiver 1 moves closer to the borehole wall. Figure 6.41 shows snapshots of horizontal motion in the x-y plane at a distance of 1.05 m from the source. The tool offset is 18 mm. The wavefields at azimuthal receivers at 0 and 180° have large differences. They would have the same amplitudes if the tool was centered.

The waveforms at the crossline receiver pair (90° and 270°) are the same but nonzero when the LWD tool is shifted along the source polarization direction.

6.3.2.2 Dipole Tool Eccentered Along 30°

The waveforms at four receiver positions for centered and eccentered dipole tools in a slow formation are shown in Fig. 6.42. The eccentered tool is shifted along 30°. The dipole characteristics of the collar wave remains unchanged. The amplitudes at azimuth of 0° (R1) and 180° (R19) increase when the tool offset increases (Fig. 6.42a, b). Formation flexural waves are contaminated by ST and formation quadrupole waves (Wang et al. 2013). The waveforms (after 2 ms) at R1 significantly change with offset, in phase and amplitude, when the eccentering increases. The waveforms at R19 only change in amplitude.

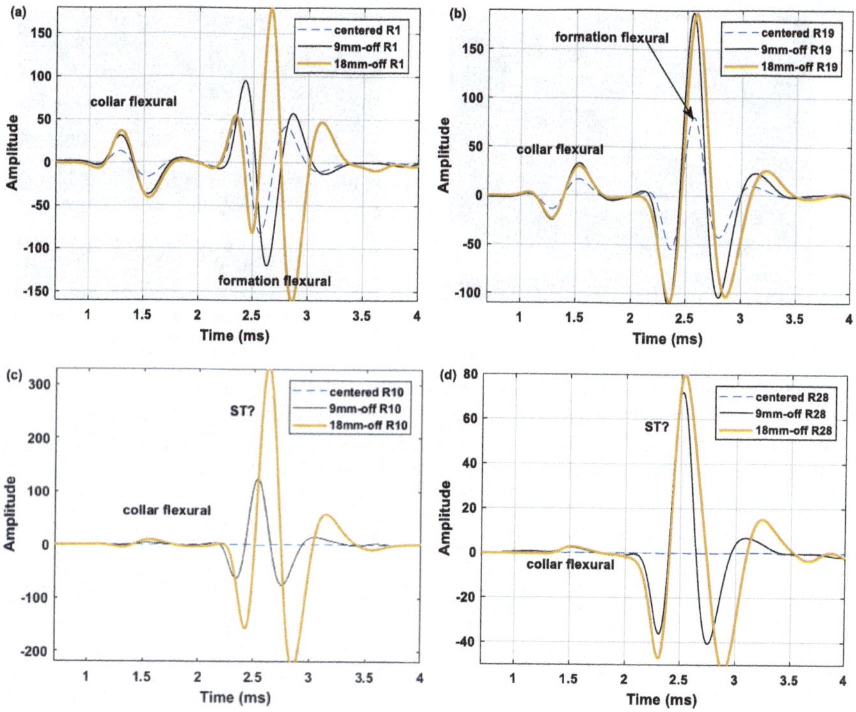

Fig. 6.42 Synthetic waveforms at four azimuthal receivers (R1, R19, R10, and R28). The source-receiver spacing is 1.5 m. The waves after 2 ms in **c** and **d** are combination of ST and formation quadrupole waves. Tool offset is at 30°

The waveforms at azimuth receivers R10 and R28 are different from those of R1 and R19. Their amplitudes become larger when the tool eccentering increases. The two waveforms are not the same because the two receivers are now not symmetric relative to the borehole center. The amplitude at R28 is smaller than at R10. The ST wave dominates the traces. These modes are substantiated by the dispersion analysis shown in Fig. 6.43, where the dispersion curves of the centered source are overlain.

6.3.2.3 Dipole Tool Eccentered Along 90°

Figure 6.44 shows the synthetic waveforms when the tool shifts along 90° (crossline direction) making R10 being closer to the borehole wall. The amplitudes at R1 and R19 increase when the tool eccentering increases. The relationship between the amplitude increments and the eccentering is not linear. The first arrival at 1 ms is a collar flexural wave. The wave after 2 ms is dominated by the formation flexural and ST waves. The amplitude at R10 and R28 (not shown here) are nearly zero (100 times smaller than that of R1 and R19).

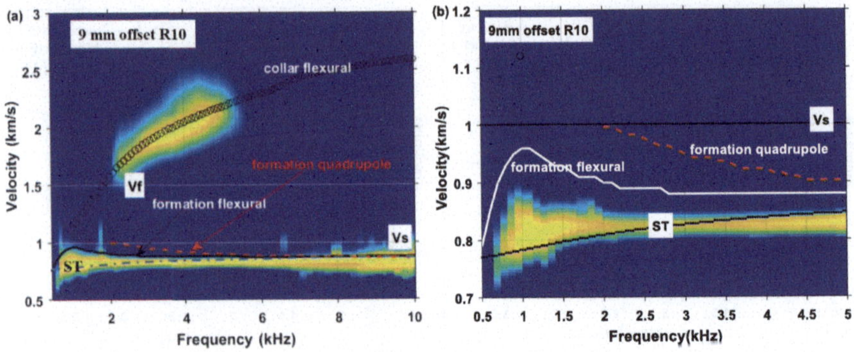

Fig. 6.43 Dispersion analysis of waveform at R10 when tool offset is 9 mm for a dipole source in a slow formation (Fig. 6.41) when tool eccentered along 30°. **b** is the same with an enlarged velocity scale from 700 to 1200 m/s. Dispersion curves of centralized monopole (ST), dipole (collar flexural and formation flexural), and quadrupole (formation quadrupole) sources are overlain

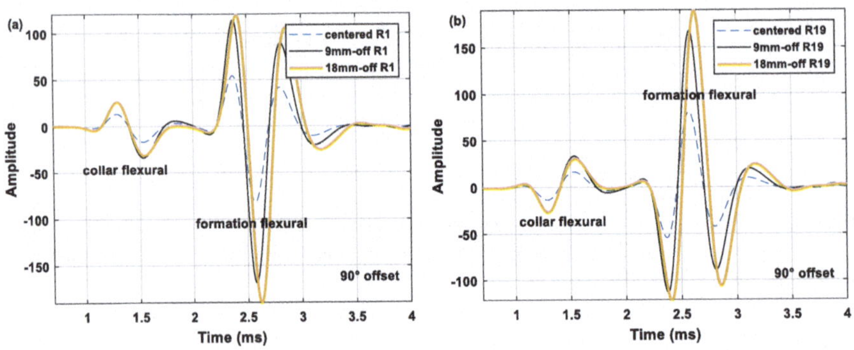

Fig. 6.44 Synthetic data for the dipole tool eccentered along the crossline direction in a slow formation (Vp = 2000 m/s, Vs = 1000 m/s, and density = 2000 kg/m^3). Collar C32 in Table 5.2 is used. **a** R1; **b** R19

6.3.3 *Quadrupole Tool*

6.3.3.1 Quadrupole Tool Eccentered Along 0°

With a centered low frequency (2 kHz) quadrupole source, there are no collar screw modes as was shown in Fig. 5.37. There is only formation screw (quadrupole) wave. This is the primary reason why the low frequency quadrupole tool is utilized in industry. To demonstrate the effect of eccentering of a quadrupole tool, we follow the procedure used for the dipole tool. First, we shift the tool along the direction of 0° making receiver 1 closer to the borehole wall.

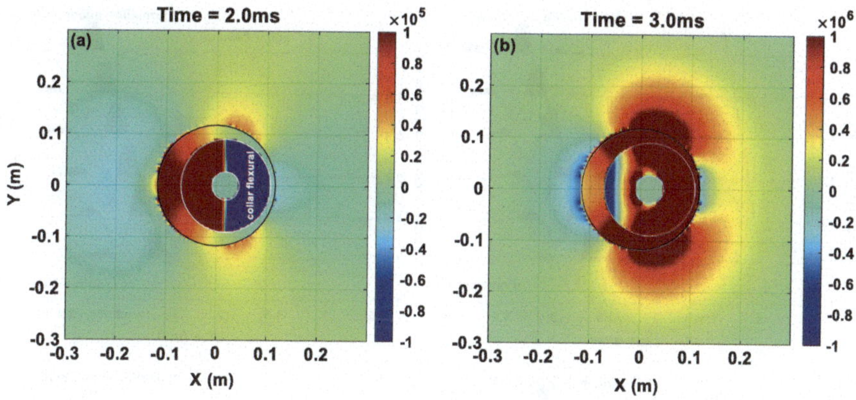

Fig. 6.45 Snapshots of x-y image of the wavefield in a slow formation (Vp = 2000 m/s, Vs = 1000 m/s, density = 2000 kg/m^3) at two different times with a 2 Hz quadrupole tool eccentered 18 mm along 0° azimuth. The source-receiver distance is 1.05 m. The two small white circles denote the inner and outer boundaries of the collar and the largest circle is the boundary of the borehole

Figure 6.45 shows top-view (x-y profile) snapshots of the pressure wavefield with the 1.5-m source-receiver spacing at times of 2 and 3 ms. Collar flexural mode appears when the tool is eccentered by 18 mm (Fig. 6.45a). The symmetry of the formation quadrupole mode is broken and an ST wave is induced (snapshot at 3 ms in Fig. 6.45b). Figure 6.46 shows the waveforms acquired by receivers at azimuths of 0, 90, 180, and 270° (relative to the x positive direction). The collar flexural wave appears as the first arrival. It is small at all receivers. The waveforms after 2 ms maintain the quadrupole characteristics.

6.3.3.2 Quadrupole Tool Eccentered Along 30°

Figure 6.47 shows synthetic waveforms at different azimuthal receivers for different amounts of tool eccentering. Clear collar flexural modes are the first arrivals. The amplitudes of collar flexural wave become larger when the tool eccentering increases.

Waveforms after 2 ms are related to the formation. Waves at different receivers are differently affected by the tool eccentering. There are prominent formation quadrupole waves. Other modes, especially formation flexural waves, appear when tool eccentering increases. Dispersion analysis, shown in Fig. 6.48a, b, demonstrates that the waveforms after 2 ms is still predominately a formation quadrupole mode. There is also some formation flexural wave.

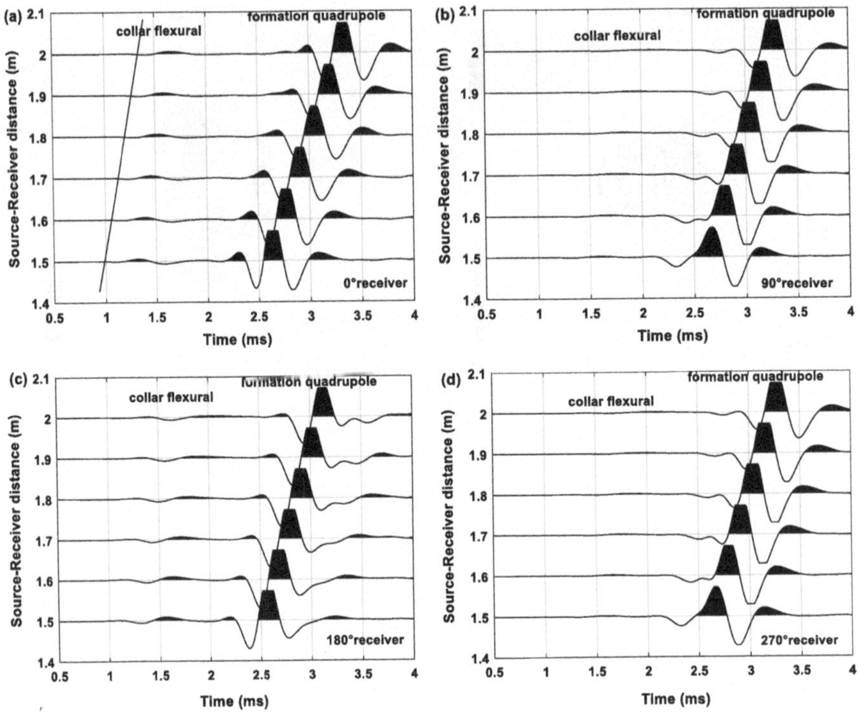

Fig. 6.46 Waveforms at 4 different azimuthal receivers when a 2 Hz quadrupole tool is eccentered 18 mm along 0°. A slow formation (Vp = 2000 m/s, Vs = 1000 m/s, and density = 2000 kg/m^3) and collar C32 in Table 5.2 are used

6.3.3.3 Quadrupole Tool Eccentered Along 45°

Figure 6.49 shows the array waveforms at different receivers for both the 9- and 18-mm cases eccentered along 45°. For the eccentered tool, small-amplitude collar flexural waves appear as first arrivals at all receivers. The formation quadrupole waves can be identified by comparing their polarities at different receivers. Waveforms at R1 are the same but with opposite polarity as the waveforms at R10. This phenomenon, which is due to symmetry, is also observed between receiver pair R19-R28, where the negative waveforms at R19 are the same as the waveforms at R28.

Waveforms calculated for other angles of tool eccentering also show that the waveforms at R1, R10, R19, and R28 for the 60° case behave in an anti-symmetrical way (the amplitude is the same and the phase is opposite) with those at R10, R1, R28 and R19 for the 30° case (Fig. 6.50). Generally, the anti-symmetry is relative to the 45° azimuth. This indicates that for a tool eccentered along two different angles: θ_1 and θ_2, if $|\theta_1 - 45°| = |\theta_2 - 45°|$, the waveforms at receiver pairs R1-R10 and R19-R28 are have same amplitude but opposite phase: R1 (θ_1) = −R10 (θ_2), R10 (θ_1) = −R1 (θ_2), R19 (θ_1) = −R28 (θ_2), and R28 (θ_1) = −R19 (θ_2). This anti-symmetry can be dealt with using a mirror transform on the LWD tool, followed by a 90°

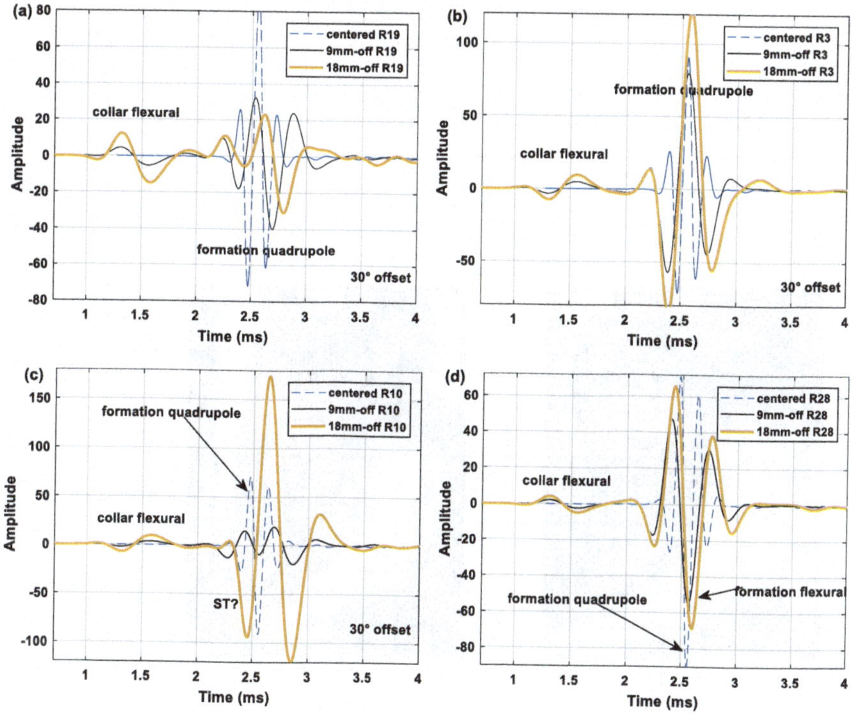

Fig. 6.47 Synthetic waveforms at azimuthal receivers for the 1.5-m source-receiver spacing when the 2 Hz quadrupole tool is eccentered 30° from the borehole center. A slow formation (Vp = 2000 m/s, Vs = 1000 m/s, and density = 2000 kg/m^3) and collar C32 in Table 5.2 are used. **a** R1. **b** R19. **c** R10. **d** R28. Note the different amplitude scales

rotation. Figure 6.51 demonstrates the anti-symmetry phenomenon. The case of tool eccentered along θ_2 (Fig. 6.51b) is shown in Fig. 6.51c after a 90° counterclockwise rotation. Then a flip transformation along horizontal direction (mirror transform) is applied (Fig. 6.51d). Comparing the tool configuration in Fig. 6.51a, d shows that the waveforms at receiver pairs R1-R10 and R19-R28 have the same amplitude but opposite phase.

6.3.4 *Estimate of the Tool Position from the ALWD Measurement*

The real-time location of the downhole drill bit is very critical for geo-steering. In wireline measurement, a multi-arm mechanical device is used to center the tool in the well and ultrasonic calipers maybe used to determine the tool location. In the

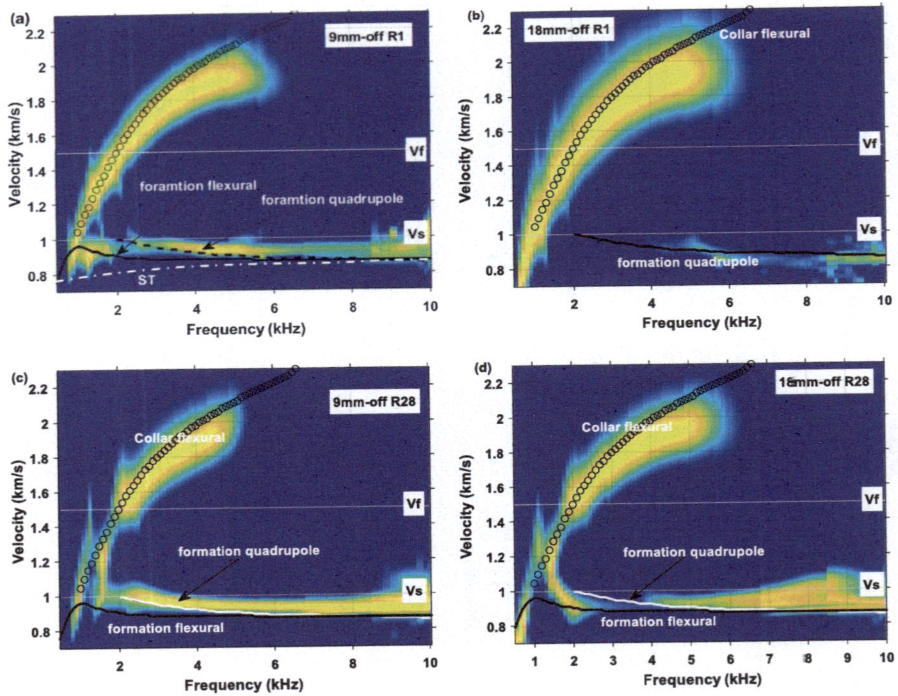

Fig. 6.48 Dispersion analysis of waveforms of the eccentered quadrupole tool along 30°. **a** R1 with tool eccentering of 9 mm. **b** R1 with tool eccentering of 18 mm. **c** R28 with tool eccentering of 9 mm. **d** R28 with tool eccentering of 18 mm

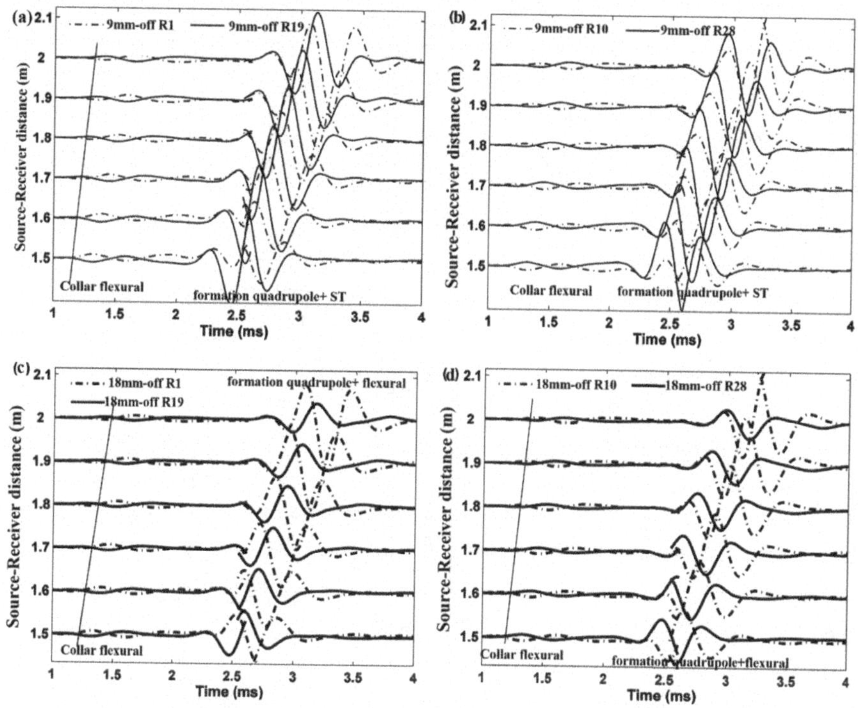

Fig. 6.49 Waveforms at four azimuthal receivers for a quadrupole tool eccentered along 45°. **a** R1 and R19 with 9 mm offset. **b** R10 and R28 with 9 mm offset. **c** R1 and R19 with 18 mm offset. **d** R1 and R19 with 18 mm offset

LWD case, mechanical calipers are not practical and an ultrasonic caliper as shown in Fig. 6.52 can be used (Market and Bilby 2011). In that method, the first arrival of the ultrasonic pulse-echo from the 4 azimuthal sensors are obtained to determine the location of the collar within the borehole.

An alternate approach is to use the waveform at sonic frequency. As shown in Sect. 6.3.2, the waveforms at different receivers in the dipole measurement are mainly affected by the tool eccentering amounts and angles. Phase difference is a function of eccentering angle θ and amount d and frequency ω. Wang et al. (2013b) used the phase differences in the waveforms at the inline receivers and at the crossline receivers to determine the tool position.

For different d and θ combinations, different relationships with phase difference can be obtained. It is possible to use the waveform phase difference to determine the tool position in LWD measurement.

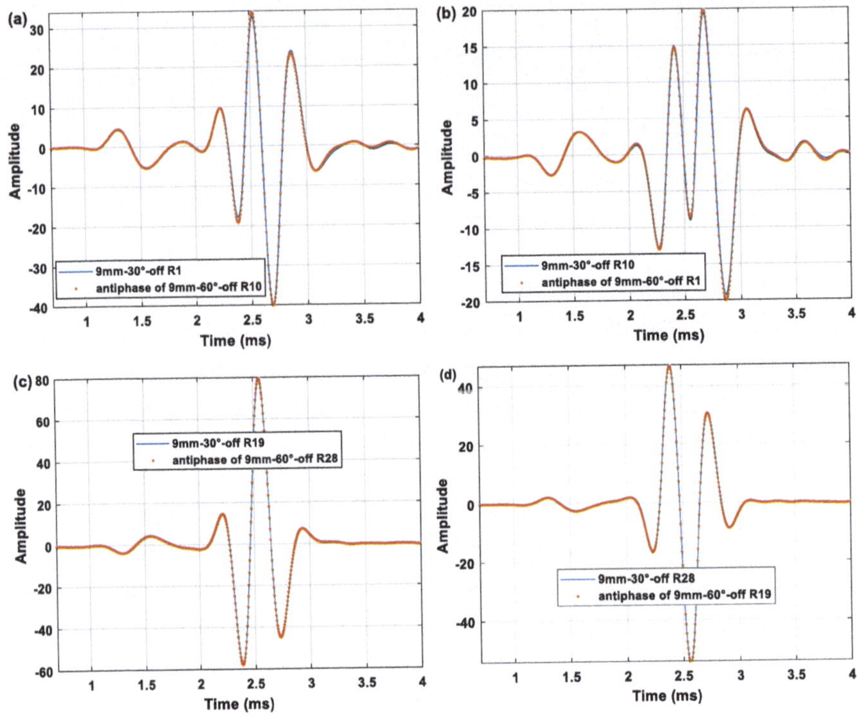

Fig. 6.50 Comparison for waveforms at different receivers when the quadrupole tool is eccentered along 30° and 60°. The amount of tool eccentering is 9 mm. Formation parameters are the same as those in Fig. 6.47. "off" in the legend means offset

6.4 Summary

This chapter covered the effects of off-centering the multi-pole wireline and LWD acoustic logging tools in wells surrounded by fast and slow formations. The summary of this chapter is as follows,

(1) The eccentered monopole wireline tool induces some asymmetric modes. The amplitudes of these modes depend on the amount of tool eccentering. The induced asymmetric modes underestimate the S-wave velocity in fast formations and affects the dispersion characteristic of ST waves in slow formations.

(2) The accuracy of formation velocity determination is affected if the tool is not centered. In monopole ALWD, the induced collar flexural waves contaminate the P- and S-waves making the velocity measurement difficult in fast formations. In the quadrupole ALWD, the induced collar flexural does affect the formation S-wave velocity measurement.

(3) Tool position can be determined using multi-azimuth waveforms from dipole measurements.

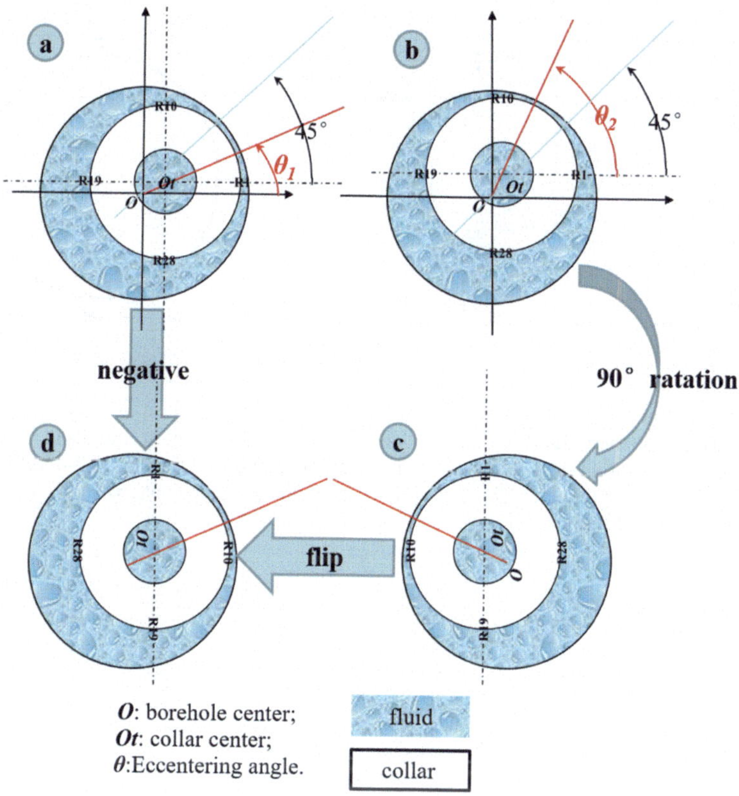

Fig. 6.51 Diagram of the anti-symmetry of the quadrupole waveforms at receiver pairs R1-R10 and R19-R28. If $|\theta_1 - 45°| = |\theta_2 - 45°|$, the tool eccentered along θ_2 (**b**) can be rotated 90° (rotating the entire picture) firstly (**c**) and then horizontally flipped to obtain a similar tool position (**d**) to the tool eccentered along θ_1 (**a**). The relationship of waveforms at different eccentering angles are R1 (θ_1) = −R10 (θ_2), R10 (θ_1) = −R1 (θ_2), R19 (θ_1) = −R28 (θ_2), and R28 (θ_1) = −R19 (θ_2)

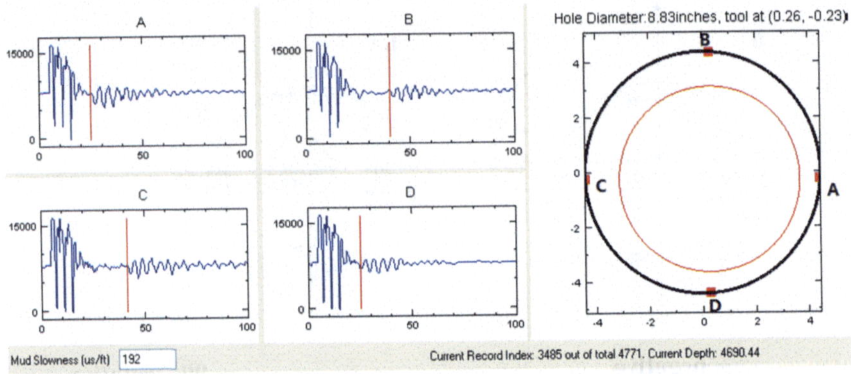

Fig. 6.52 Tool position determined from the ultrasonic pulse-echo in 4-azimuth caliper data, showing each waveform (A, B, C, and D Calipers). This is the result at one depth. The large signal at the beginning of each wave is the remnant of the transmitted signal. The red line on each trace shows the pick of the echo. The right display shows the computed hole size and tool position (Fig. 4 in Market and Bilby (2011))

Chapter 7
Peripheral Imaging Around a Borehole

Imaging around a borehole provides valuable information for high-resolution characterization of the subsurface. Surface seismic surveys provide good lateral imaging, but its capabilities for detecting small features are limited. Conventional borehole acoustic logging, described in Chap. 2, has a lateral depth of investigation of approximately one meter from the borehole. The gap between seismic surveys and conventional acoustic logging can be bridged by modifying acoustic logging for lateral imaging. Recent studies have shown that borehole acoustic reflection imaging (BARI) can detect structures as far as 20–100 m from the borehole (Tang 2004, 2017; Zhang et al. 2018). In addition, the technology can be employed in acoustic logging-while-drilling (ALWD) and in geo-steering (Tang et al. 2007). In this chapter, we describe the concepts, wave propagation modeling, and data processing methods for lateral imaging. We also show some field data examples of fractures imaged around a borehole.

7.1 Introduction

In acoustic logging, the seismic waves generated by the source are partly trapped in and around the fluid-filled borehole and partly penetrate the formation. Seismic energy from the source that goes into the formation and propagates as body waves that scatters from discontinuities may be recorded by the receivers in the borehole and used for imaging. Figure 7.1 shows a schematic of the wave propagation paths around a borehole in a formation with lateral heterogeneity.

Waves penetrating the formation propagate as P- and S-waves, and their ray paths are defined by Snell's law. Waves are reflected and converted when they encounter a discontinuity, as shown in Fig. 7.2. Therefore, the waves generated by the source produce an array of reflected and transmitted P-P (P to P), P-S (P to S) and S-P (S to P), and S-S (S to S) waves. When the geometry is favorable, these waves can be recorded by receivers in the borehole. The amplitudes and arrival times of

H. Wang et al., *Borehole Acoustic Logging—Theory and Methods*,
Petroleum Engineering, https://doi.org/10.1007/978-3-030-51423-5_7

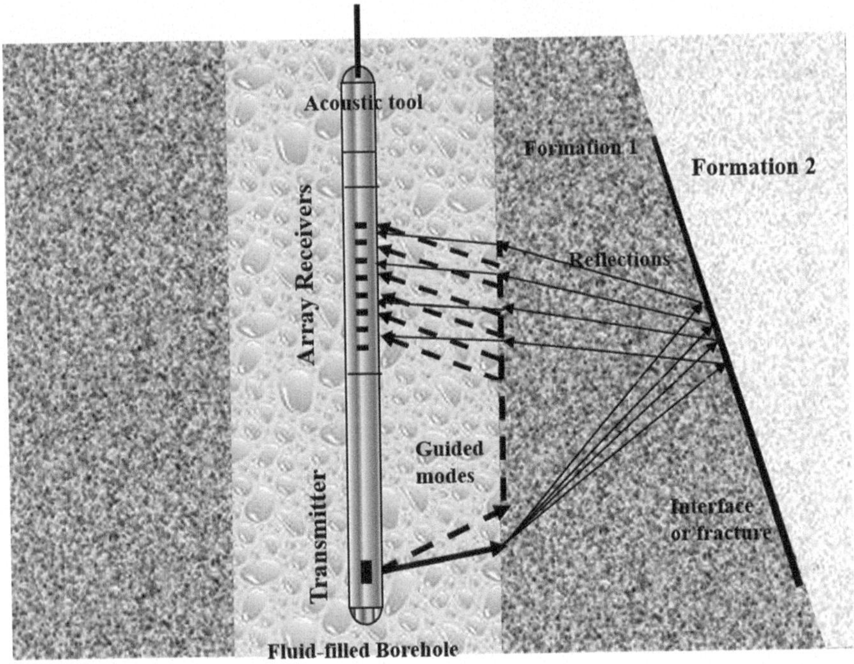

Fig. 7.1 Schematic of an acoustic reflection image from a well

received waves depend on the distance to the discontinuity, the material contrast at the discontinuity, and the geometries of the interfaces. The reflected waves can be used to image both the positions and the geometries (such as dip and strike) of the discontinuities near the borehole (Hornby 1989; Tang 2004).

In a simplified model, travel times can be determined using the ray theory. An example is shown in Fig. 7.3, where the source T and receiver R are in the borehole with a separation L and the interface intersects the borehole axis with angle θ. Let b be the distance between the receiver R and intersection point of the interface and borehole axis, V_{f} is the borehole fluid velocity, and V_{p} and V_{s} are the formation P- and S-wave velocities, respectively. R is the borehole radius.

The travel time of a P-P reflection (t_{pp}) from the transmitter to receiver is as follows,

$$t_{pp} = \frac{R}{v_f}(\sec\theta_5 + \sec\theta_1) + \frac{\sqrt{L'^2 + 4b'(L' + b')\sin^2\theta}}{v_p} \tag{7.1}$$

where L is the distance from the source to the receiver. $L' = \mathrm{L} - R(\tan\theta_1 - \tan\theta_5)$ and $b' = \mathrm{b} - R(\cot\theta - \tan\theta_5)$. θ_1 and θ_5 are incidence and arrival angles, respectively. The angles in Eq. 7.1 are related by Snell's law and geometry:

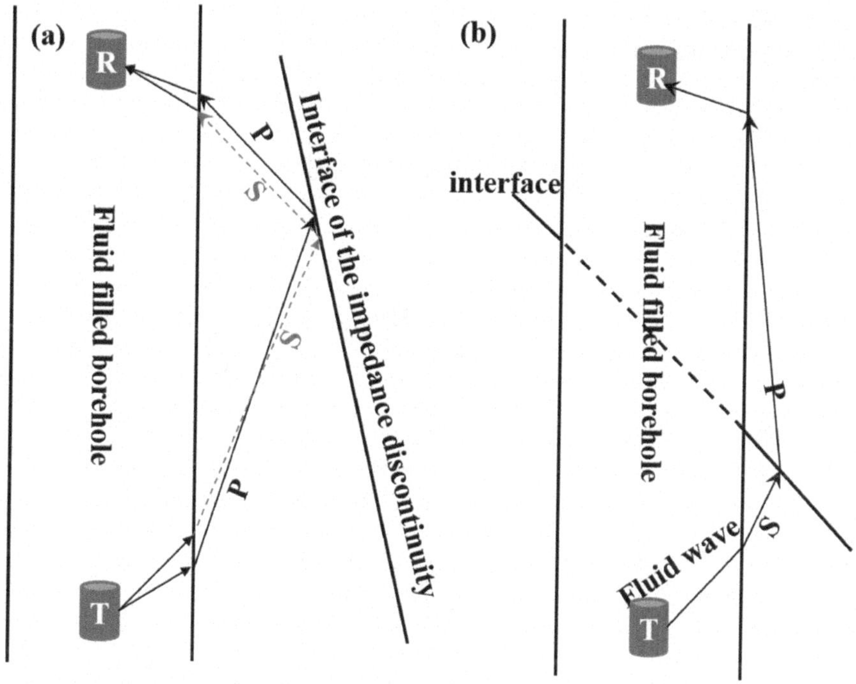

Fig. 7.2 Schematic diagram of the reflected and transmitted waves. **a** P-P and S-S reflected waves. **b** S-P converted transmitted waves

$$V_f / \sin\theta_1 = V_p / \sin\theta_2, \tag{7.2}$$

$$\theta_2 = \theta_3 + \theta, \tag{7.3}$$

$$\theta_4 = \theta_3 - \theta, \tag{7.4}$$

$$V_f / \sin\theta_5 = V_p / \sin\theta_4, \tag{7.5}$$

$$L' \cot\theta = \left(L' + 2b'\right) \tan\theta_3, \tag{7.6}$$

where θ, θ_2, θ_3, and θ_4 are shown in Fig. 7.3. The arrival times of P-S, S-P, and S-S reflections can also be obtained with a similar method.

Figure 7.4 shows the arrival times of different waves as a function of the source-receiver offset. For this example, the formation surrounding the borehole is F1, whose properties are listed in Table 2.1. The distance (L + b) from the source to the intersect point of the interface and borehole axis is 18 m. The angle $\boldsymbol{\theta}$ is 15°. The arrival times for three borehole guided waves (P, S, and ST) and four reflected waves (P-P,

Fig. 7.3 Schematics of ray paths for P-P reflection at an interface

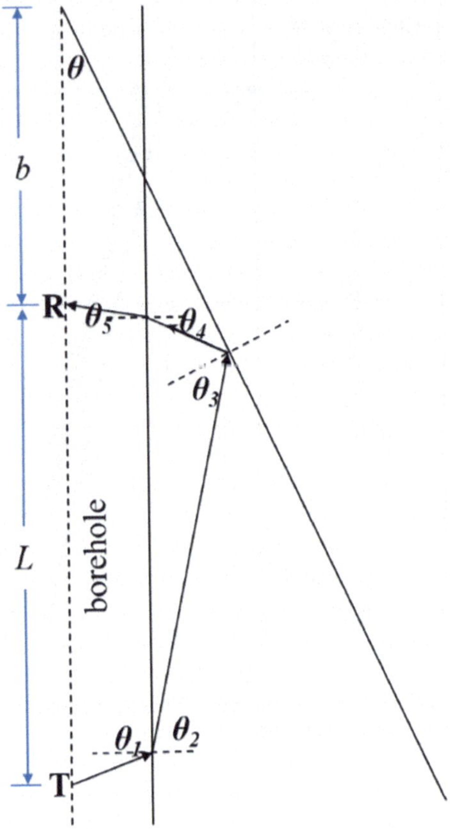

P-S, S-P, and S-S) are also shown. The arrival times of the borehole guided waves increase linearly with the increase in receiver offset. The arrival times of the reflected waves fall along hyperbolas. The reflected and borehole guided waves overlap each other for a source-receiver offset of more than 3 m. This concept illustrates one of the challenges in the application of this method. The weak reflected waves may be submerged within the strong waves that are guided along the borehole.

The first prototype of an acoustic reflection tool was called EVA (Evaluation of Velocity and Attenuation) and was developed in 1981 (Arditty et al. 1981; Fortin et al. 1991). Schlumberger called their acoustic reflection tool BARS: Borehole Acoustic Reflection Survey (Esmersoy et al. 1998). Field applications showed that acoustic imaging can detect interfaces 10–20 m from the borehole (Hirabayashi et al. 2010). Bohai Drilling of China National Petroleum Corporation also developed an acoustic reflection logging tool and data processing software in 2007 (Chai et al. 2009; Li et al. 2014). When a monopole source and omnidirectional receivers are used for imaging, the strike (i.e., azimuth) of the reflector cannot be determined. Different approaches have been introduced to obtain strike information. An example includes

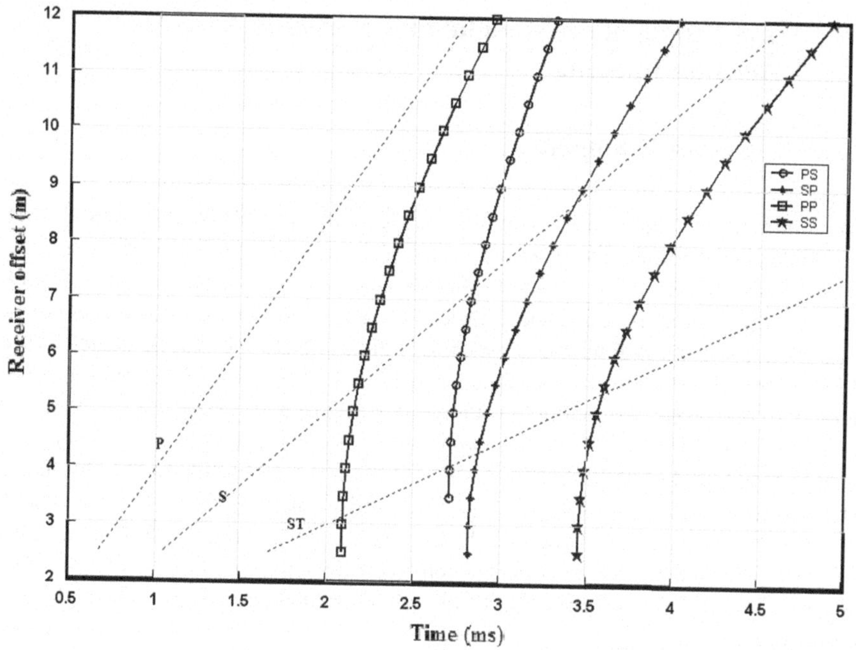

Fig. 7.4 Travel time curve of the borehole modes and reflected waves (He 2005). The formation surrounding the borehole is F1, and the properties are listed in Table 2.1. The model is shown in Fig. 7.3 with angles and dimensions given in the text

the Sonic Scanner (a Schlumberger tool) designed to detect the structure around the borehole in 3 dimensions (Pistre et al. 2005).

Tang (2004, 2016, 2017), Tang and Patterson (2009) used dipole S-wave measurements based on XMAC II (a sonic logging tool from Baker Hughes) to image interfaces and to obtain the azimuth information from the dipole data using the reflected SH and SV waves. Zhang et al. (2018) reported an improved dipole tool for imaging. The tool employs a dipole source at two frequencies of 1 and 4 kHz. The tool has two recording times of 128 and 256 ms and can detect reflectors up to 80 m from the borehole.

Although acoustic reflection imaging tools have been available for more than a decade, some problems remain. The biggest challenges are associated with extracting the relatively weak reflected waves from the dominant borehole-guided waves, and determining the dip, strike, and azimuth of the reflector. Progress is being made using methods adopted from seismic surface reflection imaging, such as migration, to detect and determine the dip of the structure using monopole logging data (e.g., Zhang et al. 2009; Li et al. 2014).

7.2 Wave Propagation Modeling for Borehole Acoustic Reflection Imaging

7.2.1 Wireline Monopole

The 3-dimensional finite difference method (see Appendix B) is used to model the wavefields in and around the borehole, including scattering from an interface outside the borehole. The model is similar to that shown in Fig. 7.1. The horizontal axis of the figure is defined as x, and the vertical axis is z. The model size is 5.5 m (x) × 1 m (y) × 15 m (z). A fluid-filled borehole with a radius of 10 cm is located at (1, 0.5) (we set it to (0, 0) in the coordinate system in the following calculations) and is parallel to the z-axis. A monopole source consisting of a 10 kHz Ricker wavelet, is located at (1, 0.5, 0.85 m). There are two formations outside the borehole. Formation F1 surrounds the borehole and formation F2 is on the other side of the interface. The elastic parameters of the borehole fluid and formations F1 and F2 are listed in Table 2.1. The angle between the borehole axis and interface is 5 degrees. The interface, with an azimuth of 0 degrees, intersects the x-axis at (4.8, 0, and 0 m). The dip (θ), azimuth (φ), and strike (ψ) of the reflection plane relative to x, y, and z are shown in Fig. 7.5.

The calculated pressure wavefields and snapshots are shown in Figs. 7.6a (x-z profile), 7.7a (y-z profile), and 7.8a (x-y profile). In these figures, the source position is set as the origin. The time of the snapshots is 2 ms. The calculation is repeated for a reference model of the same size with homogenous formation F1 but without an interface. Snapshots are shown in Figs. 7.6b (x-z profile), 7.7b (y-z profile), and 7.8b (x-y profile). When there is no interface in the model (Fig. 7.6b), no reflected waves are generated outside the borehole. The guided waves in the borehole can be identified, such as P, S, pR, ST, and pR Airy phase (marked in Fig. 7.6b), where the P-wave is the fastest arrival and the S-wave is the second arrival. The differences in the wavefields between the models with and without the reflector is shown in Fig. 7.6c. The differences in the wavefields in the y-z and x-y profiles are shown in Figs. 7.7c and 7.8c, respectively. Figures 7.9, 7.10, and 7.11 show the snapshots at 4 ms. The P-wave has nearly propagated out of the model. The differences in the wavefields between the models with and without the reflector are shown in Figs. 7.9c, 7.10c, and 7.11c for the x-z, y-z, and x-y profiles, respectively, where the reflected waves, P-S, S-P, and S-S are marked in the figures. The spatial distribution of the pressure at 4 ms and 4.2 m offset along the x direction is shown in Fig. 7.12. The maximum amplitude is that of the Airy phase of the pseudo Rayleigh (pR) wave. The reflected waves cannot be discerned without amplification. In Fig. 7.12, they are amplified by a factor of 1000. The difference between the maximum amplitudes of the guided and reflected waves is approximately 78 dB.

The waveforms collected by the centralized receiver array in the borehole are shown as a function of the offset in Fig. 7.13. The reflected waves (Fig. 7.13b) are the difference between the full waveforms and guided modes in the same model, where the guided modes are obtained from the model without the interface. Because

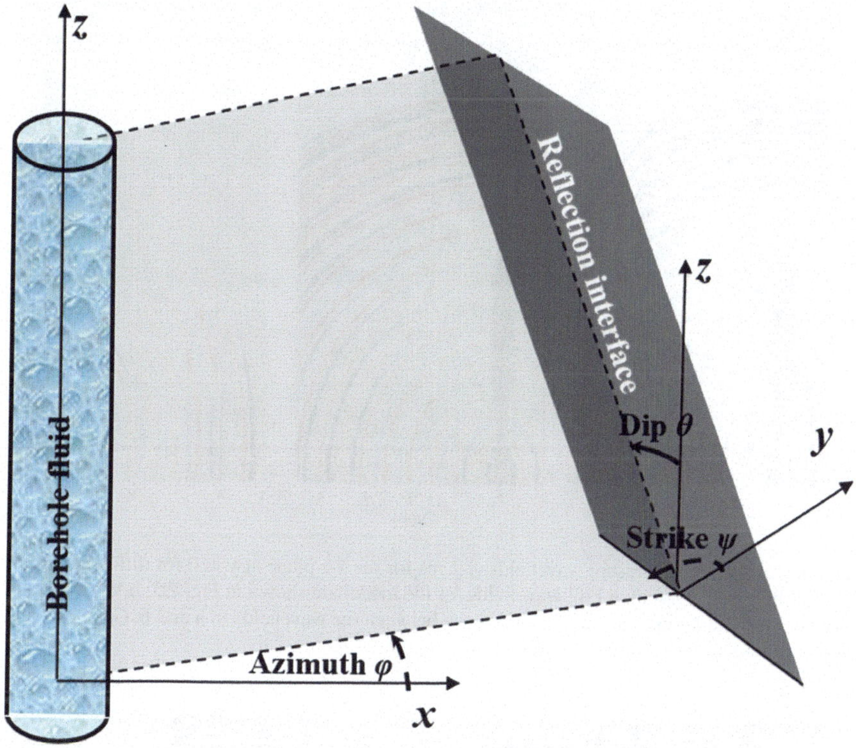

Fig. 7.5 Schematic diagram of a reflector outside of a borehole. Dip, azimuth, and strike are shown. For the modeling, $\varphi = 0°$, and $\theta = 5°$. The formation surrounding the borehole is F1. The formation on the opposite side of the interface from the borehole is F2. Note that the "dip" denotes that the angle between the interface and the z-axis, which is not a common convention, as dip is more commonly measured between the interface and horizontal plane. The geometry and interface angles are described in the text

all the receivers are below the interface, there are only four reflected waves that have a hyperbolic relationship between the arrival time and offset (shown in Fig. 7.13b). At near offset, the arrival times of different modes, calculated using ray tracing, are marked by the dashed curves. The arrival times of the P-P and S-P reflected waves are between 1.5 and 2.5 ms, and the S-S reflection arrives after 3 ms. The converted P-S and S-P waves are located between the P-P and S-S waves. The S-P reflection is partially buried in the trailing part of the P-S reflection.

The amplitude of the S-S reflected wave is very small when the offset is less than 4 m. This amplitude is small because of weak S-wave generation at angles nearly perpendicular to the borehole axis. The S-S reflected wave can be identified at large offsets.

Figure 7.14 shows the analysis for selected array waveforms with receiver offsets from 3 to 4.05 m with a receiver spacing of 0.15 m, which are the most common offsets used in commercial acoustic logging tools. In Fig. 7.14a, different reflections

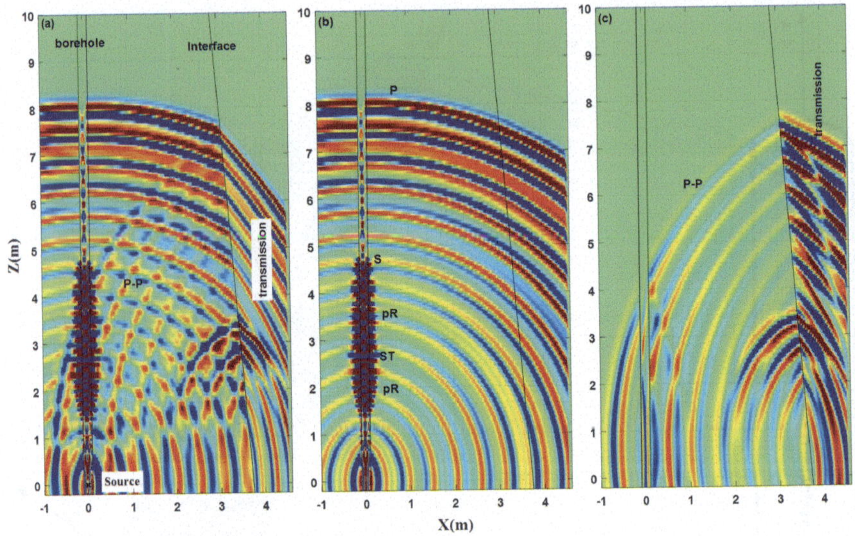

Fig. 7.6 Snapshots of pressure wavefields at 2 ms for the *x-z* plane at y = 0 for different models (10 kHz monopole source). **a** Full wavefields for the model are shown in Fig. 7.5. **b** Wavefields in the model without an interface. **c** The difference between the wavefields in **a** and **b**. Geometry and interface angles are described in the text

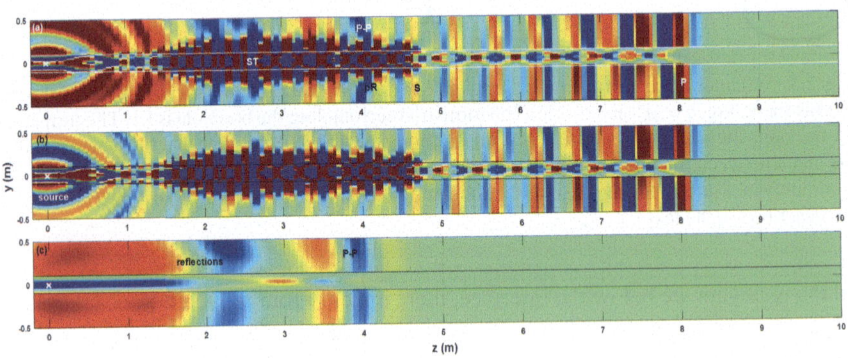

Fig. 7.7 Snapshots of pressure wavefields at 2 ms in the *y-z* plane at x = 0. **a** Full wavefields in the model (shown in Fig. 7.5). **b** Wavefields in the model without the interface. **c** The difference between wavefields in **a** and **b**

are marked with different colored lines. For these offsets, the arrivals P-S, S-P, and S-S, overlap each other. It is difficult to pick them. The velocity-time semblance plot is shown in Fig. 7.14c. The apparent velocities of reflections are large. The arrival times of reflections are marked in Fig. 7.14a, b. Because of their small amplitudes, the reflections are not easy to identify even with the arrival times marked. Only the borehole waves are observed in the velocity-time semblance plot in Fig. 7.14d. Data

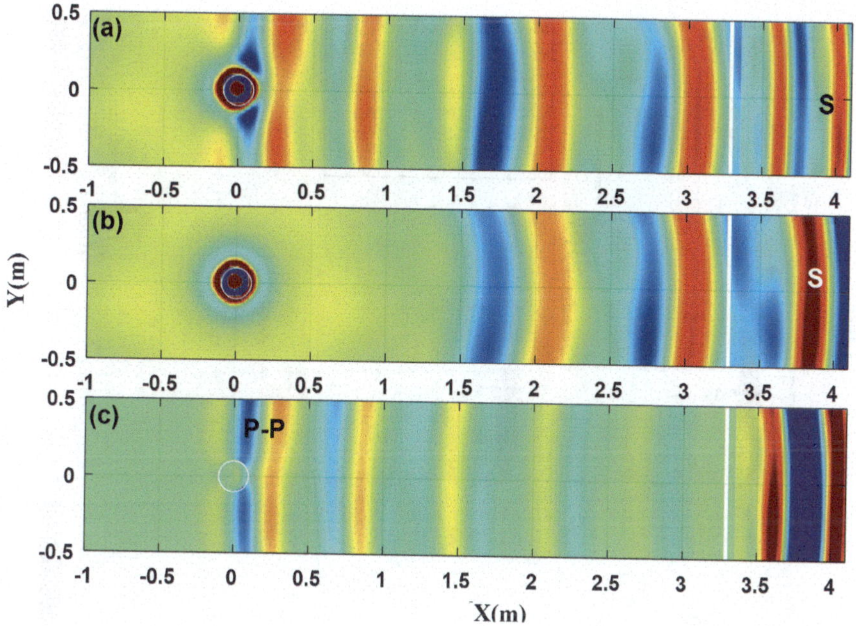

Fig. 7.8 Snapshots of pressure wavefields at 4 ms in the x-y plane at z = 5. **a** Full wavefields in the model (shown in Fig. 7.5). **b** Wavefields in the model without the interface. **c** The difference between wavefields in **a** and **b**

processing methods, described in Sect. 7.3, are needed to extract the reflections and image the reflector.

7.2.2 *Wireline Dipole*

To complement the monopole measurement, the S reflected wave from dipole measurements may be used. The polarization feature of the S-wave (SV and SH) can be used for azimuth determination (Tang 2004; He 2005). Different configurations of the source and receivers, such as dipole source and dipole receivers, dipole source and monopole receivers, or even a monopole source and dipole receivers, may be used.

Reflected waves from a dipole source include a P-wave with very small amplitude and S-waves. There are two reflected S-waves: the SH wave polarized in the horizontal plane and the SV polarized in the vertical plane. Figure 7.15 shows the polarization and propagation directions of the SV and SH waves. In the *x*-*y* plane, the amplitudes of the SH and SV waves propagating at azimuth φ from the source polarized along the x-direction are $SH \cdot \cos\varphi$ and $SV \cdot \sin\varphi$, respectively, where SH and SV are the amplitudes of the SH and SV waves propagating in the x direction. For

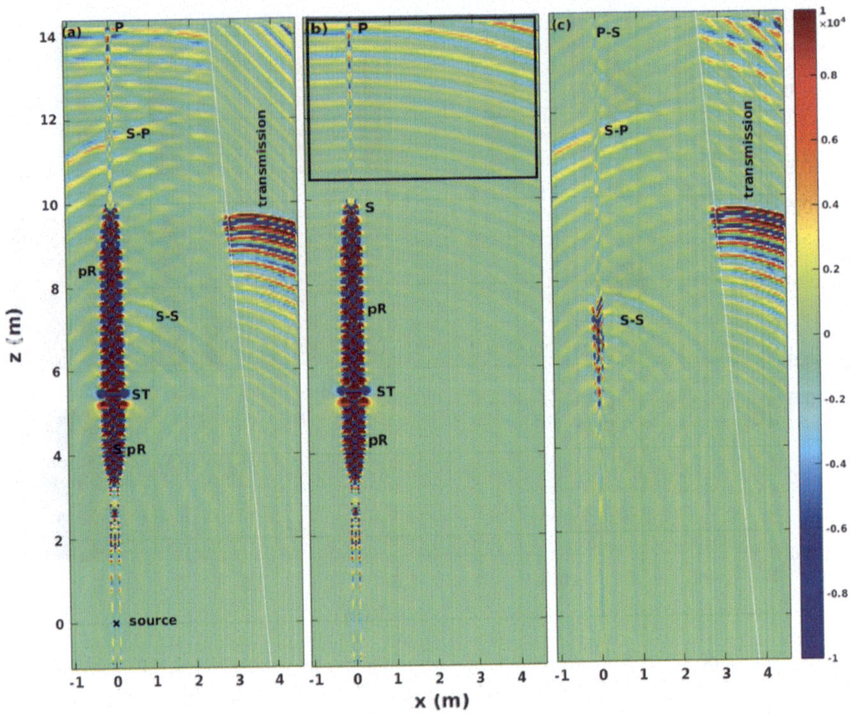

Fig. 7.9 Snapshots of pressure wavefields at 4 ms in the *x*-*z* plane at y = 0 (10 kHz monopole source). The model is shown in Fig. 7.5. The subplots are in the same order as those shown in Fig. 7.6. The direct P and P-P reflection have propagated beyond the model field at z = 14 m. Black box in **b** shows the P-wave

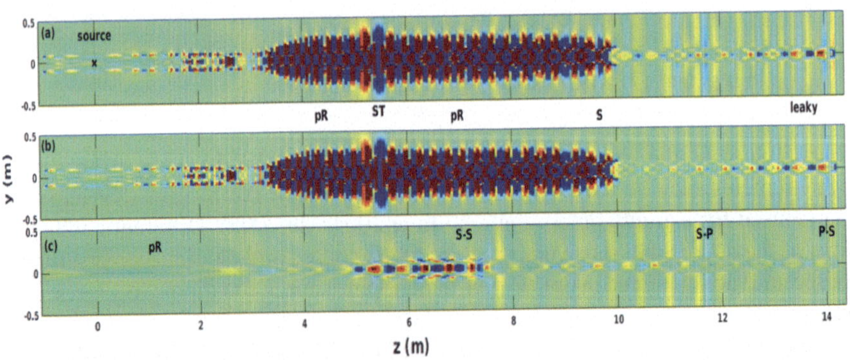

Fig. 7.10 Snapshots of pressure wavefields at 4 ms in the *y*-*z* plane at x = 0. The model is shown in Fig. 7.5. The subplots are in the same order as those in Fig. 7.7

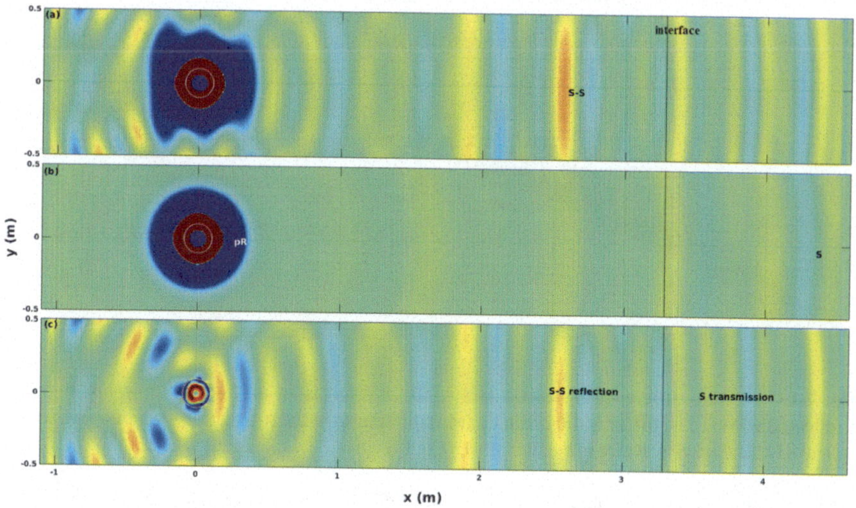

Fig. 7.11 Snapshots of pressure wavefields at 4 ms in the x-y plane at z = 5. The model is shown in Fig. 7.5. The subplots are in the same order as those in Fig. 7.8

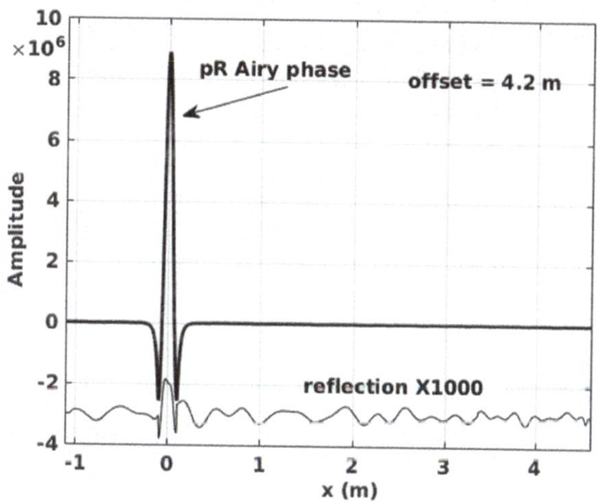

Fig. 7.12 A comparison of amplitudes of borehole guided waves and reflections as a function of radial distance at offset z = 4.2 m. The model is shown in Fig. 7.5

example, if φ is 0, the projection of SV in the x direction results in a zero amplitude. However, the projection of SV in the y direction results in a maximum amplitude. This is the method to determine the reflector azimuth using the amplitudes of the SH and SV waves. For receivers placed along the x and y axes, the borehole effect can be ignored, and the received reflected wavefields are given by:

$$xx = SH \cdot \cos^2 \varphi + SV \cdot \sin^2 \varphi, \tag{7.7a}$$

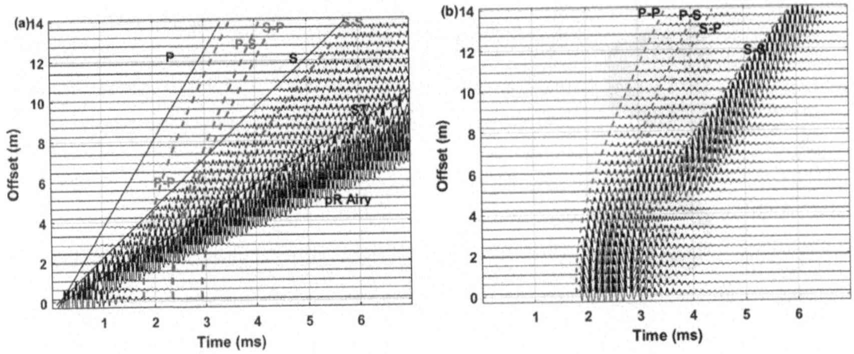

Fig. 7.13 **a** Finite difference synthetic array full waveforms in the model shown in Fig. 7.5; **b** reflected waves. The offset is the distance from the source in the z direction. A monopole source with a 10 kHz Ricker wavelet is located at (0, 0, 0.85)

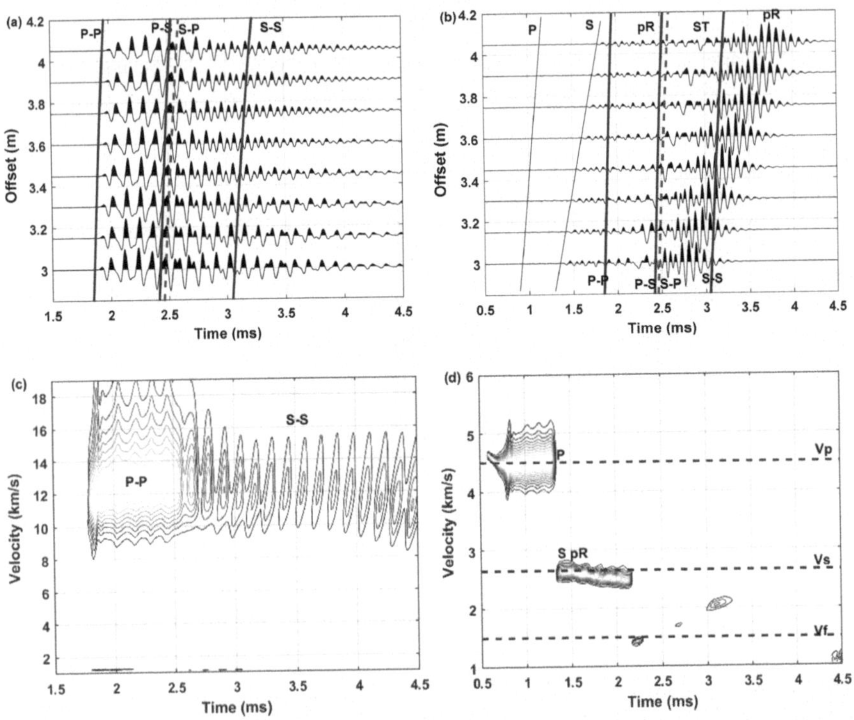

Fig. 7.14 Waveform analysis with selected receivers. A monopole source with a 10 kHz Ricker wavelet is located at (0, 0, 0.85). **a** Reflections. **b** Full waveforms. Coloured lines indicate reflections highlighted in **a**. **c** Velocity-time semblance analysis of the reflected waves in **a**. **d** Velocity-time semblance analysis of the array waveforms in **b**. The selected receivers have offsets from 3 to 4.05 m with a receiver spacing of 0.15 m

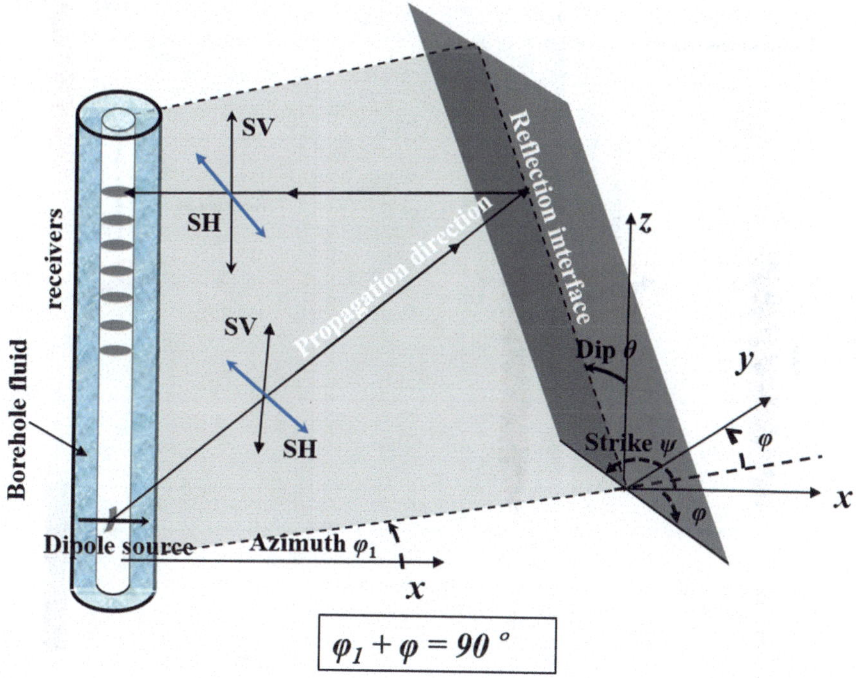

Fig. 7.15 Schematic diagram of S-wave imaging with a dipole logging tool. Note that the azimuth is defined as the complementary angle of the azimuth in Fig. 7.5

$$xy = -SH \cdot \cos\varphi \sin\varphi + SV \cdot \cos\varphi \sin\varphi. \tag{7.7b}$$

The first index x indicates that the source polarization is in the x direction and the second index is the direction of the receiver. In the same way, if the polarization of the source is along the y direction, the amplitudes of SH and SV waves are SH $\cdot \sin\varphi$ and SV $\cdot \cos\varphi$, respectively. The received reflected waves are given by:

$$yy = SH \cdot \sin^2\varphi + SV \cdot \cos^2\varphi, \tag{7.7c}$$

$$yx = -SH \cdot \sin\varphi \cos\varphi + SV \cdot \cos\varphi \sin\varphi. \tag{7.7d}$$

For the four component waveforms in the cross-dipole tool (shown in Fig. 1.6), the SH and SV reflected waves can be obtained from the data:

$$SH = xx \cdot \cos^2\varphi - (xy + yx) \cdot \sin\varphi \cos\varphi + yy \cdot \sin^2\varphi, \tag{7.8a}$$

$$SV = xx \cdot \sin^2\varphi + (xy + yx) \cdot \sin\varphi \cos\varphi + yy \cdot \cos^2\varphi. \tag{7.8b}$$

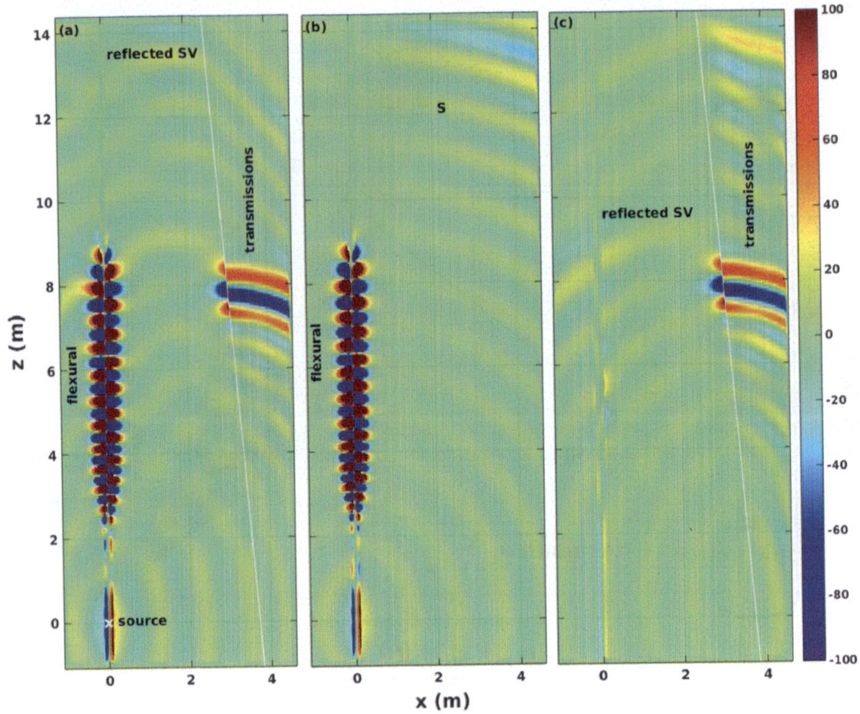

Fig. 7.16 Snapshots of wavefields at 4 ms in the *x*-*z* profile at y = 0 m (2 kHz dipole source along the *x* direction). **a** Wavefields for the model with an interface shown in Fig. 7.5. **b** Wavefields without an interface. **c** The difference between wavefields in **a** and **b**

The azimuth φ can be determined from the amplitudes of the SH and SV waveforms in the cross-dipole data.

A synthetic example of the dipole wavefields is shown as snapshots at 4 ms in Figs. 7.16 (*x*-*z* profile), and 7.17 (*x*-*y* profile), using the same model as that in Sect. 7.2.1 (Fig. 7.5). In the simulations, the polarization of the dipole source is along the *x* direction. This polarization is simulated by loading the two opposite-polarity point sources onto the normal stress grid (see Fig. B.1 in Appendix B) at (−0.005, 0, 0 m) and (0.005, 0, 0 m). The source time function is a 2 kHz Ricker wavelet. The guided modes, shown in Figs. 7.16b and 7.17b, are formation flexural waves. The body S-waves, SV in the *x*-*z* view (Fig. 7.16b) and SH in the *x*-*y* view (Fig. 7.17b), propagate ahead of the flexural wave. When the flexural waves hit the reflector, they are reflected back toward the borehole, as shown in Figs. 7.16c and 7.17c. The waves transmitted across the interface are larger than the reflected waves. Because the source polarization is along the *x* direction, there are no guided waves in the *y*-*z* plane. As shown in Eqs. 7.7a, 7.7b, 7.7c, 7.7d and 7.8a, 7.8b, the SH and SV waves can be decomposed from the horizontal components of the received waves and can be used to determine the azimuth of the reflector interface. The amplitudes of the SH and SV waves at different depths (z) are related to the dip of the reflecting interface.

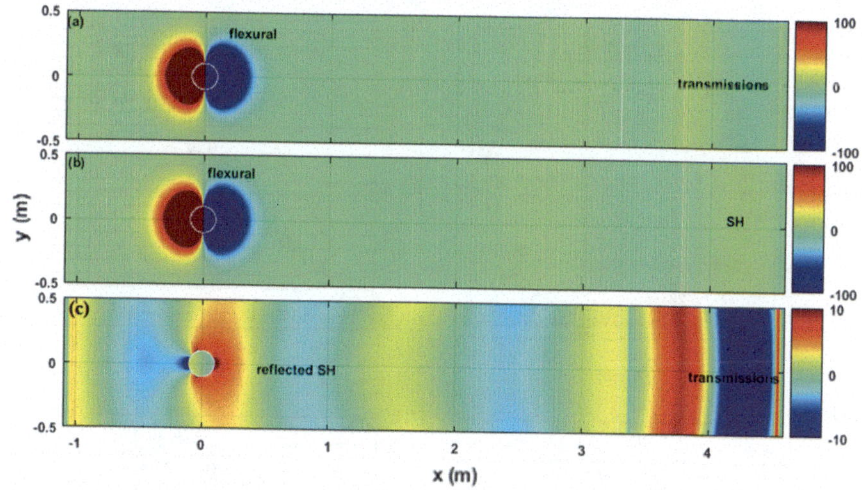

Fig. 7.17 Snapshots of wavefields at 4 ms in the x-y profile at $z = 5$ m (2 kHz dipole source along the x direction). **a** Full wavefields in the model shown in Fig. 7.5. **b** Wavefields without an interface. **c** The difference between wavefields in **a** and **b**. Small residul SH reflection, which is supposed to be zero, results from the directivity of the dipole source which consists of two point source

7.2.3 ALWD Measurements for Geo-Steering

With the increasing use of horizontal drilling, the use of reflected waves from bed boundaries provides the means to keep the borehole in the formation of interest. Data acquired ALWD is especially suitable for geo-steering.

Nakken et al. (1995) proposed a geo-steering system with a configuration of one transmitter and two receivers. The source frequency is between 1 and 4 kHz to reach a penetration depth of approximately 20 m. Here, we model the waveforms for an ALWD system. Parameters of the borehole and drill collar C32 given in Table 5.2 are used for the modeling. The location and orientation of the reflector is the same as that in the wireline case described in Sect. 7.2.1 (Fig. 7.5). A ring source (10 kHz Ricker) with 3D elements, shown in Fig. 6.7a, is used to simulate the monopole ALWD tool in the 3D finite difference code. The positions of 36 azimuthal receivers are shown in Fig. 7.18a. An array of waveforms at receiver R1 but at different offsets is shown in Fig. 7.18b. The collar, S, and Stoneley (ST) waves are clearly visible in the wavetrains. The fluid annulus in LWD is narrower than that in the wireline model. This moves the pR wave to a higher frequency range than the wireline case shown in Fig. 7.13a. pR is very small at a 10-kHz source frequency. The ST wave is very prominent. Figure 7.18c, d show the reflected waves for receivers R24 and R1, respectively. The arrival times of the reflected waves at R24 and R1 are identified with curves. Similar to the reflected waves in the wireline case (Fig. 7.13b), the P-P and S-S reflections are clear. The amplitude of reflected waves varies with offset (spacing). One challenge in ALWD reflection imaging is overcoming the noise problem.

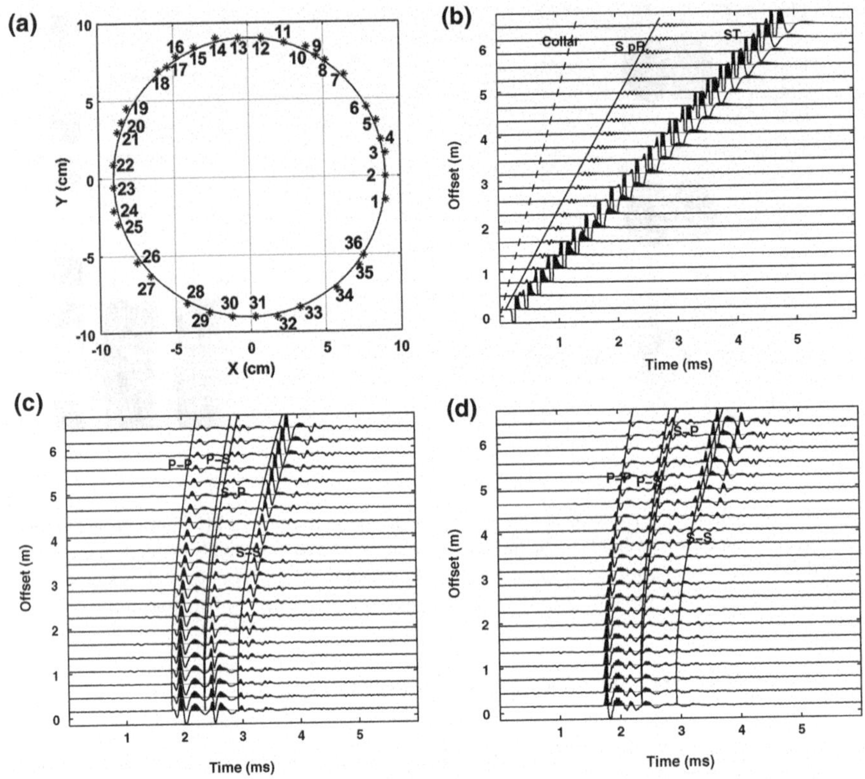

Fig. 7.18 The full waveform and reflected wave fields simulated by 3DFD for a formation interface model in an ALWD case. The model is shown in Fig. 7.5. Collar C32 in Table 5.2 is used. **a** Top view of the ring source and 36 receivers; **b** array full waveforms at receiver R1; **c** and **d** show the reflected waves at R24 and R1, respectively. Arrival times of the reflected waves at R1 and R24 are shown as lines

The use of ALWD for geo-steering is especially important in horizontal drilling to keep a borehole in the formation of interest. The reflections of interest are those from the bed boundaries. The azimuthally distributed sources and receivers shown in Fig. 7.18a make it possible to determine the collar position relative to the top and bottom of the bed.

7.3 Data Processing for Peripheral Imaging

Processing methods used in surface reflection seismology can be employed, with some modification, for reflection imaging by wireline or by LWD tools. Data collected by the receiver array can be arranged either in CSG (common source gather) or in COG (common offset gather). The geometry of the data gather determines the type of processing algorithms to be used for reflection extraction and reflector imaging.

The velocity-time semblance method, described in Chap. 3, is ideal for velocity determination. For the reflected waves, where the wave propagation directions are not along the borehole, the velocities obtained by semblance are apparent velocities and are larger than the wave propagation velocities. The velocity difference between the guided and reflected waves is dependent on the source-receiver offset, the distance to the reflector, and dip angle θ of the reflector.

Guided waves propagate parallel to the borehole, and their "ray paths" are parallel to the borehole axis. These waves appear as straight lines in the COG profile along depth. The reflected waves, which appear as hyperbolic lines, are affected by the distance from the borehole, azimuth and dip of the reflector. Amplitudes depend on impedance contrast and angle of incidence at the reflector. The ray paths of the reflected waves from different source positions provide the aperture for imaging the interface.

Figure 7.19 shows a flow diagram of data processing for BARI. Most of the data processing methods are borrowed from surface seismic reflection methods, including denoising, wavefield separation, migration, and imaging. The first step is to process the raw data into the COG format because data are acquired in CSG order. Then the apparent velocity analysis is performed on both gathers. A suitable reflection extraction method is chosen based on the difference in apparent velocities between the guided and reflected waves. After the reflection data are separated from full waveforms to obtain the image of the reflector, seismic migration is carried out using one of many time-domain migration methods (see Claerbout 1976; Stolt 1978; Beylkin 1985; Miller et al. 1987; Bancroft et al. 1998; Whitmore 1983; McMechan 1983; Baysal et al. 1983; Yilmaz 2001; Bleistein et al. 2001; Op't Root et al. 2012; Brytik et al. 2012; Ikelle and Amundsen, 2018). Then the azimuth of the reflector is determined. Finally, the time domain results are converted into the space domain where the distance to the interface is determined. The desired outputs are dip, strike, azimuth, impedance contrast and distance to the interface.

Borehole imaging faces challenges different from those of surface seismic surveys. In boreholes, data acquisition is confined to a single line (the borehole). There are fewer sensors. The weak reflected waves are buried in the strong guided waves. It is a challenge to extract the weak reflected signals whose amplitude are often down more than 70 dB from guided waves. In addition, the borehole effects need to be considered.

In this section, we demonstrate the data processing methods, using synthetic data generated from the two models (one is for fast formations, and the other is for slow formations) shown in Fig. 7.20. The model dimensions are 9 m, 1 m, and 14.5 m in the x, y, and z directions, respectively. The borehole is along the z direction and is located at (1, 0.5 m) in the x-y plane. The dip of the reflecting interface is 80° and the interface intersects the x-axis at 5.8 when y and z are 0. Both strikes are 0°. The elastic parameters of the formations and borehole dimensions are listed in Table 2.1. A 10 kHz Ricker wavelet is used as the source for generating monopole synthetics. Synthetic data used in this section are calculated by the 3D finite difference (see Appendix B).

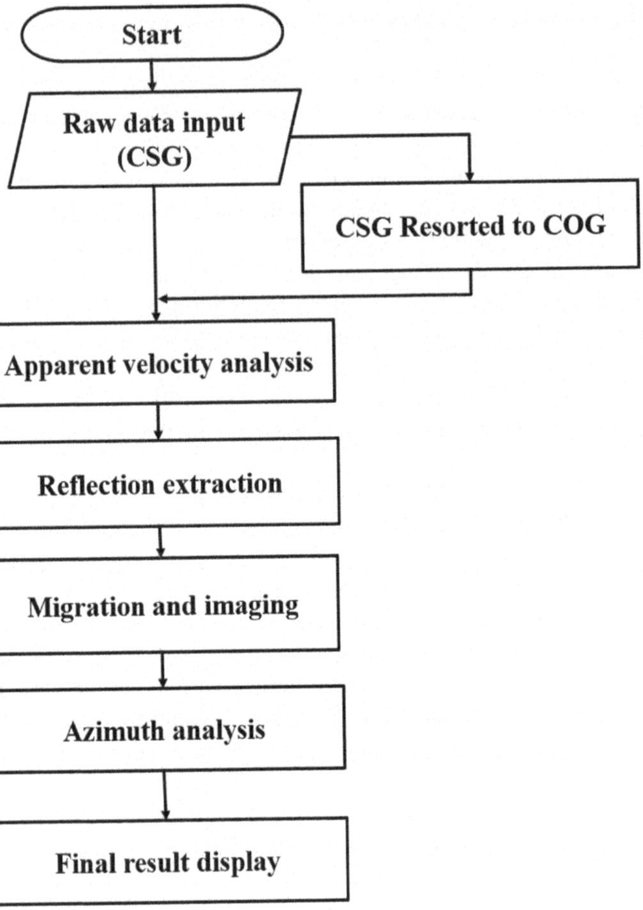

Fig. 7.19 A data processing flow for BARI

7.3.1 *Extraction of Weak Reflections*

Different methods, such as frequency-wavenumber filtering (Hornby 1989), parameter estimation (Tang 1997), median filtering (Song and Toksöz 2010), Adaptive Interference Canceling (AIC) (Al Rougha et al. 2005; Haldorsen et al. 2006), multi-scale velocity-time semblance (Tao et al. 2008a), Radon transform (Li and Yue 2017), and others may be used to extract the weak reflections. Here we describe and compare results from the frequency-wavenumber filtering, median filtering, and Radon transform methods for extraction of weak reflections.

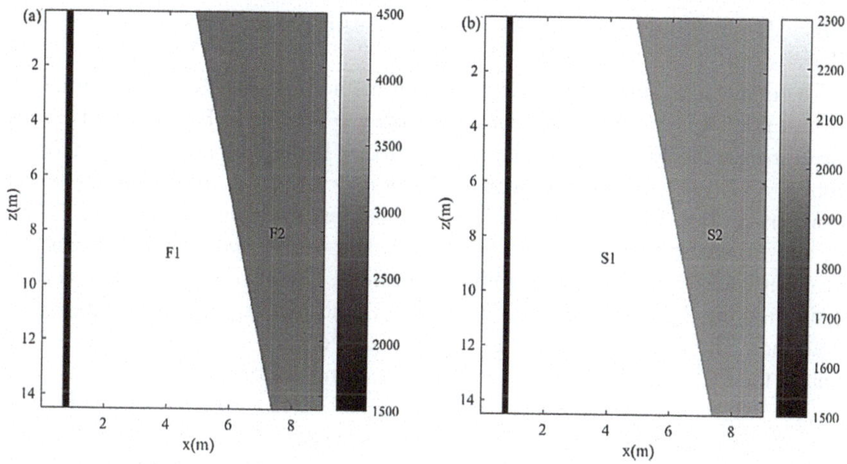

Fig. 7.20 Schematic diagram of the reflectors and velocities used for models to calculate synthetic waveforms. F1, F2 and S1, S2 are fast and slow formations (Table 2.1). The vertical black column is the borehole

7.3.1.1 Frequency-Wavenumber Filtering

One can obtain the velocities of different modes using a frequency-wavenumber transform, described by Embree et al. (1963) and Waters (1978), on the array waveforms. Using a filter in the f-k domain, reflected waves can be extracted given their different apparent velocities from those of the guided waves. The f-k transform can be used to transform the COG data to the f-k domain and the reflected waves are extracted by f-k filtering. An inverse transform converts the data back to the time-space domain for further processing.

The guided waves in the COG profile are located near the frequency axis f, as shown in Fig. 7.21. The reflected waves are mapped away from the f axis in the f-k domain. The upgoing reflected waves are in the positive f-k half plane. The downgoing reflections are located in the negative half plane. Using a bandpass filter in the f-k domain, the guided waves are suppressed. Then the reflected waves are

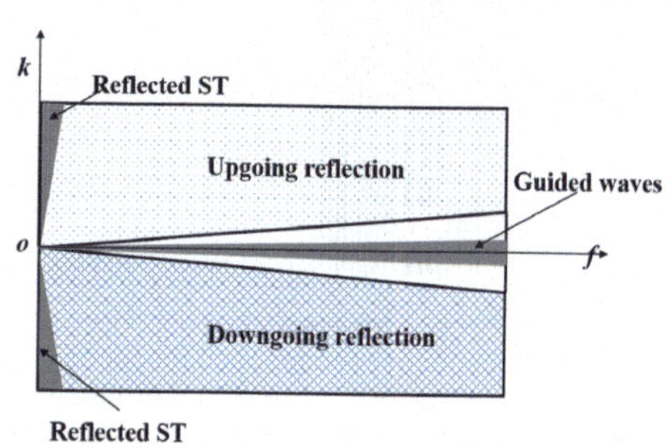

Fig. 7.21 Different waves of the COG profile in the frequency-wavenumber domain. The reflected ST waves are from an interface intersecting the borehole

obtained by inverting the filtered data back into the time-space domain with an inverse f-k transform.

To demonstrate application of the f-k filter method for reflected wave extraction, we use the two synthetic COG data sets from the models in Fig. 7.20, where the waveforms at the 3 m-offset receivers are collected by moving the monopole source from $z = 13.8$ m to $z = 4.8$ m, with a 0.15-m interval between the adjacent source positions. In Fig. 7.22a, the COG data from the fast formation model (Fig. 7.20a) are shown, where the black and white colors are positive and negative phases and the color intensity defines the amplitude. The vertical axis of the plot is the receiver position along the borehole axis. The P, S, ST, and pR Airy waves are marked in the figure. The reflected waves can be observed between 2 and 3 ms. The arrival time of the P-P reflection, marked with a dashed line, is obtained from the modeled reflected waves. Figure 7.22b shows the modeled reflected waves, which are obtained by subtracting the guided waves from the total wavefield. Comparing the full waves with the modeled reflected waves, it is obvious that the P-P reflections overlap with pR waves (Airy phase) at all depths. The guided waves are straight lines without moveout. The reflected waves are oblique with a distinct moveout.

The f-k transform is applied to the COG profiles in Fig. 7.22. The f-k transform of the modeled guided waves, which is obtained from the subtraction between Fig. 7.22a, b, is shown in Fig. 7.23b. The guided waves are located in a region where the wavenumbers are close to zero. The grayscale in the plot designates the relative magnitudes of the modes, which correspond to the amplitudes in the time domain. A darker color indicates a lager amplitude. The dashed line in Fig. 7.23c marks the upper limit of the reflected wave's apparent velocity where a large slope results in a lower velocity. Below this line, there is no reflected wave in the f-k domain of Fig. 7.23c. This dashed line is also plotted in Fig. 7.23a, b. The reflected waves, which are extracted by the inverse f-k transform, are shown in Fig. 7.24. Although some residual guided waves still exist before the arrival of the reflected P-P wave (marked with a dashed line), reflected waves are clearly visible.

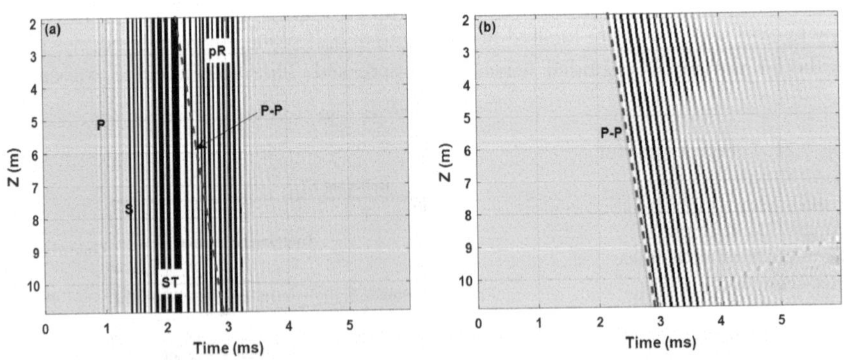

Fig. 7.22 Modeled COG (3 m offset) profile for a 10 kHz monopole source in a fast formation model (model in Fig. 7.20a). **a** Normalized full waveforms. **b** Normalized reflected waveforms

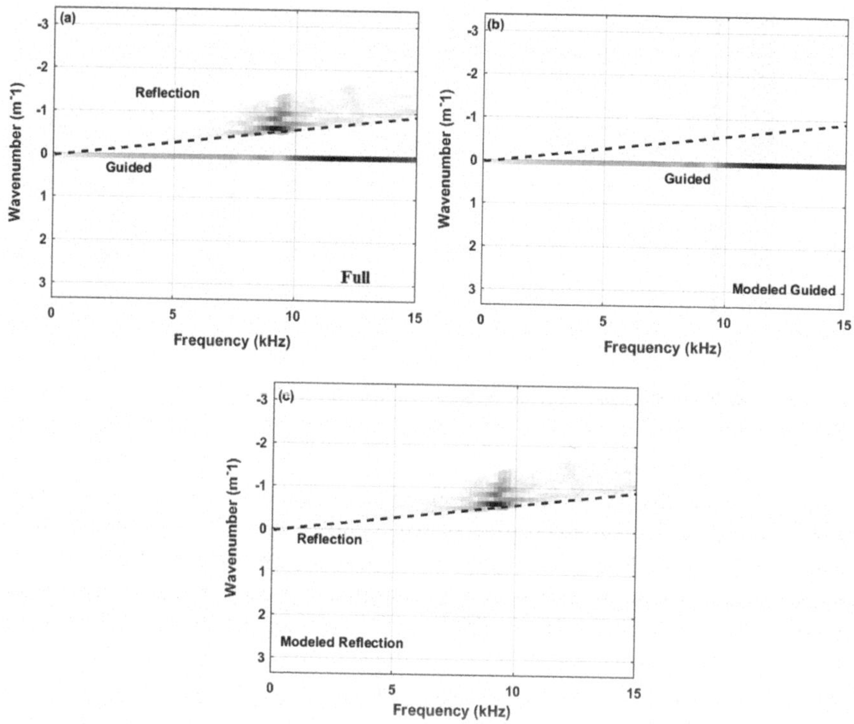

Fig. 7.23 Different waves in the f-k domain of the COG profile collected from the fast formation model shown in Fig. 7.20a. **a** Full waves. **b** Modeled guided waves. **c** Modeled reflected waves

Fig. 7.24 Reflected waves in the COG profile extracted by the f-k filter method in the fast formation model shown in Fig. 7.20a. Anomalies at both ends (z) are due to insufficient aperture

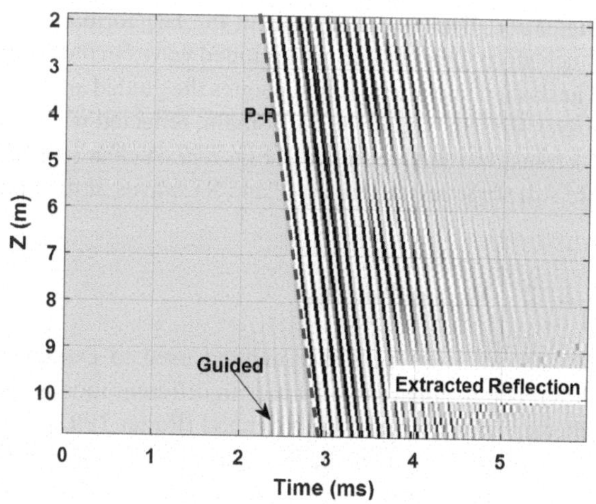

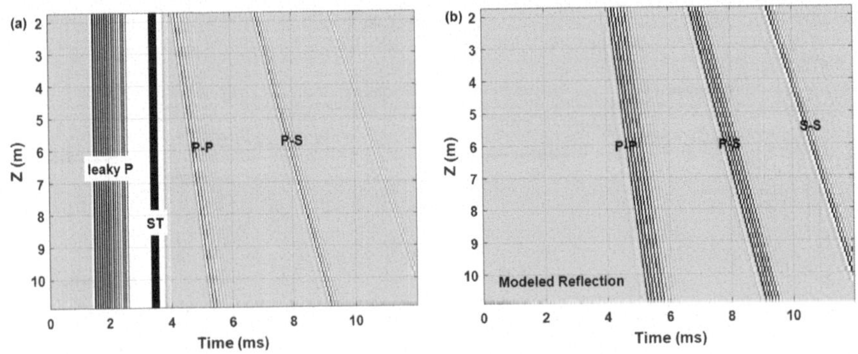

Fig. 7.25 Modeled COG profile in a slow formation model (model in Fig. 7.20b). **a** Full waveforms. **b** Reflected waves

Another example is the slow formation model (model in Fig. 7.20b). In this case, there are no pR waves and no overlaps between guided and reflected waves. Similar to Fig. 7.22, the modeled waves are collected as a 3 m offset COG profile and shown in Fig. 7.25. In the full wave COG profile, the leaky P and ST modes are straight lines without any moveout. The leaky P-waves are dispersive and have large amplitudes. The reflected waves have an oblique moveout, and there is no overlap between these waves and guided waves. From the f-k transform of the modeled guided waves, which is shown in Fig. 7.26b, there is the large magnitude of leaky P at approximately 10 kHz. The clear slopes of different reflected phases can be observed in Fig. 7.26c. Figure 7.26a shows the full waves in the f-k domain. A dashed line marks the upper limit of the apparent velocity of the reflected waves obtained from Fig. 7.26c. This is different from the fast formation case. The reflected waves are much more separated from the guided waves in the f-k domain as shown in Fig. 7.26a. The dashed line effectively separates the guided and reflected waves. After the guided waves are removed in the f-k domain, reflected waves can be obtained by the inverse f-k transform and are shown in Fig. 7.27. A clear reflection is obtained, although there are still some residual leaky P and ST waves at the bottom and top of the profile.

7.3.1.2 Median Filtering

Median filtering is another method used to extract reflected waves based on the difference in apparent velocities in different modes. Median filtering is a nonlinear filtering and data smoothing method (Reiter 1991).

Median filtering can be illustrated simply using a 1D signal, such as a one wave trace as a function of time. In each window, the amplitude in an array with a length of N is sorted in an increasing order. The output of the window is the median number of the ordered array. The same operation is repeated sample-by-sample over the entire waveform. For example, take an input array {3, 2, 6, 11, 4, 1, 5}, and apply a 3-point length window median filter. The first and last values are repeated to obtain enough

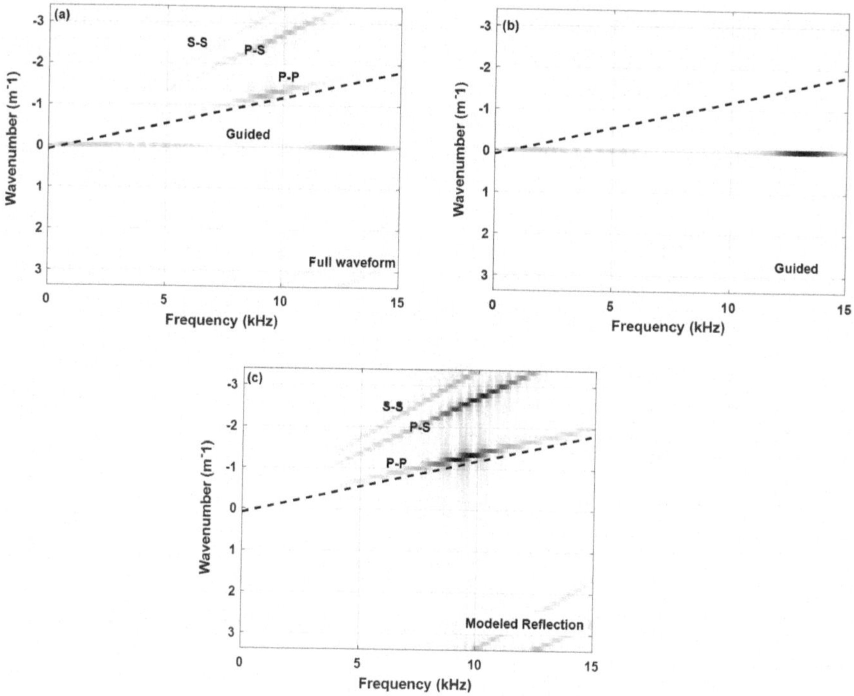

Fig. 7.26 Different waves in the f-k domain of the COG profile collected from the slow formation model (shown in Fig. 7.20b). **a** Full waves. **b** Modeled guided waves. **c** Modeled reflected waves

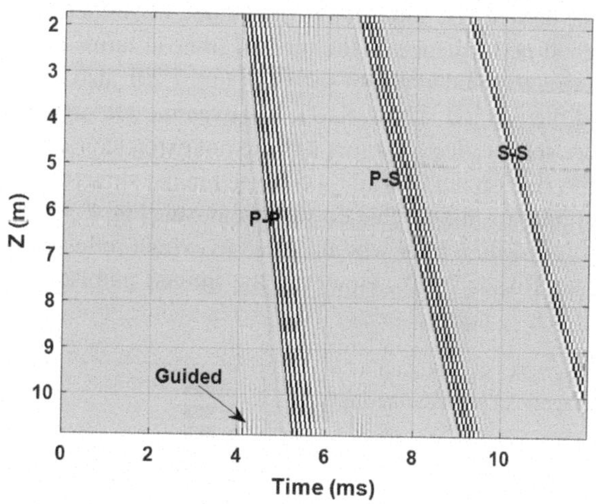

Fig. 7.27 Extracted reflected waves in the COG profile obtained by the f-k filtering method in the slow formation model (shown in Fig. 7.20b). The residual leaky P and ST waves at the bottom and top of the profile do not affect the reflection extraction

entries to fill the window. The first window consists of an array with 3 numbers {3, 3, 2} and corresponding reordered array is {2, 3, 3}. The output is 3. In the same way, the array in the second window is {3, 2, 6}, and the output of the filter is 3. By repeating the operation, the final output of the entire array is {3, 3, 6, 6, 4, 4, 5}.

For 2D data such as data in the time-space domain, more complex window patterns are applied. The simplest method is to apply the median filter in a specific direction. As shown in Fig. 7.28, the median filter can be applied in the waveforms at z_0, and the main characteristics of the waveforms will be kept along the time direction. This can be applied in both COG and CSG acoustic logging datasets (Li et al. 2002; Tang 2004). Here, the application of the median filter to COG and CSG data is shown.

The total waveform W_{tol} in the COG profile consists of guided (W_g) and reflected (W_r) waves. By applying the median filter to W_{tol}, we can extract the guided waves by $median\{W_{tol}\}$. The median filter is applied to the time direction. For each depth, the reflected waves can be extracted by the following equation,

$$W_r(i,t) = W_{tol}(i,t) - median\left\{W_{tol}^{j}(i,t),\ j = i-L, \ldots, i, i+L\right\},$$
$$i = 1, 2, \ldots, N_{dep}, \tag{7.9}$$

where L is the length of the filter window in time samples and N_{dep} is the number of depths.

To demonstrate the application of median filtering, we use synthetic waveforms shown in Fig. 7.22. Figure 7.29a shows the extracted reflected waves. The arrival time of the P-P, obtained from modeling the reflected waves, is overlain with a dashed red line. The first arrival of the extracted reflected waves perfectly matches the dashed line and there are no guided waves before the dashed line. This shows the good performance of the median filter in terms of extracting reflected waves. The extracted guided waves shown in Fig. 7.29b also demonstrate the good performance of the method. The P, S, ST, and even the pR Airy phases are clearly visible. There are some reflected waves that are not extracted (after 3 ms). The example of median filtering on the COG data collected in the slow formation model (Fig. 7.20b) shows similar results for guided wave removal (Fig. 7.30).

A median filter was also used to extract reflected waves in a CSG data set (Song and Toksöz 2010). However, the limited number of channels, usually 8 channels,

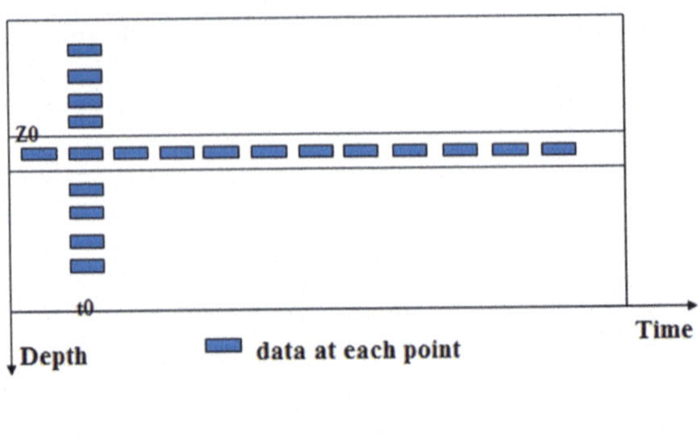

Fig. 7.28 Schematic diagram of median filtering

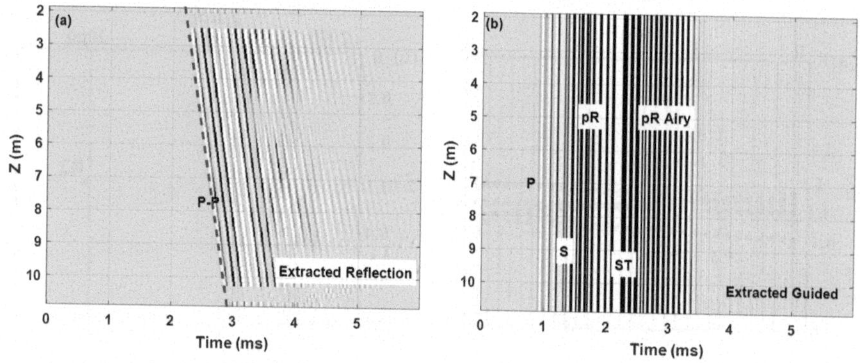

Fig. 7.29 Application of median filtering to reflected wave extraction from the synthetic COG data in a fast formation model (model in Fig. 7.20a). **a** Extracted reflected waves. **b** Extracted guided waves

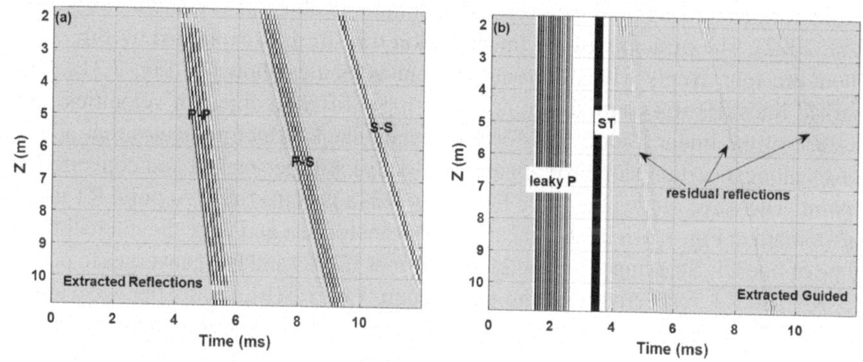

Fig. 7.30 Application of median filtering to reflected wave extraction from the synthetic data in a slow formation model (model in Fig. 7.20b). **a** Extracted reflected waves. **b** Extracted guided waves

limits the application. The reason is that median filtering is based on the differences in apparent velocities between guided and reflected waves. The zero moveout of the guided waves in the COG profile easily distinguishes them from the reflected waves. The difference in the apparent velocities between the guided and reflected waves in the CSG profile is not as large as that in the COG profile.

7.3.1.3 Linear Radon Transform

The linear Radon transform is another method of separating different modes by employing the difference in their apparent velocities. The linear Radon transform was initially proposed by the Austrian mathematician Radon in 1917. In this method, the time domain data collected in space (i.e., depth for COG and offset for CSG profile)

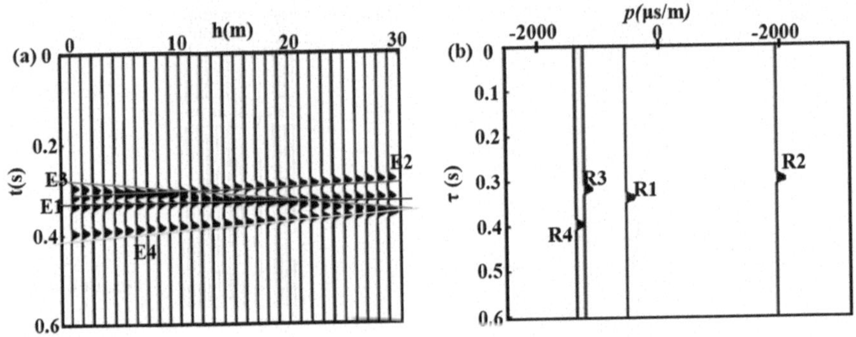

Fig. 7.31 The principle of the linear Radon transform. **a** Data set in the time-space domain including 4 different modes (marked with different lines). **b** Data in the Radon domain (Modified from Fig. 2 in Li and Yue 2017)

is transformed into the time-slowness (τ-p) domain (Kabir and Verschuur 1995; Trad et al. 2002). The principle of the linear Radon transform is explained by Fig. 7.31. There are four events with different apparent velocities shown in Fig. 7.31a. The marked lines (E1–E4) with different slopes have different apparent velocities. The essence of the linear Radon transform is a slant stack, which indicates that all the waves along a certain slope (or apparent velocity) will be stacked and converted to a point. The wave mode along line E1 in Fig. 7.31a is converted as a point R1 in the τ-p domain in Fig. 7.31b by the linear Radon transform. τ and p are the intercept and slope of line E1. Stacking other modes along lines E2, E3 and E4, converges to points R2, R3, and R4, respectively, in the τ-p domain (Fig. 7.31b). If only the wave along line E1 is desired, other modes can be easily removed in the τ-p domain, and then, the waveform of the mode along E1 can be obtained by the inverse Radon transform. The traditional linear Radon transform is a nonorthogonal operator. This results in equal energy in the forward and inverse transforms (Trad et al. 2003), resulting in a poor resolution. As shown in Fig. 7.32a, there is an edge effect for each point, and the anticipated focus points appear as "scissors". Thus, the inverse transform of R1 marked with a circle will have multiple modes rather than a single mode.

A high-resolution Radon transform method was proposed that combines maximum entropy and the Bayes principle (Sacchi and Ulrych 1995), where the data in the τ-p domain are collapsed to points. The method has been widely applied in VSP and surface seismic processing (Sacchi and Ulrych 1995; Sacchi et al. 1999), as well as in global seismology (Wang et al. 2006; de Hoop et al. 2009; Shang 2014). Recently, this method has been used to extract the reflected waves from the guided waves in BARI (Li and Yue 2015, 2017). Here, we briefly discuss the use of the high-resolution Radon transform for extracting reflected waves.

A set of traces $W(z, t)$ can be transformed into the Radon domain as $m(p, \tau)$:

$$m(p, \tau) = \int_{-\infty}^{+\infty} W(z, t = \tau + pz)dz, \tag{7.10}$$

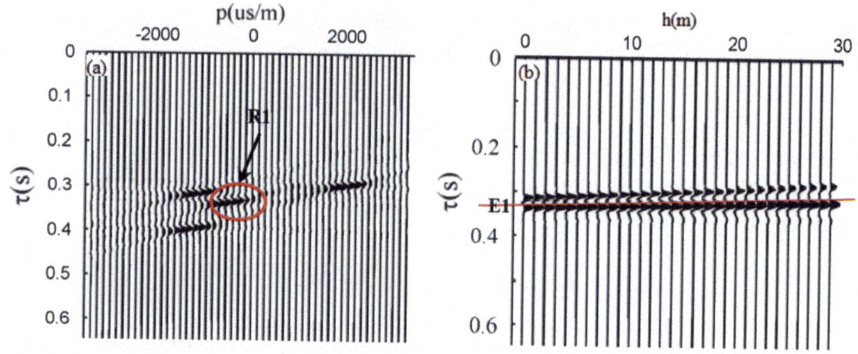

Fig. 7.32 Edge effect of the conventional linear Radon transform. **a** Data in the Radon domain for the data set shown in Fig. 7.31a using the conventional linear Radon transform. **b** Inversed waveform of R1 shown in **a** (Modified from Fig. 2 in Li and Yue 2017)

where z is the offset from the source to receiver.

If only the mode with apparent slowness p is chosen, the waveform in the time-space domain can be obtained by the inverse transform:

$$\hat{u}(z, t) = \int_{-\infty}^{+\infty} m(p, \tau) dp, \tag{7.11}$$

where $\tau = t - pz$.

The corresponding form in the frequency domain is as follows:

$$M(p, \omega) = \int_{-\infty}^{+\infty} W(z, \omega) \exp(i\omega pz) dz, \tag{7.12}$$

$$\widehat{U}(z, \omega) = \frac{|\omega|}{2\pi} \int_{-\infty}^{+\infty} M(p, \omega) \exp(-i\omega pz) dp, \tag{7.13}$$

where, $M(p, \omega)$, $\widehat{U}(z, \omega)$, and $W(z, \omega)$ are the frequency domain forms of $m(p, \tau)$, $\hat{u}(z, t)$, and $W(z, t)$, respectively. The matrix forms of Eqs. (7.12) and (7.13) are as follows:

$$\mathrm{m} = \mathrm{L^H W}, \tag{7.14}$$

$$\hat{u} = Lm, \tag{7.15}$$

Here $L = \exp(-i\omega p_j z_i)$ is the forward transform operator, and the subscripts i and j range from 1 to N and 1 to Lp, respectively. N is the number of receivers. Lp is the number of slowness values. $\mathbf{L^H}$ is the conjugate transpose matrix of $\mathbf{L}$. In data processing, it is common practice to eliminate part of the apparent slowness and offset, and then Eq. (7.15) cannot reconstruct the original signal. A generalized inverse is required to obtain the solution. Because N is less than L_p in array acoustic logging data, the forward transform can be changed to the following equation:

$$\mathrm{M} = \mathrm{L^H}\left(\mathrm{LL^H}\right)^{-1}\mathrm{W}. \tag{7.16}$$

To stabilize the calculation, a damping factor is usually added, and Eq. 7.16 is rewritten as follows:

$$\mathrm{M} = \mathrm{L^H}\left(\mathrm{LL^H} + \lambda\mathrm{I}\right)^{-1}\mathrm{W}. \tag{7.17}$$

Equations (7.15) and (7.17) form a forward and inverse transform pair. λ is the damping factor, which is usually set to 0.01 times the leading diagonal value of LL^H. This is the dampened least square Radon transform (LSRT).

Because the damping factor is usually set as a constant, the images in the Radon domain are usually contaminated with a trailer or "scissors". To overcome these problems, the high-resolution Radon transform (HRRT) was developed (Sacchi and Ulrych 1995). One of the common implementations employs Bayes principle combined with a priori solutions in the iterations and the Cauchy distribution to obtain a weighted matrix, making the following objective function reach its minimum:

$$J_c = \sum_k \ln\left[1 + \frac{|v_k|^2}{\sigma_c^2}\right] + (LV - W)C_n^{-1}(LV - W), \tag{7.18}$$

where V is the result of the HRRT and v_k is the kth element. σ_c is a parameter of the Cauchy distribution. C_n is the covariance matrix of noise. W is the data in the frequency domain. The minimization of J_c leads to the following solution:

$$\mathrm{V} = Q_c^{-1}L^H\left(C_n + LQ_c^{-1}L^H\right)^{-1}W, \tag{7.19}$$

where Q_c is the weight matrix, which is a diagonal matrix and the diagonal element Q_{cii} in the matrix is as follows:

$$Q_{cii} = \frac{1}{\sigma_c^2\left[1 + \frac{|v_i|^2}{\sigma_c^2}\right]}. \tag{7.20}$$

If the initial solution is obtained by the LSRT, the improved results can be obtained by 3–5 iterations of Eq. 7.19. This improved method is called the HRRT. Using the HRRT, the data in Fig. 7.31a are converted to the τ-p domain as shown in Fig. 7.33a, where the "scissors" of the data, shown in Fig. 7.32a, are highly suppressed. Using the inverse transform, the wave mode E1 (Fig. 7.33b) can be obtained without interference from other modes or noise.

For reflected wave extraction, we use the COG data shown in Fig. 7.22a. The full and modeled reflected waves in the τ-p domain are shown in Fig. 7.34a and b,

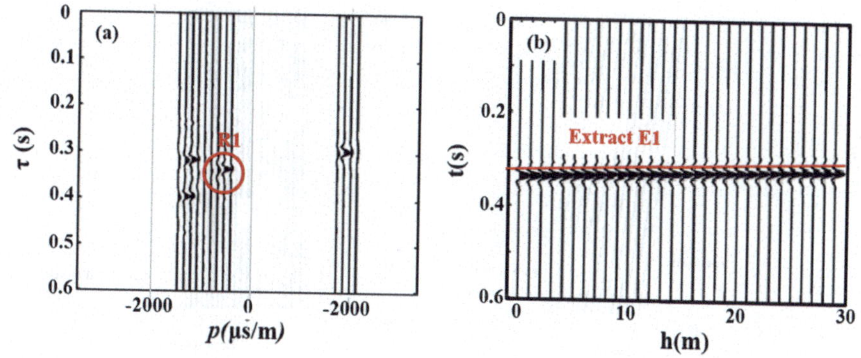

Fig. 7.33 The performance of the forward and inverse HRRT. **a** Forward HRRT for the data in Fig. 7.31a. **b** Extracted mode using the inversed HRRT

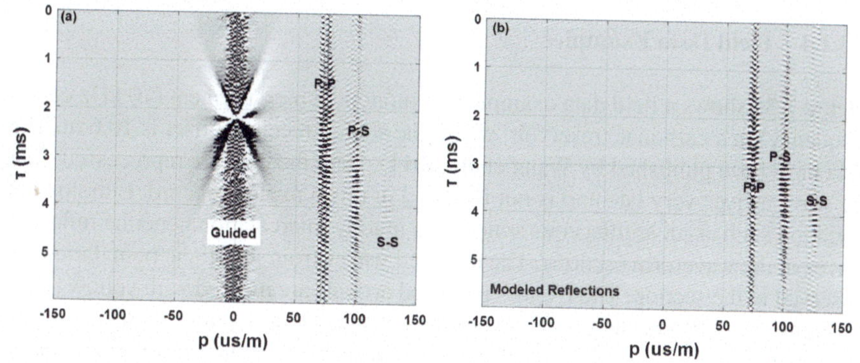

Fig. 7.34 The Radon transform of the model data shown in Fig. 7.22a. **a** Full wavefield. **b** Reflected waves

respectively. The guided waves in the COG profile, which are marked in Fig. 7.34a, have zero apparent velocity and are in the region around $p = 0$. Most of the apparent velocities of the modes are zero in the rectangle, except approximately $\tau = 2$ ms, where the ST and pR modes are located. The reflected waves in the τ-p domain are clearly visible and separated from the guided waves. This makes separation of the reflected waves from the strong guided waves possible. After removing the guided waves from the τ-p domain, the remaining information can be inverted into the time-space domain as reflected waves. Figure 7.35a shows the extracted reflected waves and they are similar to the modeled reflected waves shown in Fig. 7.22b. The arrival times of the modeled P-P reflected wave (red dashed line) exactly match the first arrival of the extracted reflected waves. Figure 7.35b shows the extracted guided waves obtained using the HRRT. The main guided modes are clear and match those shown in Fig. 7.22a.

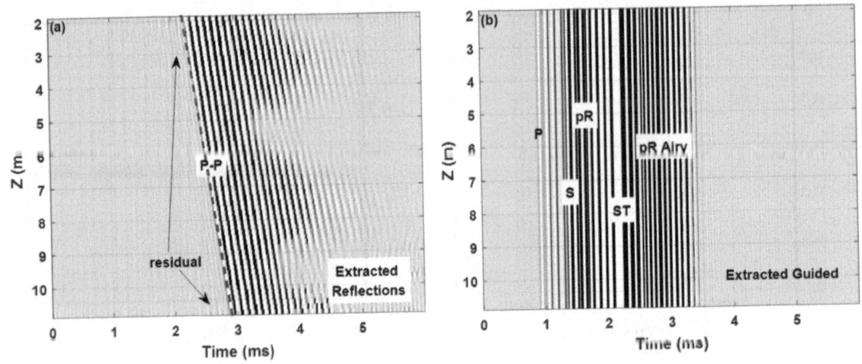

Fig. 7.35 Results of applying the HRRT to the COG data shown in Fig. 7.22a. **a** Extracted reflected waves. **b** Extracted guided waves

7.3.1.4 Field Data Examples

Figure 7.36 shows a field data example of a monopole measurement (10 kHz center frequency) in a carbonate reservoir, where the source-receiver offset is 10.6 m. The data have been published by Wang et al. (2011). Here the data are reprocessed. The ST wave arrives very late and is not included in this 6 ms long record. Lithological variation such as an argillaceous zone and a muddy intercalation generate reflected waves in the waveform sections. The reflected waves from lithologic boundaries are observed in the section. The P and S refracted arrivals are marked with yellow lines.

Figure 7.37 shows the array waveforms collected at a depth of 6718 m and its velocity time semblance plot. The spacing between receivers is 0.15 m. The P and S arrivals can be easily picked from the array waveforms in Fig. 7.37a. The reflected waves, between the P and S arrivals, are marked by a dashed rectangle. The velocities of the P and S waves obtained from the velocity-time semblance analysis, shown in Fig. 7.37b, are 6000 and 3200 m/s, respectively. The apparent velocity of the reflected waves, arriving after the P- and S-waves, match those of the P and S waves. These are reflections from horizontal bed boundaries and not from lateral interfaces.

Using a linear Radon transform, the data in Fig. 7.36 were converted into the τ-p domain. The upgoing and downgoing reflections were identified by their apparent slownesses. Using the HRRT, the guided waves were removed. The result is shown in Fig. 7.38. In the figure, the arrival times of the P- and S-waves obtained from Fig. 7.36 are marked on the waveforms. The arrival times of the P- and S-waves match well with the filtered waveforms. A comparison of the raw data and one with reflections removed is shown in Fig. 7.39. The solid and dashed lines are the raw data and the same trace with the reflected waves removed, respectively. After the guided waves are removed, the reflected waves are obtained and are shown in Fig. 7.40. In the left panel (a), the extracted reflected waves, including the upgoing and downgoing waves, are shown. These can be further separated into the upgoing (middle panel) and

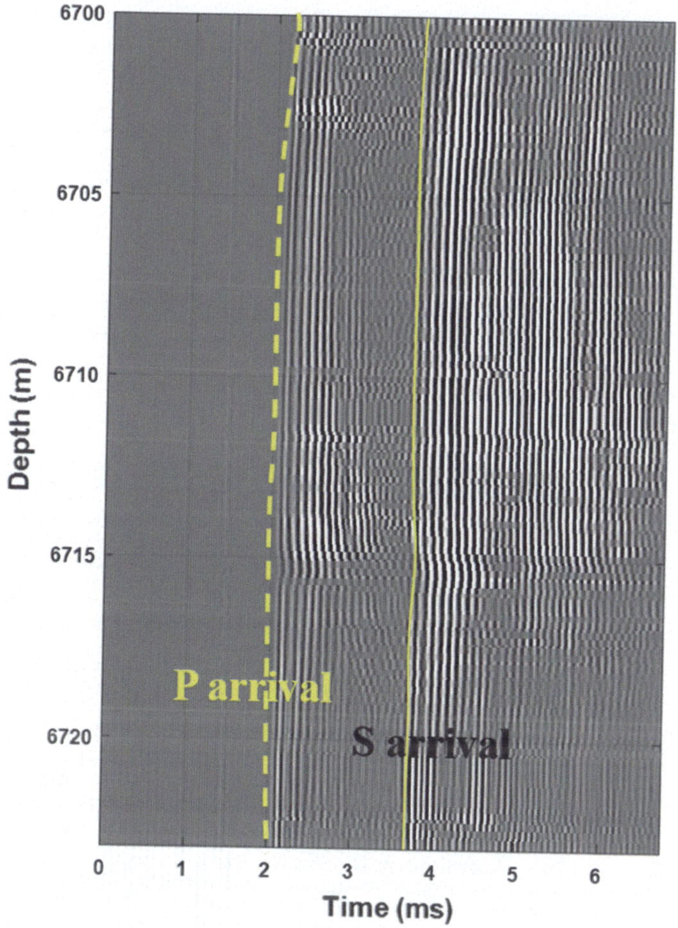

Fig. 7.36 Field data (COG profile) of a monopole measurement in a carbonate reservoir (Wang et al. 2011). The source-receiver offset is 10.6 m

downgoing (right panel) waves. A bed boundary (interface) at the depth of 6715 m appears to be a major reflector.

Another field example is a dipole log (Li and Yue 2017), which is shown in Fig. 7.41. The figure shows the COG profile of inline dipole data in a carbonate reservoir. The source-receiver offset is 3.048 m. The Chirp source frequency sweeps from 300 to 8 kHz (Pistre et al. 2005). Although the flexural waves are strong, one can still see some upgoing and downgoing reflected waves. The data are converted to the τ-p domain by a linear Radon transform, as shown in Fig. 7.42. The guided waves are marked with a rectangle having a slowness of approximately zero. The upgoing and downgoing reflected waves can be identified by the signs (+ and −) of their apparent slowness. The largest reflections are located in the lower right and upper left regions outside the rectangle. Using HRRT, the guided waves are suppressed

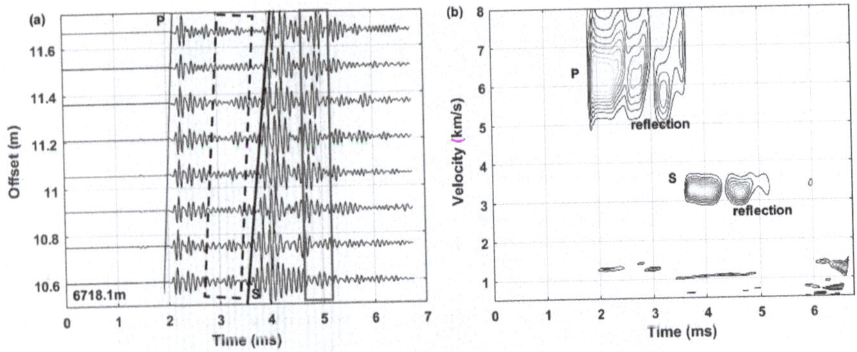

Fig. 7.37 **a** Field data monopole array waveforms at depth of 6718.1 m. **b** Velocity time semblance plot of the waveforms in **a**

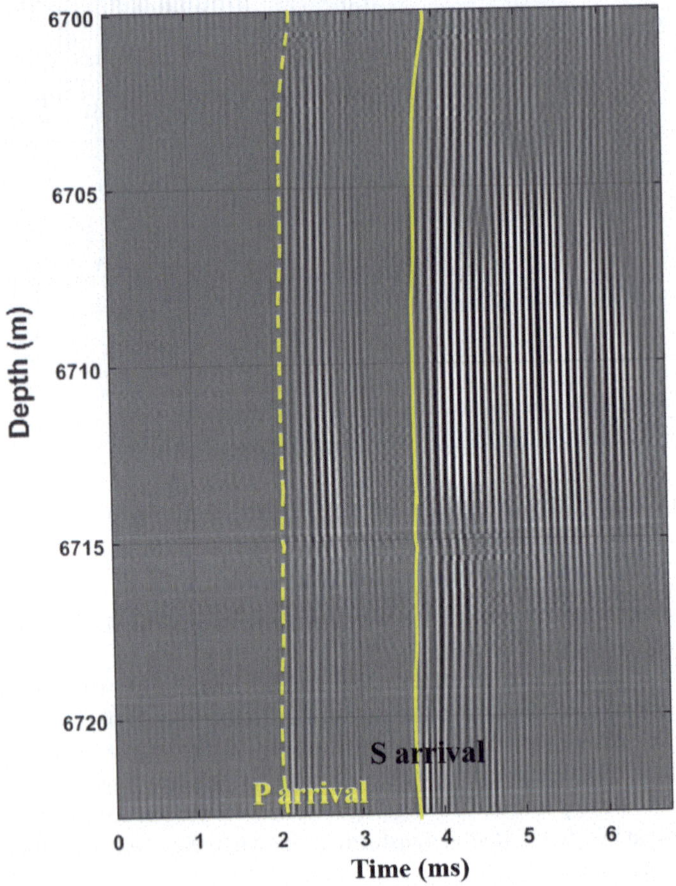

Fig. 7.38 Data, shown in Fig. 7.36, after the reflections were removed using the HRRT

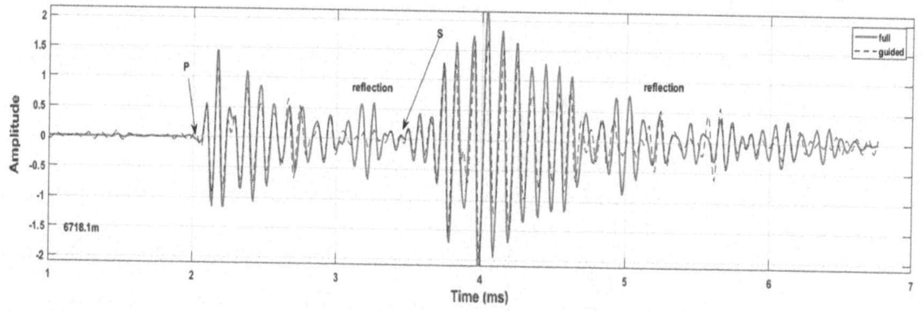

Fig. 7.39 Comparisons between the field data and extracted guided waves by HRRT. The waveform is at a depth of 6718.1 m. The tool offset is 10.6 m. The solid and dashed lines are the raw data and extracted guided waves, respectively. Large amplitude arrival after the S are pR waves

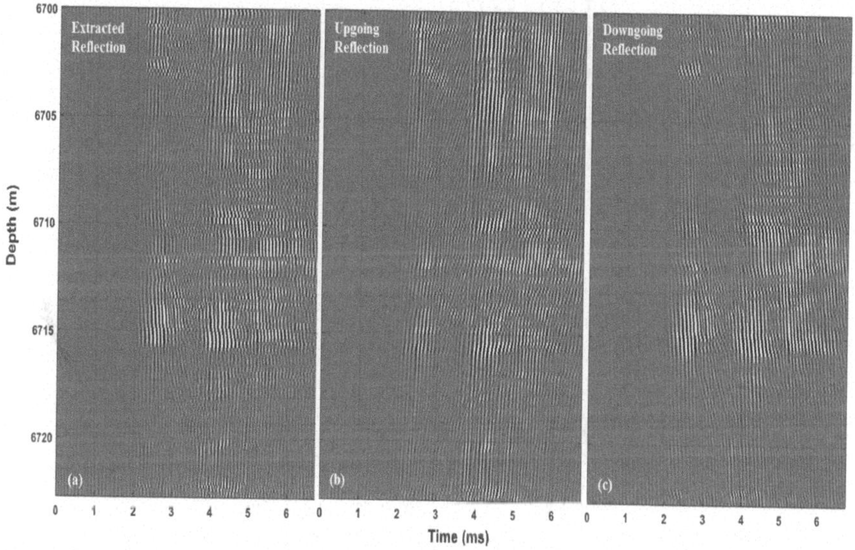

Fig. 7.40 Reflected waves extracted from Fig. 7.36 using the HRRT. **a** Total reflected waves. **b** Upgoing reflected waves. **c** Downgoing reflected waves

and then the upgoing and downgoing reflected waves are extracted. The results are shown in Fig. 7.43.

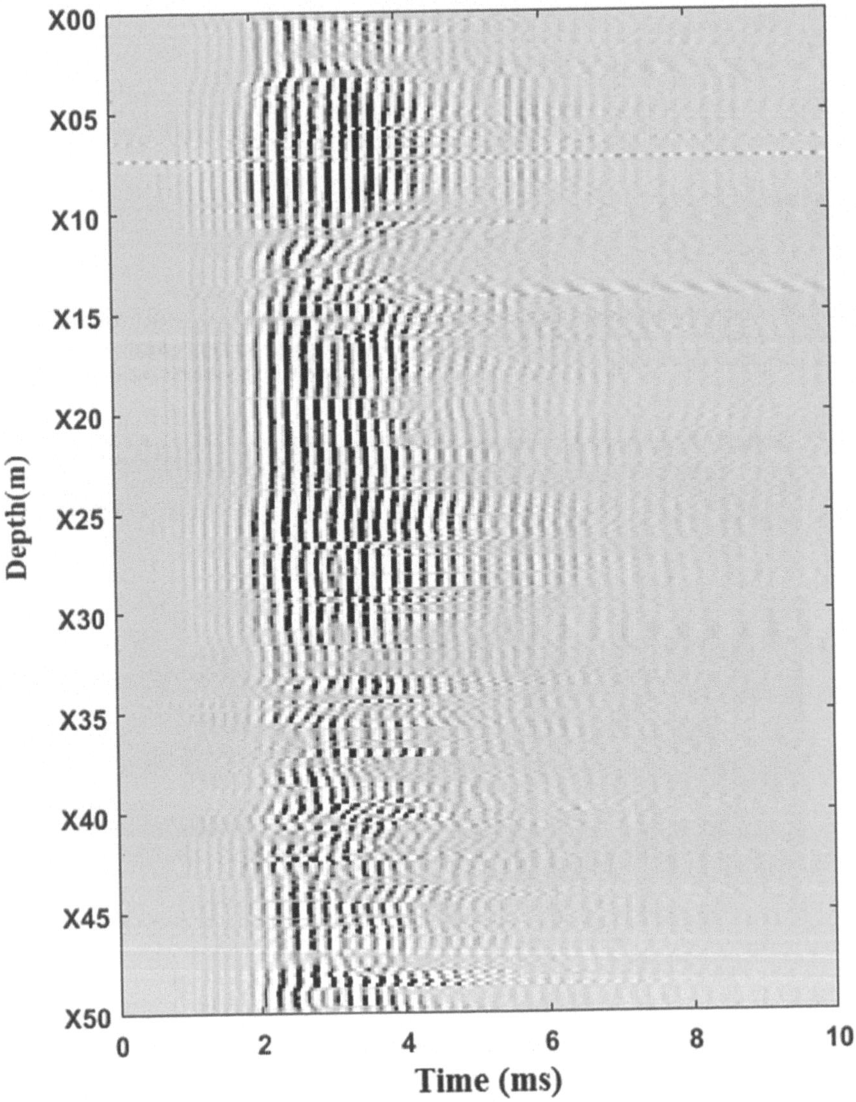

Fig. 7.41 The COG profile (3.048 m offset) of inline dipole acoustic logging data in a carbonate reservoir

7.3.2 *Migration Methods for Borehole Peripheral Imaging*

7.3.2.1 General Overview

Seismic migration has been very important for subsurface imaging in exploration seismology. Migration can also be used to map reflectors around a borehole. The

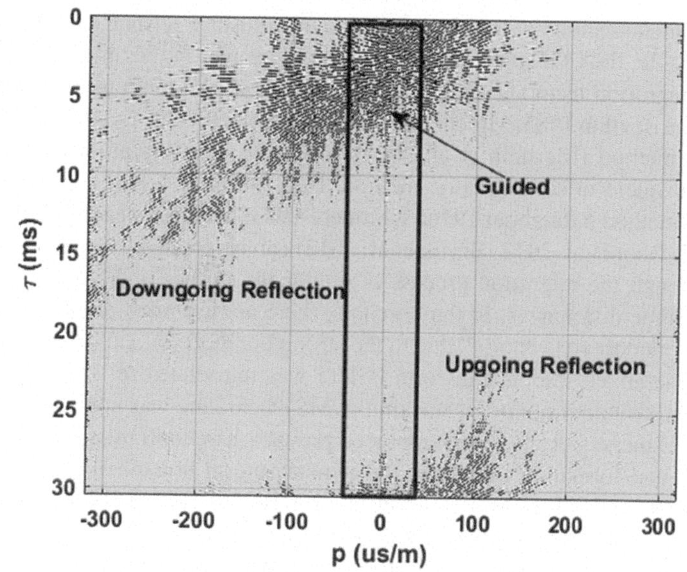

Fig. 7.42 The HRRT of the COG profile in Fig. 7.41. The guided and reflected waves are marked

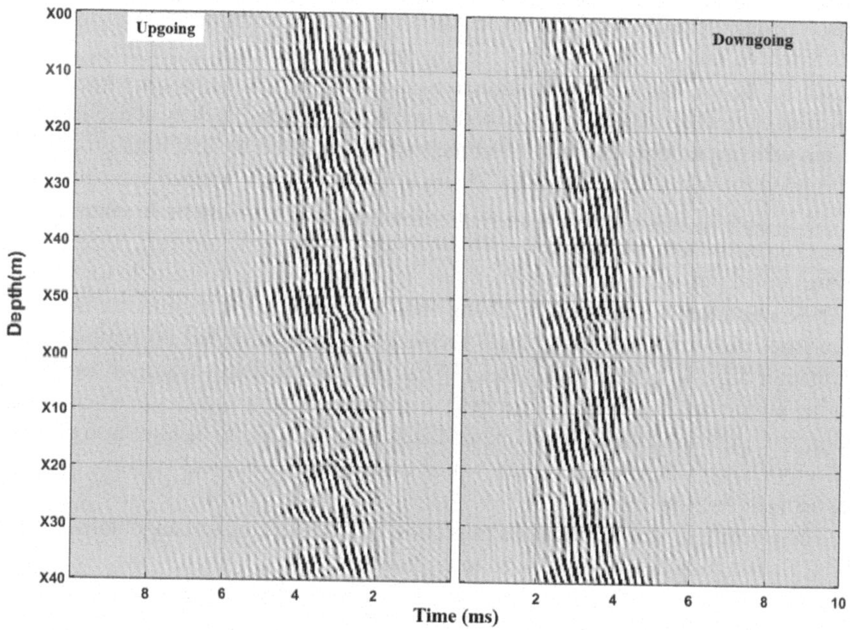

Fig. 7.43 Extracted upgoing and downgoing reflected waves from the dipole data in Fig. 7.41. The figure shows that there are many reflections

migration methods used in BARI are adopted from surface seismic data. Similar to a seismic profile, the COG profile includes reflected waves in the time-space domain.

Most migration methods, including the generalized Radon transform migration method (see Beylkin 1985; Miller et al. 1987, Bleistein et al. 2001), Kirchhoff integral migration method (Bleistein et al. 2001), improved F-K migration method (Stolt 1978), equivalent offset migration method (Bancroft et al. 1998) and reverse time migration method (Claerbout 1976; Whitmore 1983; McMechan 1983; Baysal et al. 1983; Op't Root et al. 2012; Brytik et al. 2012) can be applied to image with BARI data. Although the migration process is similar for surface and borehole seismic, there are some differences. In the borehole, there are a limited number of sensors (channels), and there is "noise" due to the strong borehole guided waves.

The generalized Radon transform (GRT) was introduced to BARI by Hornby (1989). The common midpoint stacking (CMS) technique was applied to suppress the noise. Esmersoy et al. (1998) employed prestack Kirchhoff migration to map the well track and formation interfaces for geo-steering in horizontal drilling. Coates et al. (2000), Li et al. (2002), Tang et al. (2007), and Tang and Patterson (2009) obtained borehole images utilizing Kirchhoff migration.

Chabot et al. (2001) used a waveform processing procedure, similar to that of surface seismic, to image using borehole acoustic logging data. Their processing formed common scatter point gathers based on the dipole S-wave image logging data for equivalent offset migration and then stacked the calibrated gathers to obtain the borehole images. This method was further improved by Chabot et al. (2002), including the use of trace equalization to equalize the amplitude for each trace, optimization of equivalent parameters to better accommodate the borehole and refinement of the velocity model. The improved method was applied to field data processing with good results in inclined wells. Zhang et al. (2009) also applied the equivalent offset migration method to image an interface outside of a borehole using BARI data. The equivalent offset migration method has a higher stacking fold compared to conventional poststack migration methods, thus making it more suitable for BARI.

F-K migration was adopted by Zheng and Tang (2005) to accommodate the characteristics of borehole acoustic data. Prestack F-K migration has no limitation for formation dips. In addition, this method can avoid wavefield separation before or during migration. The method separates the upgoing and downgoing wave fields after migration. A disadvantage of this method is that it is not able to handle strong lateral velocity changes (surface seismic), which correspond to vertical velocity changes for acoustic logging.

The velocity model is critical for migration. In BARI, velocities are obtained from the P, S, and guided waves and used in migration.

7.3.2.2 Reverse Time Migration for Borehole Acoustic Reflection Imaging

If the borehole effect is ignored, acoustic reflection logging can directly employ seismic exploration migration methods. Wave-equation-based migration methods,

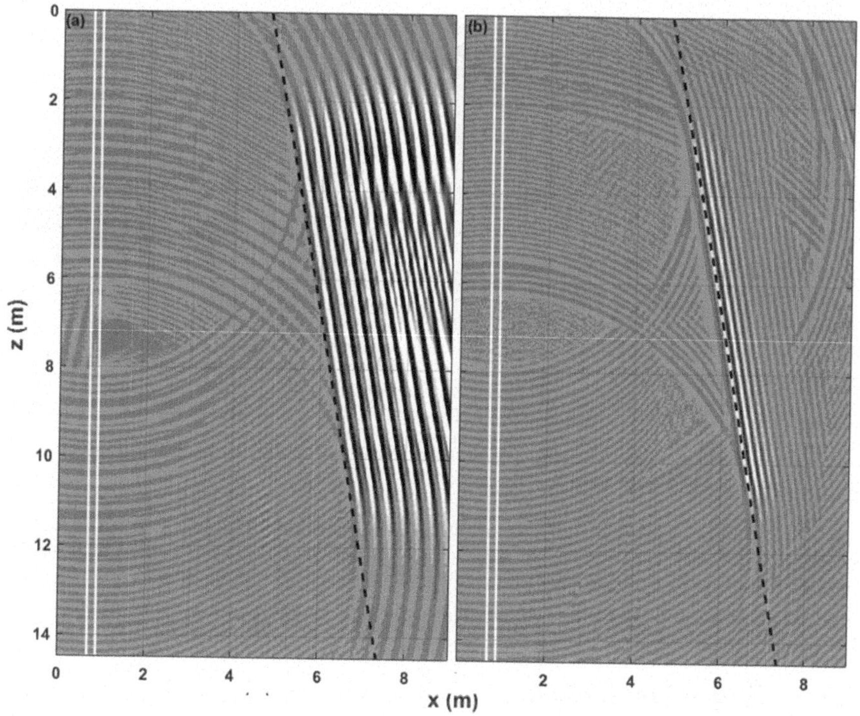

Fig. 7.46 Imaging results obtained using the RTM method. **a** Fast formation model in Fig. 7.20a; **b** Slow formation model in Fig. 7.20b. The interface between formations in the two models are marked with dashed lines. Borehole walls are marked with white lines

Data were acquired in a fast formation, Vp = 5790 m/s and Vs = 3048 m/s, by a monopole tool. The tool has an array of 13 receiver stations. At each station, there are 8 azimuthal receivers with 45° intervals. The minimum source-receiver offset is 3.35 m (11 ft), and the receiver interval is 0.15 m. For data acquisition, the tool was moved at steps of 0.15 m along the horizontal borehole from 0 to 200 m. Figure 7.47a shows a COG profile (3.35 m offset) of the full waveform recordings along the borehole. The reflections are not easily identifiable. Figure 7.47b shows the same section after the borehole guided waves are removed with a median filter. Both "upgoing" and "downgoing" reflections can be observed in Fig. 7.47b. These reflections were separated into updip and downdip events for final processing. Figure 7.47c shows the fractures obtained from imaging.

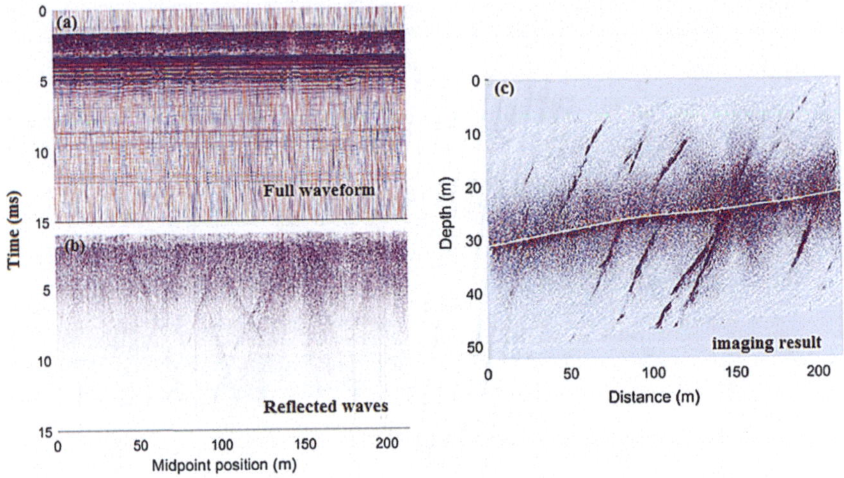

Fig. 7.47 Example of borehole reflection imaging. **a** Full waveform recording in the COG (3.35 m offset) profile. **b** Reflected waves after the borehole guided waves were removed using a median filter. **c** Fracture imaging result. The line in the middle of the image is the well trajectory (Modified from Figs. 3 and 5 in Hirabayashi et al. 2016)

7.3.3 Dipole Log Imaging

In dipole measurements, the directional source generates directional S-waves. The azimuth of the reflector can be determined from the amplitudes of the reflected S-waves in different directions (Tang 2004). Figure 7.48 illustrates an example of dipole S-wave imaging in a horizontal well (Fig. 1 in Tang 2017). A dipole source in the borehole generates SH and SV waves. In Fig. 7.48, the S-waves are shown in the vertical plane for SH and horizontal plane for SV. Figure 7.49 shows the imaging results in a deviated well in a shale gas reservoir (modified from Fig. 3 in Tang et al. [2017]). The shale is overlain by caprock. Figure 7.49a shows the caprock imaged by SH waves. The SV waves are used for fracture imaging (Fig. 7.49b). Two fractures, marked with red arrows, intersect the borehole.

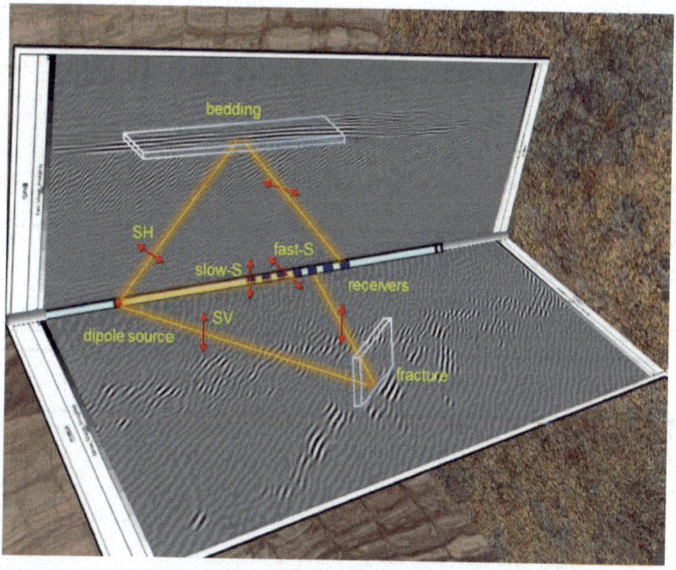

Fig. 7.48 Schematic diagram of S-wave imaging using a dipole acoustic tool in a horizontal well. The SH and SV are two S-waves in the formation. These two S-waves can be used to image the formation bedding and fractures (Fig. 1 in Tang [2017])

7.4 Summary

In this chapter, methods for imaging around a borehole are described. Data processing techniques for BARI were covered. Imaging of reflectors in 3D space using monopole, dipole, and cross-dipole measurements were presented. The important developments resulting from this chapter include obtaining improved images of reflectors, determining reflector azimuths using a monopole tool with receivers placed at multiple azimuths around the tool, dipole SV and SH measurements, cross-dipole logging tools, and bed boundary and fracture mapping.

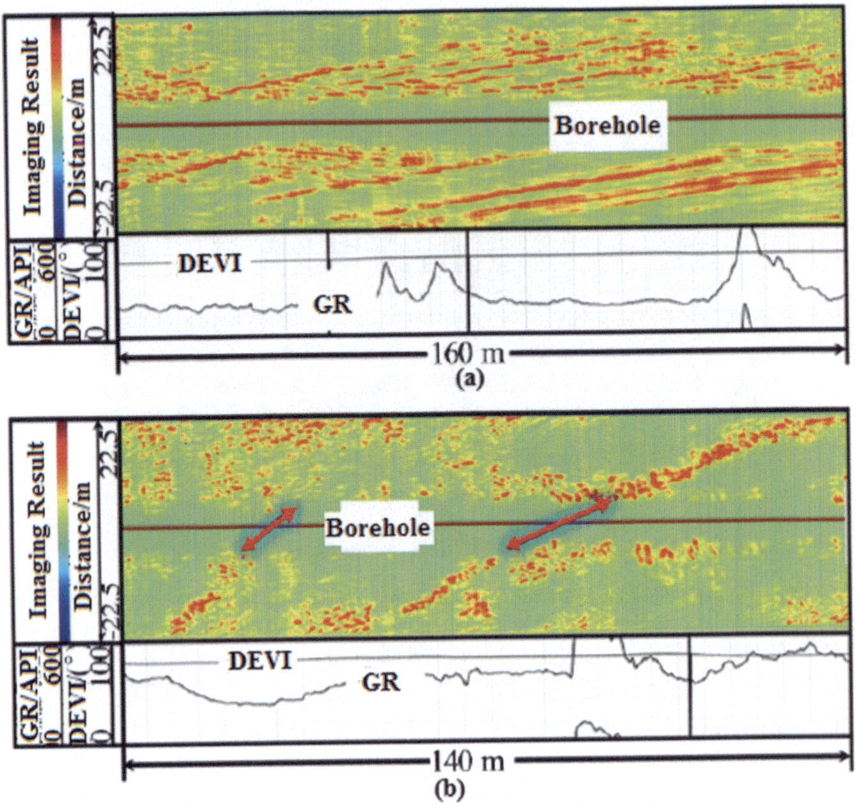

Fig. 7.49 Dipole S-wave imaging results in a deviated well in a shale gas reservoir. **a** Formation bedding imaged from SH waves. **b** Two fractures imaged by SV waves Figure is modified from Fig. 3 in Tang et al. (2017)

Appendix A
Elements of the Matrices for Openhole Logging and for Logging-While-Drilling

A.1 Openhole Logging

Assuming the radius of the borehole is a. Elements in $[\mathrm{m_{ij}}]_{4\times4}$ in Eq. 2.15 for the openhole case are,

$m_{11}=-\left[\frac{n}{a}I_n(fa)+fI_{n+1}(fa)\right]$,	$m_{12}=\frac{n}{a}K_n(pa)-pK_{n+1}(pa)$,
$m_{13}=\frac{n}{a}K_n(sa)$,	$m_{14}=\mathrm{i}k_z\left[\frac{n}{a}K_n(sa)-sK_{n+1}(sa)\right]$,
$m_{21}=\rho_f\omega^2I_n(fa)$,	$m_{22}=\mu\left\{\left[\frac{2n(n-1)}{a^2}+2k_z^2-k_s^2\right]K_n(pa)+\frac{2p}{a}K_{n+1}(pa)\right\}$,
$m_{23}=2\mu\frac{n}{a}\left[\frac{n-1}{a}K_n(sa)-sK_{n+1}(sa)\right]$,	$m_{24}=2\mathrm{i}k_z\mu\left\{\left[\frac{n(n-1)}{a^2}+s^2\right]K_n(sa)+\frac{s}{a}K_{n+1}(sa)\right\}$,
$m_{32}=2\mu\frac{n}{a}\left[-\frac{n-1}{a}K_n(pa)+pK_{n+1}(pa)\right]$,	$m_{33}=\mu\left\{-\left[\frac{2n(n-1)}{a^2}+s^2\right]K_n(sa)-\frac{2s}{a}K_{n+1}(sa)\right\}$,
$m_{34}=2\mathrm{i}k_z\mu\frac{n}{a}\left[\frac{1-n}{a}K_n(sa)+sK_{n+1}(sa)\right]$,	$m_{42}=2\mathrm{i}k_z\mu\left[\frac{n}{a}K_n(pa)-pK_{n+1}(pa)\right]$,
$m_{43}=\mathrm{i}k_z\mu\frac{n}{a}K_n(sa)$,	$m_{44}=\mu\left(k_s^2-2k_z^2\right)\left[\frac{n}{a}K_n(sa)-sK_{n+1}(sa)\right]$.

Elements not listed are zero.

H. Wang et al., *Borehole Acoustic Logging—Theory and Methods*, Petroleum Engineering, https://doi.org/10.1007/978-3-030-51423-5

Nomenclature

a	Borehole radius
n	The order of the multipole source, 0 for monopole, 1 for dipole, and so on
I, K	Bessel functions of the first kind and second kind, respectively
k_z	Axial wavenumber
k_f	Wavenumber of the fluid, $k_f = \omega/V_f$
k_p	Wavenumber of the P-wave in the collar, $k_p = \omega/V_p$
k_s	Wavenumber of the S-wave in the collar, $k_s = \omega/V_s$
f	Radial wavenumber in the fluid, $f = \sqrt{k_z^2 - k_f^2}$
p	Radial wavenumber of the P-wave in the collar, $p = \sqrt{k_z^2 - k_p^2}$
s	Radial wavenumber of the S-wave in the collar, $s = \sqrt{k_z^2 - k_s^2}$
ρ_f	Fluid density
ω	Angular frequency
μ	Shear modulus of the formation

A.2 Acoustic Logging-While-Drilling

The elements of $[m_{ij}]_{12\times 12}$ in Eq. 5.2 are listed as follows, where a and b are inner and outer radii of the collar, and c is the borehole radius. Coefficient n designates different source configurations, where n = 0, 1, 2, 3 are monopole, dipole, quadrupole, and hexapole, respectively. Definitions of other terms follow the equations.

$m_{11} = -\left[\frac{n}{c} I_n(fc) + f I_{n+1}(fc)\right]$,	$m_{12} = \frac{n}{c} I_n(pc) + p I_{n+1}(pc)$,
$m_{13} = \frac{n}{c} K_n(pc) - p K_{n+1}(pc)$,	$m_{14} = \frac{n}{c} I_n(sc)$,
$m_{15} = \frac{n}{c} K_n(sc)$,	$m_{16} = \mathrm{i}k_z\left[\frac{n}{c} I_n(sc) + s I_{n+1}(sc)\right]$,
$m_{17} = \mathrm{i}k_z\left[\frac{n}{c} K_n(sc) - s K_{n+1}(sc)\right]$,	$m_{21} = \rho_f \omega^2 I_n(fc)$,
$m_{22} = \mu\left\{\left[\frac{2n(n-1)}{c^2} + 2k_z^2 - k_s^2\right] I_n(pc) - \frac{2p}{c} I_{n+1}(pc)\right\}$,	$m_{23} = \mu\left\{\left[\frac{2n(n-1)}{c^2} + 2k_z^2 - k_s^2\right] K_n(pc) + \frac{2p}{c} K_{n+1}(pc)\right\}$,
$m_{24} = 2\mu \frac{n}{c}\left[\frac{n-1}{c} I_n(sc) + s I_{n+1}(sc)\right]$,	$m_{25} = 2\mu \frac{n}{c}\left[\frac{n-1}{c} K_n(sc) - s I_{n+1}(sc)\right]$,
$m_{26} = 2\mathrm{i}k_z\mu\left\{\left[\frac{n(n-1)}{c^2} + s^2\right] I_n(sc) - \frac{s}{c} I_{n+1}(sc)\right\}$,	$m_{27} = 2\mathrm{i}k_z\mu\left\{\left[\frac{n(n-1)}{c^2} + s^2\right] K_n(sc) + \frac{s}{c} K_{n+1}(sc)\right\}$,
$m_{32} = 2\mu \frac{n}{c}\left[\frac{1-n}{c} I_n(pc) - p I_{n+1}(pc)\right]$,	$m_{33} = 2\mu \frac{n}{c}\left[\frac{1-n}{c} K_n(pc) + p K_{n+1}(pc)\right]$,
$m_{34} = \mu\left\{-\left[\frac{2n(n-1)}{c^2} + s^2\right] I_n(sc) + \frac{2s}{a} I_{n+1}(sc)\right\}$,	$m_{35} = \mu\left\{-\left[\frac{2n(n-1)}{c^2} + s^2\right] K_n(sc) - \frac{2s}{a} K_{n+1}(sc)\right\}$,
$m_{36} = 2\mathrm{i}k_z\mu \frac{n}{c}\left[\frac{1-n}{c} I_n(sc) - s I_{n+1}(sc)\right]$,	$m_{37} = 2\mathrm{i}k_z\mu \frac{n}{c}\left[\frac{1-n}{c} K_n(sc) + s K_{n+1}(sc)\right]$,
$m_{42} = 2\mathrm{i}k_z\mu\left[\frac{n}{c} I_n(pc) + p I_{n+1}(pc)\right]$,	$m_{43} = 2\mathrm{i}k_z\mu\left[\frac{n}{c} K_n(pc) - p K_{n+1}(pc)\right]$,
$m_{44} = \mathrm{i}k_z\mu \frac{n}{c} I_n(sc)$,	$m_{45} = \mathrm{i}k_z\mu \frac{n}{c} K_n(sc)$,
$m_{46} = \mu\left(k_s^2 - 2k_z^2\right)\left[\frac{n}{c} I_n(sc) + s I_{n+1}(sc)\right]$,	$m_{47} = \mu\left(k_s^2 - 2k_z^2\right)\left[\frac{n}{c} K_n(sc) - s K_{n+1}(sc)\right]$,
$m_{52} = \frac{n}{b} I_n(pb) + p I_{n+1}(pb)$,	$m_{53} = \frac{n}{b} K_n(pb) - p K_{n+1}(pb)$,
$m_{54} = \frac{n}{b} I_n(sb)$,	$m_{55} = \frac{n}{b} K_n(sb)$,
$m_{56} = \mathrm{i}k_z\left[\frac{n}{b} I_n(sb) + s I_{n+1}(sb)\right]$,	$m_{57} = \mathrm{i}k_z\left[\frac{n}{b} K_n(sb) - s K_{n+1}(sb)\right]$,
$m_{58} = -\left[\frac{n}{b} I_n(fb) + f I_{n+1}(fb)\right]$,	$m_{59} = -\left[\frac{n}{b} K_n(fb) - f K_{n+1}(fb)\right]$,
$m_{62} = \mu\left\{\left[\frac{2n(n-1)}{b^2} + 2k_z^2 - k_s^2\right] I_n(pb) - \frac{2p}{b} I_{n+1}(pb)\right\}$,	$m_{63} = \mu\left\{\left[\frac{2n(n-1)}{b^2} + 2k_z^2 - k_s^2\right] K_n(pb) + \frac{2p}{b} K_{n+1}(pb)\right\}$,
$m_{64} = 2\mu \frac{n}{b}\left[\frac{n-1}{b} I_n(sb) + s I_{n+1}(sb)\right]$,	$m_{65} = 2\mu \frac{n}{b}\left[\frac{n-1}{b} K_n(sc) - s K_{n+1}(sb)\right]$,

(continued)

(continued)

$m_{66} = 2\mathrm{i}k_z\mu\left\{\left[\frac{n(n-1)}{b^2} + s^2\right]I_n(sb) - \frac{s}{b}I_{n+1}(sb)\right\}$,	$m_{67} = 2\mathrm{i}k_z\mu\left\{\left[\frac{n(n-1)}{b^2} + s^2\right]K_n(sb) + \frac{s}{b}K_{n+1}(sb)\right\}$,
$m_{68} = \rho_f\omega^2 I_n(fb)$,	$m_{69} = \rho_f\omega^2 K_n(fb)$,
$m_{72} = 2\mu\frac{n}{b}\left[\frac{1-n}{b}I_n(pb) - pI_{n+1}(pb)\right]$,	$m_{73} = 2\mu\frac{n}{b}\left[\frac{1-n}{b}K_n(pb) + pK_{n+1}(pb)\right]$,
$m_{74} = \mu\left\{\left[-\frac{2n(n-1)}{b^2} - s^2\right]I_n(sb) + \frac{2s}{b}I_{n+1}(sb)\right\}$,	$m_{75} = \mu\left\{\left[-\frac{2n(n-1)}{b^2} - s^2\right]K_n(sb) - \frac{2s}{b}K_{n+1}(sb)\right\}$,
$m_{76} = 2\mathrm{i}k_z\mu\frac{n}{b}\left[\frac{1-n}{b}I_n(sb) - sI_{n+1}(sb)\right]$,	$m_{77} = 2\mathrm{i}k_z\mu\frac{n}{b}\left[\frac{1-n}{b}K_n(sb) + sK_{n+1}(sb)\right]$,
$m_{82} = 2\mathrm{i}k_z\mu\left[\frac{n}{b}I_n(pb) + pI_{n+1}(pb)\right]$,	$m_{83} = 2\mathrm{i}k_z\mu\left[\frac{n}{b}K_n(pb) - pK_{n+1}(pb)\right]$,
$m_{84} = \mathrm{i}k_z\mu\frac{n}{b}I_n(sb)$,	$m_{85} = \mathrm{i}k_z\mu\frac{n}{b}K_n(sb)$,
$m_{86} = \mu\left(k_s^2 - 2k_z^2\right)\left[\frac{n}{b}I_n(sb) + sI_{n+1}(sb)\right]$,	$m_{87} = \mu\left(k_s^2 - 2k_z^2\right)\left[\frac{n}{b}K_n(sb) - sK_{n+1}(sb)\right]$,
$m_{98} = -\left[\frac{n}{a}I_n(fa) + fI_{n+1}(fa)\right]$,	$m_{99} = -\left[\frac{n}{a}K_n(fa) - fK_{n+1}(fa)\right]$,
$m_{910} = \frac{n}{a}K_n\left(p^t a\right) - p^t K_{n+1}\left(p^t a\right)$,	$m_{911} = \frac{n}{a}K_n\left(s^t a\right)$,
$m_{912} = \mathrm{i}k_z\mu\left[\frac{n}{a}K_n\left(s^t a\right) - s^t K_{n+1}\left(s^t a\right)\right]$,	$m_{108} = \rho_f\omega^2 I_n(fa)$,
$m_{109} = \rho_f\omega^2 K_n(fa)$,	$m_{1010} = \mu^t\left\{\left[\frac{2n(n-1)}{a^2} + 2k_z^2 - k_s^2\right]K_n\left(p^t a\right) + \frac{2p^t}{a}K_{n+1}\left(p^t a\right)\right\}$,
$m_{1011} = 2\mu^t\frac{n}{a}\left[\frac{n-1}{a}K_n\left(s^t a\right) - s^t K_{n+1}\left(s^t a\right)\right]$,	$m_{1012} = 2\mathrm{i}k_z\mu^t\left\{\left[\frac{n(n-1)}{a^2} + s^{t^2}\right]K_n\left(s^t a\right) + \frac{s^t}{a}K_{n+1}\left(s^t a\right)\right\}$,
$m_{1110} = 2\mu^t\frac{n}{a}\left[\frac{1-n}{a}K_n\left(p^t a\right) + p^t K_{n+1}\left(p^t a\right)\right]$,	$m_{1111} = \mu^t\left\{\left[\frac{2n(1-n)}{a} - s^{t^2}\right]K_n\left(s^t a\right) - \frac{2s^t}{a}K_{n+1}\left(s^t a\right)\right\}$,
$m_{1112} = 2\mathrm{i}k_z\mu^t\frac{n}{a}\left[\frac{1-n}{a}K_n\left(s^t a\right) + \frac{s^t}{a}K_{n+1}\left(s^t a\right)\right]$,	$m_{1210} = 2\mathrm{i}k_z\mu^t\left[\frac{n}{a}K_n\left(p^t a\right) - p^t K_{n+1}\left(p^t a\right)\right]$,
$m_{1211} = \mathrm{i}k_z\mu^t\frac{n}{a}K_n\left(s^t a\right)$,	$m_{1212} = \mu^t\left(k_s^{t^2} - 2k_z^2\right)\left[\frac{n}{a}K_n\left(s^t a\right) - s^t K_{n+1}\left(s^t a\right)\right]$.

Elements not listed are zero.

Nomenclature

a	Borehole radius
b	Outer radius of the collar
c	Inner radius of the collar
n	The order of the multipole source, 0 for monopole, 1 for dipole, and so on
I, K	Bessel functions of the first kind and second kind, respectively
k_z	Arial Wavenumber
k_f	Wavenumber of the fluid, $k_f = \omega/V_f$
k_p	Wavenumber of the P-wave in the collar, $k_p = \omega/V_p$
k_s	Wavenumber of the S-wave in the collar, $k_s = \omega/V_s$
k_p^t	Wavenumber of the P-wave in the formation, $k_p^t = \omega/V_p^t$
k_s^t	Wavenumber of the S-wave in the formation, $k_s^t = \omega/V_s^t$
f	Radial wavenumber in the fluid, $f = \sqrt{k_z^2 - k_f^2}$
p	Radial wavenumber of the P-wave in the collar, $p = \sqrt{k_z^2 - k_p^2}$
s	Radial wavenumber of the S-wave in the collar, $s = \sqrt{k_z^2 - k_s^2}$
p^t	Radial wavenumber of the formation P-wave, $p^t = \sqrt{k_z^2 - k_p^{t2}}$
s^t	Radial wavenumber of the formation S-wave, $s^t = \sqrt{k_z^2 - k_s^{t2}}$
ρ_f	Fluid density
ω	Angular frequency
μ	Shear modulus of the collar
μ^t	Shear modulus of the formation

The explicit difference scheme in Eqs. B.19–B.33 can be iteratively solved and is very suitable for parallel computation.

The source is implemented by assigning the source time function at normal stress nodes (Coutant et al. 1995).

B.2 Dispersion and Stability Analysis

The essence of numerical simulation for the wave equation using the finite difference method is to discretize the continuous media to obtain a numerical solution. This generates numerical dispersion. Cheng (1994) suggested that at least 6–10 grids are used per wave length to avoid numerical dispersion in staggered grid finite difference simulation.

Because the staggered grid finite difference in the time domain is explicit, certain stability conditions should be satisfied to ensure the convergence of the computation results. Assuming V_{max} is the maximum velocity for all the grids in the medium, the stability condition (Courant et al. 1967) on the discrete time and spatial steps should satisfy the following relationship,

$$\Delta t < \frac{\min(\Delta x, \Delta y, \Delta z)}{\sqrt{3}V_{\max}(|\eta_1|+|\eta_2|)}. \tag{B.34}$$

B.3 Absorbing Boundary Conditions

The numerical solution of the physical problems usually involves the solution in an infinite domain as shown in Fig. B.2a, although the region of the interest is finite. Some problems can be treated by some simple solution. For example, for a periodic problem, a simple coordinate transformation from infinite domain $(-\infty, \infty)$ to finite domain, such as $(-1, 1)$, can be used. In most applications, including seismic wave propagation, the infinite space is usually truncated into a limited computational domain (as shown in Fig. B.2b). The numerical FDTD implementations include boundaries that inevitably bring reflected energy back into the computational domain, that reflected energy contaminates the signal.

Many methods have been developed to avoid the artificial reflections from computational domain boundaries: nonreflecting plane boundary condition (Smith 1974), absorbing boundary conditions (ABCs) (Clayton and Engquist 1977; Higdon 1990), absorbing boundary layers (Cerjan et al. 1985), and transparent boundary (Zhu 1999). The most commonly used method is the perfectly matched layer (PML) (Berenger 1994) that was initially developed for Maxwell's equation. The idea of PML is to add layers of absorbing material outside of the computational domain (as shown in Fig. B.2). This material exponentially attenuates the incident energy and attenuates

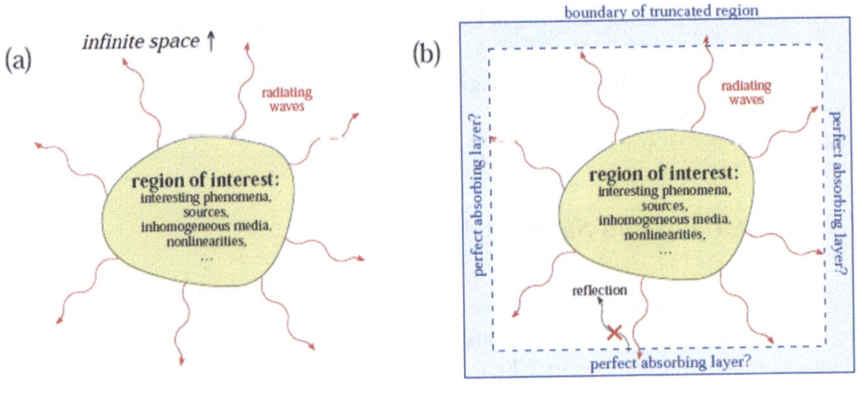

Fig. B.2 The effect of artificial boundary (Fig. 1 in Johnson [2007])

it again when it is reflected back from the outer boundary of the PML. If the layer is large enough, the energy will be absorbed completely in the layer.

PML was introduced into seismic wave propagation simulation by Collino and Tsogka (2001) and to borehole wave propagation by Wang et al. (2009, 2013a). The PML method later evolved into several different types from field-splitting PML (SPML) (Berenger 1994; Collino and Tsogka 2001) to complex frequency shifted PML (CFS-PML) (Kuzuoglu and Mittra 1996; Komatitsch and Martin 2007). In this book, the CFS-PML method is included in the FDTD to simulate the acoustic/seismic wave propagation in the borehole environment.

In the PML method, a complex stretch factor $S_j = \beta + d_j/(\alpha + i\omega)$ is used (Roden and Gendney 2000) to stretch the original coordinate in the j direction in the PML region by jS_j, where j can be x, y, or z in the Cartesian coordinate system, and α and β are frequency-shifted factor and scaling factor, respectively. d_j is a damping function, $d_J = \partial\gamma_j/\partial j$ $(\gamma_j > 0)$, which is the function of space in the j direction. ω is angular frequency. Assuming that the plane wave solution of the wave equations B.5a, B.5b is $\exp\left[-i\sum k_j j\right]$, the solution can be transformed as $\exp\left[-i\sum k_j j\right]\exp\left[-\sum k_j\gamma_j/\omega\right]$, if α and β are 0 and 1, respectively (Chew and Weedon 1994; Chew and Liu 1996). In such a configuration, the incident plane wave along the j direction can be exponentially attenuated in the PML region (Fig. B.3d). To implement this product in the time domain requires a convolution, which is computationally expensive.

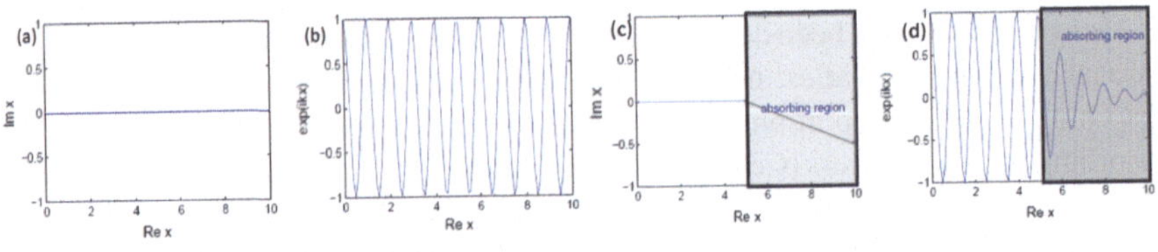

Fig. B.3 The realization and function of PML (modified from Fig. 2 in Johnson [2007]). **a** The original coordinates. **b** The original solution. **c** The new coordinates with the complex stretch factor. **d** The new solution with the complex stretch factor

B.3.1 Split-Field Perfectly Matched Layer (SPML)

SPML is one of the methods which can avoid the convolution operation. Using the x direction velocity in governing equation B.6, the expression in the frequency domain is as follows:

$$i\omega\rho V_x = \frac{\partial t_{xx}}{\partial x} + \frac{\partial t_{yz}}{\partial y} + \frac{\partial t_{zx}}{\partial z}, \tag{B.35}$$

where the terms V_x, t_{xx}, t_{yx}, and t_{zx} are the expressions in the frequency domain of v_x, τ_{xx}, τ_{yx}, and τ_{zx}. With the complex stretch factor, the space derivatives $\partial/\partial x$, $\partial/\partial y$ and $\partial/\partial z$ are replaced by $\partial/\partial x' = \partial/\partial x \cdot 1/S_x$, $\partial/\partial y' = \partial/\partial y \cdot 1/S_y$, and $\partial/\partial \mathbf{z}' = \partial/\partial \mathbf{z} \cdot 1/S_z$, in the complex stretch plane, respectively. Equation B.35 can then be expressed as follows,

$$i\omega\rho V_x = \frac{1}{s_x(x)}\frac{\partial t_{xx}}{\partial x} + \frac{1}{s_y(y)}\frac{\partial t_{yx}}{\partial y} + \frac{1}{s_z(z)}\frac{\partial t_{zx}}{\partial z}, \tag{B.36}$$

To avoid a convolution operation in the time domain, each velocity and stress component is split further into parallel and perpendicular components with respect to the coordinate directions (Wang et al. 2009; Tao et al. 2008b; Berenger 1994; Collino and Tsogka 2001). For example, V_x can be split into three parts: $V_x = V_{xx} + V_{xy} + V_{xz}$, and Eq. B.36 is expressed as,

$$\begin{aligned} i\omega\rho V_{xx} &= \frac{1}{S_x(x)}\frac{\partial t_{xx}}{\partial x}, \\ i\omega\rho V_{xy} &= \frac{1}{S_y(y)}\frac{\partial t_{yx}}{\partial y}, \\ i\omega\rho V_{xz} &= \frac{1}{S_z(z)}\frac{\partial t_{zx}}{\partial z}. \end{aligned} \tag{B.37}$$

The transformations into the time domain to become, ($\alpha = 0$, $\beta = 1$)

$$\begin{aligned} \frac{\partial v_{xx}}{\partial t} + d_x v_{xx} &= \frac{1}{\rho}\frac{\partial \sigma_{xx}}{\partial x}, \\ \frac{\partial v_{xy}}{\partial t} + d_y v_{xy} &= \frac{1}{\rho}\frac{\partial \sigma_{yx}}{\partial y}, \\ \frac{\partial v_{xz}}{\partial t} + d_z v_{xz} &= \frac{1}{\rho}\frac{\partial \sigma_{zx}}{\partial z}. \end{aligned} \tag{B.38}$$

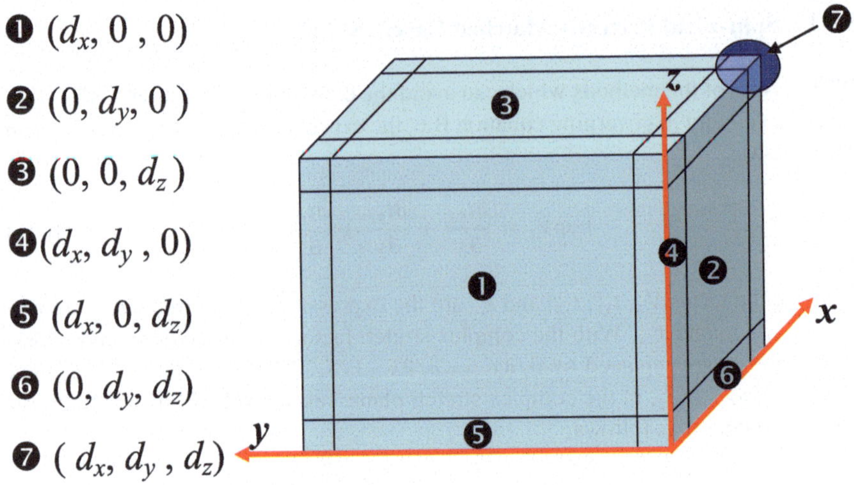

Fig. B.4 Parameter setting of PML in different positions of the model

The discrete form in the time domain of Eq. B.38 is as follows,

$$
\begin{aligned}
v_{xx}^{i+1/2} &= \frac{1 - 1/2d_x\Delta t}{1 + 1/2d_x\Delta t} v_{xx}^{i-1/2} + \frac{\Delta t}{\rho(1 + 1/2d_x\Delta t)} \frac{\partial \sigma_{xx}}{\partial x}|^i, \\
v_{xz}^{i+1/2} &= \frac{1 - 1/2d_z\Delta t}{1 + 1/2d_z\Delta t} v_{xz}^{i-1/2} + \frac{\Delta t}{\rho(1 + 1/2d_z\Delta t)} \frac{\partial \sigma_{xz}}{\partial z}|^i, \\
v_{x}^{i+1/2} &= v_{xx}^{i+1/2} + v_{xz}^{i+1/2}.
\end{aligned} \tag{B.39}
$$

In the implementation of SPML in FDTD, the d_x, d_y, and d_z are given different values according to the different PML domains, and only the components normal to the axis are used while others are set to zero except at the corner of the PML domain (Collino and Tsogka 2001). Figure B.4 shows the distribution of the damping factors in different regions.

B.3.2 Multi-axial Perfectly Matched Layer (M-PML)

Meza-Fajardo and Papageorgiou (2008) analyzed the numerical stability of SPML and introduced a modified version, M-PML, in which damping in different directions is coupled. Again, taking the x direction as an example, the damping profile consists of 3 parts, $d_x^x(x)$, which can be derived from SPML, $d_y^x(x)$ and $d_z^x(x)$, which are the corrections in the y and z directions, respectively. Here the subscript x, y, and z are the normal directions and the superscript x is the damping direction, where $d_x^y(x) = p_{yx}d_x^x(x)$, $d_x^z(x) = p_{zx}d_x^x(x)$, p_{yx} and p_{zx} are correction coefficients that can be tuned according to specific cases. According to Meza-Fajardo and Papageorgiou (2008), a p_0 ($p_0 \in [0,1]$) can be found where for all $p > p_0$, the MPML is stable.

However, reflectivity increases when the stability is improved. In other words, the wave in the x direction will be damped in the x direction and will also be damped in the other two directions (y and z). Therefore, the damping coefficients of M-PML are:

$$\begin{aligned} d_x &= d_x^x(x) + d_x^y(y) + d_x^z(z), \\ d_y &= d_y^x(x) + d_y^y(y) + d_y^z(z), \\ d_z &= d_z^x(x) + d_z^y(y) + d_z^z(z). \end{aligned} \tag{B.40}$$

In fact, Martin et al. (2010) made the case that the M-PML should not be considered as PML because the theoretical reflection coefficient for an infinite PML is not exactly zero in this approach. It is just a modification of sponge and the reflection coefficients are not zeros even for differential formulation (Dmitriev and Lisitsa 2011). The M-PML is a brute-force approach that works well with media having anisotropy and high material property contrasts (Meza-Fajardo and Papageorgiou 2008).

B.3.3 Non-split Perfectly Matched Layer (NPML)

To simplify the implementation of classic PML, Wang and Tang (2003) introduced the non-split PML (NPML), in which a trapezoidal rule is applied to calculate the convolutions in the PML formulation. For example, Eq. B.36 can be transformed into the time domain using inverse Fourier transforms,

$$\rho\frac{\partial v_x}{\partial t} = F^{-1}\left(\frac{1}{S_x(x)}\right) * \frac{\partial \tau_{xx}}{\partial x} + F^{-1}\left(\frac{1}{S_y(y)}\right) * \frac{\partial \tau_{yx}}{\partial y} + F^{-1}\left(\frac{1}{S_z(z)}\right) * \frac{\partial \tau_{zx}}{\partial z}, \tag{B.41}$$

where $F^{-1}\left(\frac{1}{S_x(x)}\right) * \frac{\partial \tau_{xx}}{\partial x} = \frac{\partial \tau_{xx}}{\partial x} - d_x(x)\int_0^T \exp[-d_x(x)(T-t)]\frac{\partial \tau_{xx}}{\partial x}dt$. Therefore, the formulation of velocity in the x direction will be,

$$\begin{aligned} \rho\frac{\partial v_x}{\partial t} = {} & \frac{\partial \sigma_{xx}}{\partial x} - d_x(x)\int_0^T exp[-d_x(x)(T-t)]\frac{\partial \sigma_{xx}}{\partial x}dt \\ & + \frac{\partial \sigma_{yx}}{\partial y} - d_y(y)\int_0^T exp[-d_y(y)(T-t)]\frac{\partial \sigma_{yx}}{\partial y}dt \\ & + \frac{\partial \sigma_{zx}}{\partial z} - d_z(z)\int_0^T exp[-d_z(z)(T-t)]\frac{\partial \sigma_{zx}}{\partial z}dt. \end{aligned} \tag{B.42}$$

Taking the time step as Δt, the time for step i is $T = i\Delta t$. Then the formulation B.42 becomes,

$$\rho\frac{\partial v_x}{\partial t} = \frac{\partial \sigma_{xx}}{\partial x} - d_x(x)\int_0^{i\Delta t} exp[-d_x(x)(i\Delta t - t)]\frac{\partial \sigma_{xx}}{\partial x}dt$$
$$+ \frac{\partial \sigma_{yx}}{\partial y} - d_y(y)\int_0^{i\Delta t} exp[-d_y(y)(i\Delta t - t)]\frac{\partial \sigma_{yx}}{\partial y}dt$$
$$+ \frac{\partial \sigma_{zx}}{\partial z} - d_z(z)\int_0^{i\Delta t} exp[-d_z(z)(i\Delta t - t)]\frac{\partial \sigma_{zx}}{\partial z}dt. \tag{B.43}$$

The time discrete form of Eq. B.43 is

$$v_x^{i+1/2} = v_x^{i+1/2} + \frac{\Delta t}{\rho}\left(\frac{\partial t_{xx}}{\partial x} + \frac{\partial t_{yx}}{\partial y} + \frac{\partial t_{zx}}{\partial z}\right),$$
$$v_x^{i+1/2} = v_x^{i+1/2} - P_{xx}^i - P_{yx}^i - P_{zx}^i. \tag{B.44}$$

where $P_{xx}^i = d_x(x)\int_0^{i\Delta t} \exp[-d_x(x)(i\Delta t - t)\frac{\partial t_{xx}}{\partial x}]dt$, $P_{yx}^i = d_y(y)\int_0^{i\Delta t} \exp[-d_y(y)(i\Delta t - t)\frac{\partial t_{yx}}{\partial y}]dt$, and $P_{zx}^i = d_z(z)\int_0^{i\Delta t} \exp[-d_z(z)(i\Delta t - t)\frac{\partial t_{zx}}{\partial z}]dt$.

The trapezoidal rule can be used for numerical approximation of the integrations above. For example, $P_{xx}^i = d_x(x)P_{xx}^{i-1} + \frac{1}{2}\Delta t d_x(x)\left\{exp[-\Delta t d_x(x)]\frac{\partial t_{xx}}{\partial x}\Big|^{(i-1)} + \frac{\partial t_{xx}}{\partial x}\Big|^{i}\right\}$, in which the auxiliary function is introduced to do the integration with second order time accuracy.

B.3.4 Complex Frequency-Shifted Perfectly Matched Layer (CFS-PML)

Poor damping of evanescent waves and instability for long time duration simulations have been reported in electromagnetic wave simulation (Berenger 1997) by FDTD with conventional PML (Berenger 1994). To reduce the limitations of conventional PML, many researchers have devoted a great deal of effort on the theory and practice of modifying PML. Kuzuoglu and Mittra (1996) analyzed the causality of conventional PML and found that the conventional stretch factor does not preserve causality. They introduced the Complex-Frequency-Shifted (CFS) PML where they use a modified factor $S = 1 + d/(1 + i\omega)$.

The conventional PML method does not work if the wave-number is a pure imaginary number, such as for evanescent waves or guided waves. For example, if k_x is a negative imaginary number (Skelton et al. 2007), it can be replaced by $k_x = -ik$ (k is a real number). The plane wave solution in the x direction becomes

$\exp(-ixk_x)\exp(ikxd_x/\omega)$. The factor $\exp(ikxd_x/\omega)$ makes the wave oscillate without attenuation. On the contrary, the solution of modified factor used by Kuzuoglu and Mittra (1996) is $\exp(-ixk_x)\exp\left(ik\omega xd_x/\left(1+\omega^2\right)\right)\exp\left(-kxd_x/\left(1+\omega^2\right)\right)$, in which the factor $\exp\left(-kxd_x/\left(1+\omega^2\right)\right)$ can exponentially attenuate the energy with increasing distance.

In order to absorb guided waves and evanescent waves efficiently, Roden and Gendney (2000) proposed a general stretch factor $S = \beta + d/(\alpha + i\omega)$ for CFS-PML, where α is a frequency-shift factor and β is a scaling factor. Komatitch and Martin (2007) used a recursive convolutional method to implement the CFS-PML with FDTD. Taking Eq. B.41 as an example, the inverse Fourier transform of 1/S is expressed as,

$$\hat{S} = F^{-1}(1/S) = \frac{\delta(t)}{\beta} - \frac{d}{\beta^2}F^{-1}\left[\frac{1}{\alpha + d/\beta + i\omega}\right] = \frac{\delta(t)}{\beta} - \frac{d}{\beta^2}H(t)e^{-(\alpha+d/\beta)t}, \tag{B.45}$$

Then Eq. B.41 becomes,

$$\rho\frac{\partial v_x}{\partial t} = \hat{S}_x * \frac{\partial \sigma_{xx}}{\partial x} + \hat{S}_y * \frac{\partial \sigma_{yx}}{\partial y} + \hat{S}_z * \frac{\partial \sigma_{zx}}{\partial z}, \tag{B.46}$$

The recursive convolutional method, which is used to obtain Eq. (B.46), is of only second-order accuracy in space and time (Martin et al. 2010). To keep PML time accuracy the same as that of computational domain, Zhang and Shen (2010) used the auxiliary differential equations (ADE) method to attain higher-order time accuracy.

Table B.1 compares the different PML methods to illustrate where the CFS-PML differs from the others. In general, the damping profile is chosen as a polynomial function. Here we follow the formulation by Collino and Tsogka (2001), for damping along the x direction,

$$d_x = d_0(l_x/L)^n, \tag{B.47}$$

where l_x is the distance from the PML-interior interface for the location in the PML domain, n is 2, d_0 is the maximum value of d which can be obtained from Collino and Tsogka (2001, and L is the thickness of the PML layer.

Table B.1 Summary of PML methods (Wang et al. 2013)

	SPML	M-PML	NPML	CFS-PML
α	0	0	0	Non-zero
β	1	1	1	Variable
d_x	$d_x^x(x)$	$d_x^x(x), d_x^y(x), d_x^z(x)$	$d_x^x(x)$	$d_x^x(x)$
Convolution	No	No	Yes	Yes

The values of α and β in CFS-PML are usually given by the following polynomials (Komatitsch and Martin 2007),

$$\beta_x = 1 + (\beta_0 - 1)(l_x/L)^m, \tag{B.48}$$

$$\alpha_x = \alpha_0\left[1 - (l_x/L)^p\right], \tag{B.49}$$

where m and p are 2 and 1, respectively, and α_0 and β_0 are the maximum values of α and β.

B.3.5 PML Methods Used in the Numerical Simulation of Elastic Wave Propagation in a Borehole

Unlike the open hole logging models, cased-hole and acoustic logging while drilling (ALWD) FDTD simulation requires very fine grids because of the thin casing or fluid annulus. Small spatial grid size requires small time step and calculation times increase which increases the cumulative numerical error. In addition, the high impedance contrast (often more than 30) between the fluid and steel casing requires a high efficiency method to capture the subtle features in the late arrivals. Because of these challenges, efficient computational algorithms are needed to make realistic simulations of cased-hole and ALWD problems. Given the high material contrasts in our models, Wang et al. (2013a) evaluated the applicability of non-reflective boundary conditions for ALWD simulations. Different PML implementations (SPML, M-PML, NPML, and CFS-PML) were used for large material contrast models (ALWD model).

Here we show results with four different PML methods in a 2D LWD case with a monopole source. The model is shown in Fig. B.5, where only the x and z directions are considered and y direction is assumed as infinite. Vp, Vs and density of the formation are 3927 m/s, 2455 m/s, and 2300 kg/m^3, respectively. The outer radius of the inner fluid, collar, and position of the boundary between the outer fluid and formation are 27 mm, 90 mm, and 117 mm, respectively. Collar C32 in Table 5.2 is used.

The staggered grid FDTD scheme used for testing has a fourth order accuracy in space and 2nd order accuracy in time (Cheng 1994; Tao et al. 2008b). The model is discretized into 123 by 334 grids along the x- and z- directions, respectively. The grid spacing is 9 mm, and time step is 0.9 μs. The PML layer thickness is 20 grids. The source time function of the monopole source is a Ricker wavelet with central frequency f_c of 10 kHz. SPML parameters d_0 and α_0 are chosen as 1 and πf_c, respectively. The total recording time of the simulated waveform is 14 ms. The array waveform for the entire simulation time and the first 4 ms are shown in Figs. B.6 and B.7, respectively. For the case of SPML (Fig. B.7a), the drill collar wave, shear (S)- wave, and Stoneley (ST) wave can be identified from their arrival times. Also, the artificial reflection from the top boundary of the model (dashed black line) is visible. This arrival is a ST wave reflected from the interface where the borehole

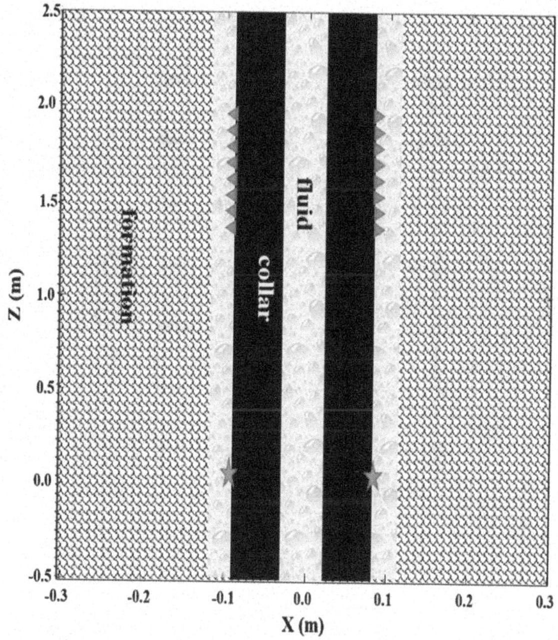

Fig. B.5 A typical large material contrast model (LWD model). Stars and triangles mark the source and receiver locations, respectively. Only x and z directions are considered in the 2D model, where the y direction is assumed as infinite

intersects the model boundary. The ST-wave velocity is 1389 m/s as calculated by the velocity-time semblance method (Kimball and Marzetta 1984; More detail can also be found in Chap. 3). The simulation becomes unstable after 10 ms (shown in Fig. B.6a), indicating the ill-posed nature of the SPML scheme for a model with a large material contrast. Figures B.6b and B.7b show the result of M-PML with correction coefficients p_{zx} and p_{xz} taken as 0.1. The reflection artifacts are still visible as in the case of SPML (shown in Fig. B.6b), where the instability is reduced to some extent, as shown in Fig. B.6b, and it appears after 13 ms and without the high-frequency component. This indicates that the M-PML can be used to simulate a longer signal if the correction coefficients are properly chosen. We find a suitable p_0 to be between 0 and 1. The simulations are stable when p values are greater than p_0. However, the reflectivity is increased. For a good simulation, we must find a good coefficient with a good tradeoff between stability and reflectivity. Figures B.6c and B.7c show the result of NPML. Comparing Figs. B.7a to B.7b and B.7c, we note that NPML (Fig. B.7c) is superior to SPML and M-PML in suppressing the reflected ST wave. However, the instability of NPML (after 12 ms in Fig. B.6c) in the ALWD case indicates that it is not suitable for long simulation times. From Figs. B.6d and B.7d, we find that CFS-PML can attenuate the reflected wave from the model boundary and remain stable for long simulation times. CFS-PML appears to be the best of the four implementation methods for both stability and absorption efficiency.

Figure B.8 shows the wavefield snapshots at time 14.4 ms of FDTD simulations for the 2D ALWD model shown in Fig. B.5 with four different PML methods. The PML domains are also shown on the boundaries of the snapshots. The computational

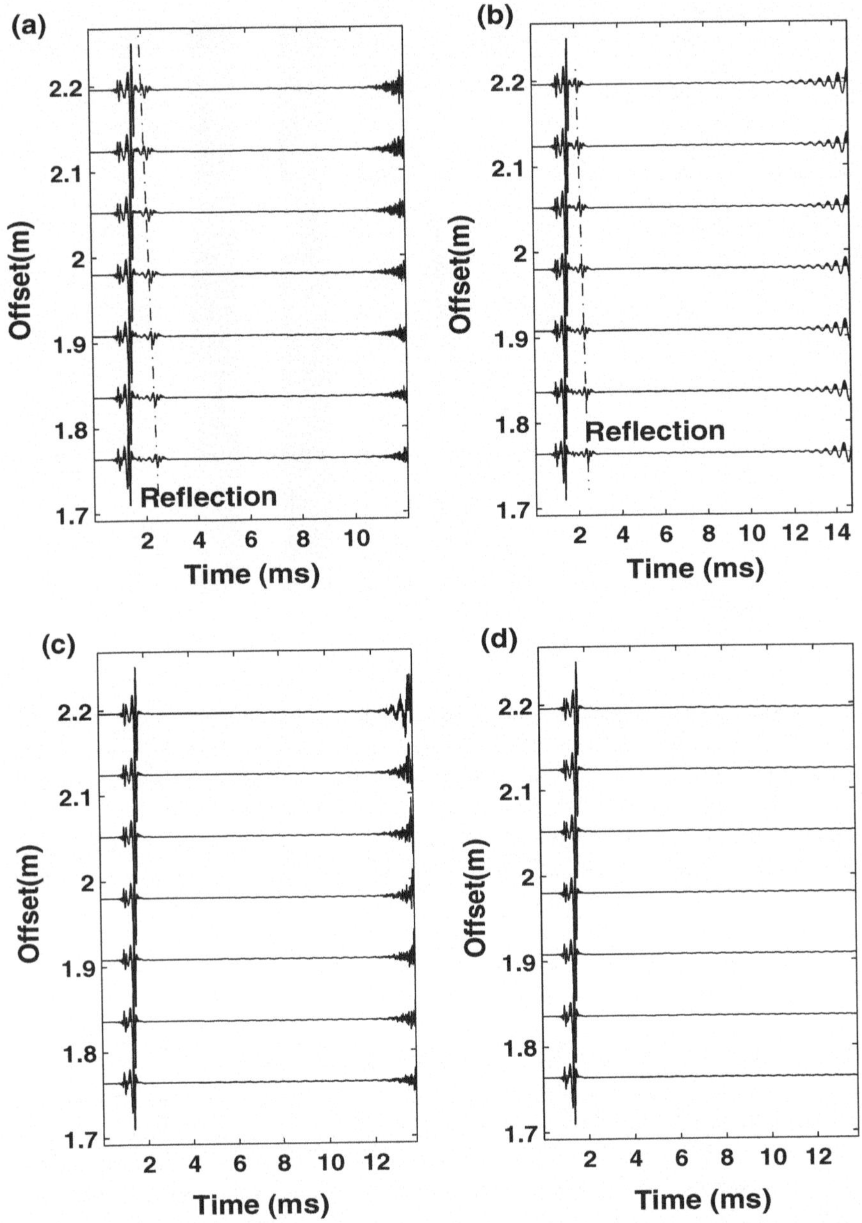

Fig. B.6 Waveform for an array of receivers in a 2D LWD case shown in Fig. B.5. **a** Result for SPML. **b** M-PML. **c** NPML. **d** CFS-PML. Material properties are given in the text. Dashed black lines in **a** and **b** show reflections from the top of the model

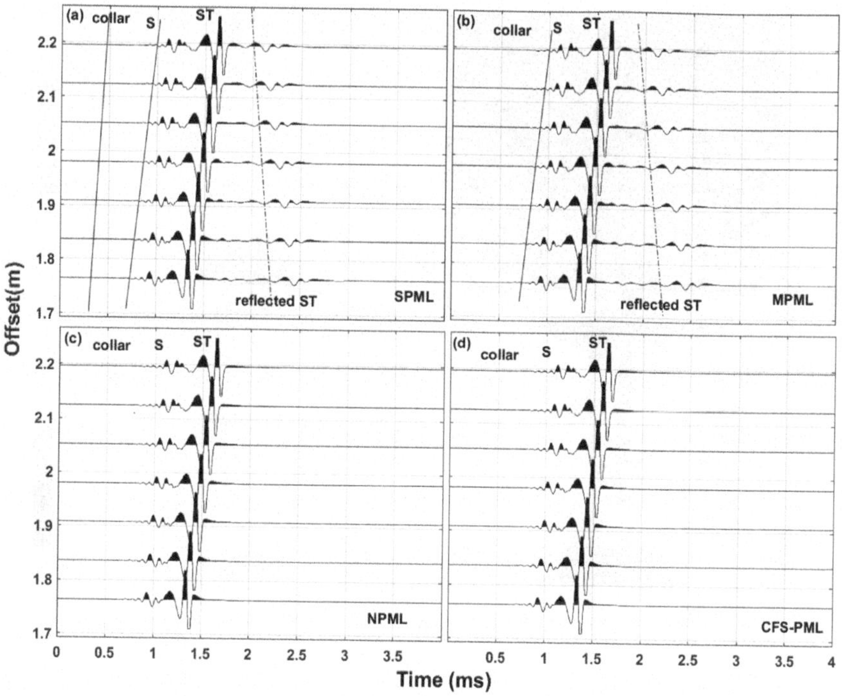

Fig. B.7 Waveform for an array of receivers in a 2D LWD case shown in Fig. B.5. **a** Result for SPML. **b** M-PML. **c** NPML. **d** CFS-PML. Material properties are given in the text

domain is marked with a rectangle. In addition, there are six solid lines from the bottom to top of the model along z direction. The two innermost lines at approximately $x = 0.55$ are the boundaries of the inner fluid column, and the two outermost lines are the boundaries of the borehole wall. The other two lines are the outer boundaries of the collar. The source location in the z direction is 0.45 m. The 10 kHz monopole source with Ricker wavelet is used here.

Here, one can observe that the numerical instability and reflected waves that are generated the fluid-steel interface intersect the model boundary. The CFS-PML gives the best results, although there are still some unabsorbed guided waves that can be clearly observed on the lower part of the snapshot in Fig. B.8d. M-PML is the second best of the four methods.

Our simulation results indicate that NPML and CFS-PML absorb more efficiently the guided wave reflections from the computation boundaries than SPML and M-PML. For long duration simulations, numerical instability is observed in SPML, M-PML and NPML, though the stability in M-PML may be improved by fine tuning of the non-reflecting layer parameters. Among all methods, CFS-PML is the best choice for large material contrast, for both efficient absorption of waves incident on the model boundaries and for numerical stability. The CFS-PML may

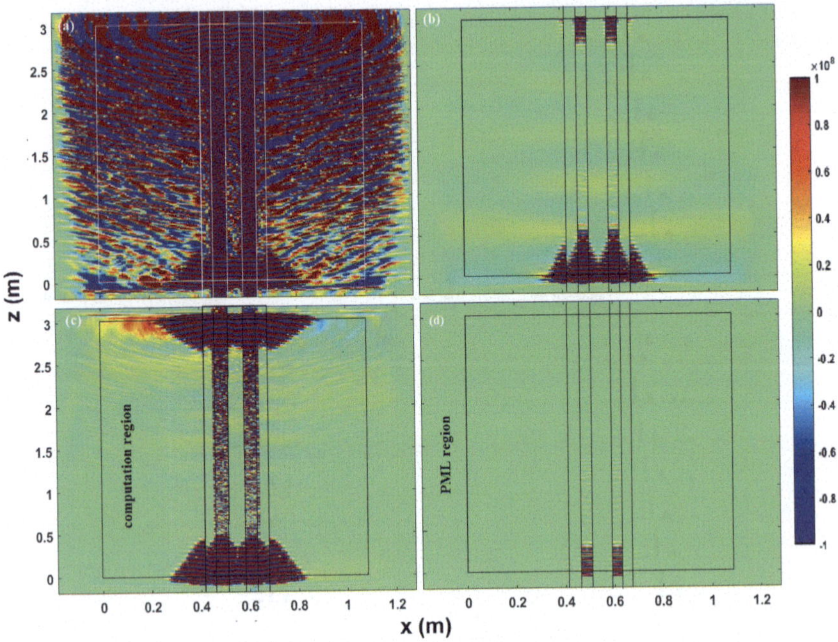

Fig. B.8 Snapshot of wave field at time 14.4 ms for the 2D ALWD sonic case shown in Fig. B.5. **a** Result for SPML. **b** M-PML. **c** NPML. **d** CFS-PML. The computational domain is marked with a rectangle. Region outside the rectangle contains the absorbing boundary. There are six solid lines from the bottom to top of the model along z direction. The two innermost lines approximately $x = 0.55$ are the boundaries of the inner fluid column, and the two outermost lines are the boundaries of the borehole wall. The other two lines are the outer boundaries of the collar

be combined with MPML as a hybrid PML in more challenging environments such as a borehole surrounded by an anisotropic formation.

B.4 *Treatment on the Fluid-Solid Boundary in FDTD*

There are significant accuracy and efficiency issues when conducting finite difference elastic wave simulation in a model with large material contrasts like those that occur at steel-water boundaries (cased-hole and ALWD models). Several methods have been proposed to handle those problems. Here, we discuss the effects on the results of the FDTD having different accuracies for spatial derivatives and for different spatial discretization intervals. As a typical high contrast medium model, the ALWD case can be easily extended to cased-hole models. Here the 2D ALWD model in Fig. B.5 and formation S1 in Table 2.1 and Collar C32 in Table 5.2 are used.

The results from the 2D discrete wavenumber integration method (DWM) (see Sect. 2.1.3) are used to benchmark the simulation results for the 2D FD with different grid sizes. A 10 kHz Ricker wavelet monopole source is used and the axial positions of receivers relative to the source range from 1.305 to 1.935 m with an interval of 0.09 m. We compare the results from fourth order accurate in space 2D FD with different grids and the DWM result in Fig. B.9a and find that the comparison between FD and DWM is better when the grid is smaller. However, there is still a significant difference between results from the 2D FD and DWM even for smaller grid size. However, the result when using a 2nd order accuracy in space FD method matches the result from DWM quite well (even though the grid is large) as shown in Fig. B.9b. The corresponding spatial snapshots of the FD results using different accuracies and grid sizes are shown in Fig. B.10, where Figs. a, b, and c are the results for 4th order space accuracy with a 3 mm grid, 4th order space accuracy with a 1 mm grid and 2nd order space accuracy with a 3 mm grid in space. The wavefield calculated using the fourth order spatial accuracy scheme have considerable noise between z = −0.5 m and z = 1.0 m. The results show that we obtain a better result with the 2nd order spacing accuracy FD than with the 4th order spacing accuracy FD when the model has high contrasts.

An explanation for the counter-intuitive result about the reliability of FD schemes is given below. Taking the differential of τ_{xx} at grid (m, k) along the x direction—$D_x \tau_{xx_{m+1/2,k}}$ in Eq. B.20 as an example, the spatial derivatives of $D_x \tau_{xx_{m+1/2,k}}$ in second and fourth orders accuracy are given in Eqs. B.50 and B.51, respectively.

$$\left[\tau_{xx}\left(m+\frac{1}{2},k\right)-\tau_{xx}(m-1/2,k)\right]/dx, \tag{B.50}$$

$$\left\{9\left[\tau_{xx}\left(m+\frac{1}{2},k\right)-\tau_{xx}\left(m-\frac{1}{2},k\right)\right]/8-\left[\tau_{xx}\left(m+\frac{3}{2},k\right)-\tau_{xx}\left(m-\frac{3}{2},k\right)\right]/24\right\}/dx. \tag{B.51}$$

More grid points are involved in the calculation for the 4th order spacing accuracy FD. If the properties of the medium change a lot, as they do for a high contrast medium

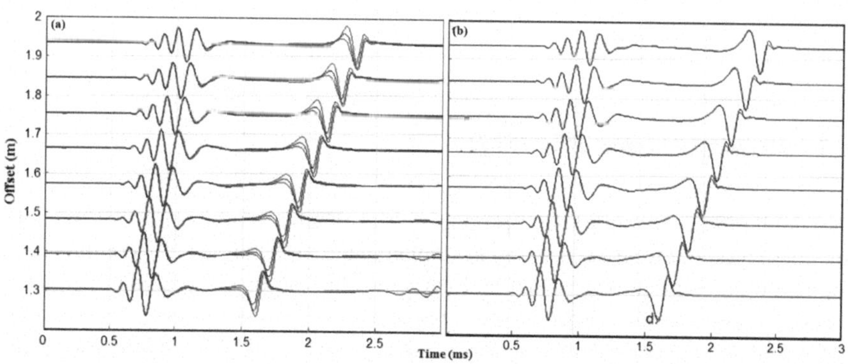

Fig. B.9 FD simulations with different accuracies by DWM (red lines). **a** Results from DWM and fourth order in space finite difference using different grid sizes. The black and blue lines are for grid size of 3 mm and 1 mm, respectively. **b** Results for DWM and second order finite difference. The blue lines are for grid size of 3 mm (Modified from Figs. 5 and 6 in Wang and Zhang [2019])

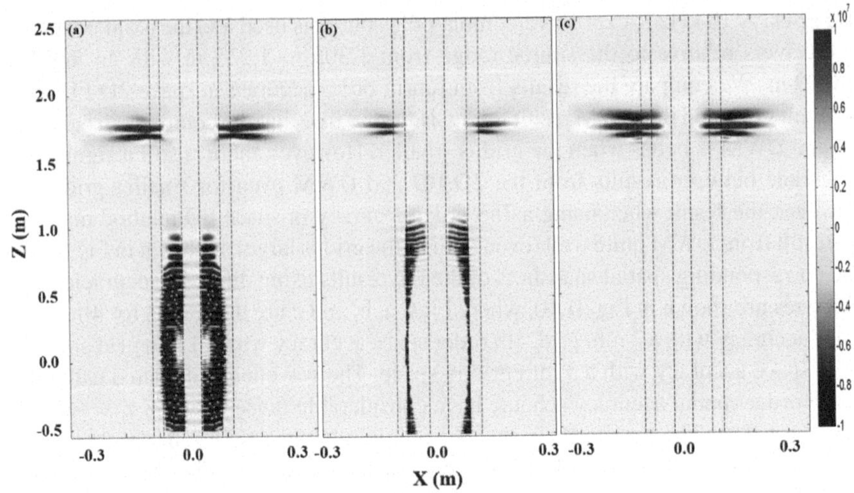

Fig. B.10 Snapshots of the wavefield calculated using FD with different grid spacings and spacing accuracies. **a** and **b** are results found using 2D FD with fourth order accuracy in space and grid spacing of 3 mm and 1 mm, respectively. **c** Result obtained using 2D FD with second order accuracy in space and a grid spacing of 3 mm

model, a finer grid is required to obtain a reliable description of the model. When using the second order spacing accuracy FD, the use of only two grid points during the updating can be better for describing the sharp contrast in material properties of the medium (as shown in Fig. B.11). Generally, higher-order finite difference schemes are used because they can provide good results for models lacking high contrast, and they require fewer grid points per wavelength for accuracy. Fewer grid points lead to significant reductions in computational cost. However, when the model has high contrasts, a lower order scheme is better. In this book, the FDTD with 2nd order spacing accuracy is used.

B.5 Parallel Design of FDTD

The simulation of seismic wave propagation problems requires massive memory and computational time, so it is rarely implemented on a single CPU computer. To improve the computational speed and carry out the large-scale simulations, large memory and the parallel processors are required.

Current parallel computation structure mainly consists of SMP (symmetric multi-processor), MPP (Massively Parallel Processor array) and COW (Cluster of Workstations). Although COW has high cost-performance advantage, its slow computation speed limits the application. SMP, which is able to share memory, is fit for fine-grained parallel computing. For instance, the variables in Eqs. B.20–B.34 at the

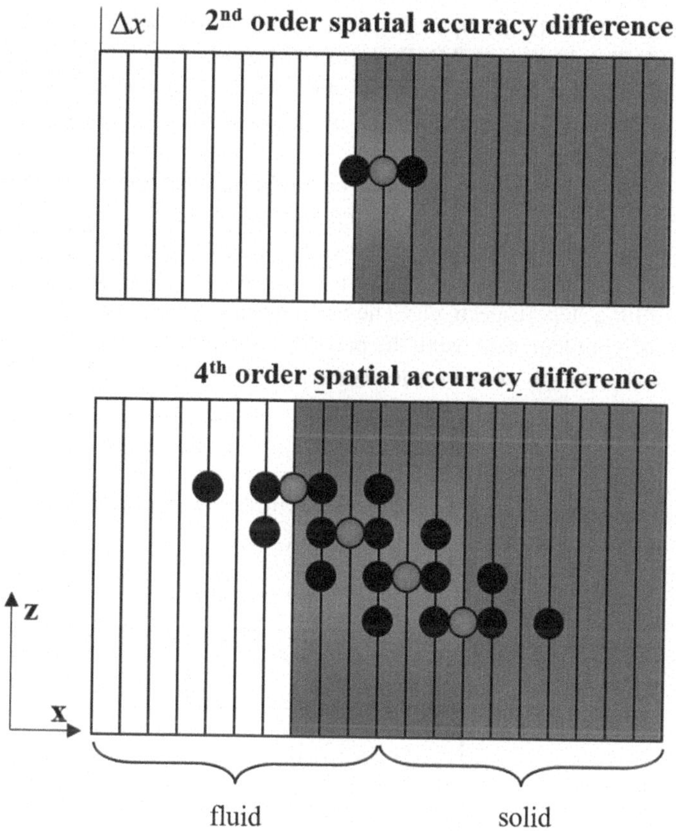

Fig. B.11 Grids involved in the FD with different order spacing accuracies. The black points are the location of discrete grids and the gray points are the positions of the obtained derivatives by different spatial accuracy difference methods (Mortified from Fig. 3 in Zhang [2016])

current time step are only related to the previous time step, and the variable in a certain spatial grid is only related to the surrounding grids. Based on the recursive feature in time domain, all the variables at a given time step can be simultaneously updated.

In addition, for models requiring massive memory that SMP is not able to provide, the coarse-grained parallel computation based on MPP can be applied. In MPP, each node has an independent processor and memory. The nodes communicate within a special network with high passband and low delay. The system is easy to expand and can run with multiple threads. Each thread has an independent address and can communicate with others via message delivery. The coarse-grained parallel computation divides the computation area into several sub area (2D) or volume (3D) regions based on spatial characteristics of the model. Independent computations and data are exchanged only at the border of the regions.

The library of SMP is OpenMP (Open Multi-Processing), which is easily programmed. The library of MMP is MPI (Message Passing Interface), which requires the effective separation of the computational domain into several sub domains. In addition, the management of message delivery makes it relatively complex to program.

The simulations in this book are done by the FDTD with the OpenMP structure. Figure B.12 shows the parallel program structure based on the OpenMP. The first step is to allocate and initialize the memory for each variable in the shared memory stack. Then, perform the time loop, and update the velocity and stress variable in the first and last half time step, respectively. The update of all velocity or stress variables is simultaneously implemented using the parallel computation. Synchronization must be performed before the stress or velocity is updated to ensure that the calculation of

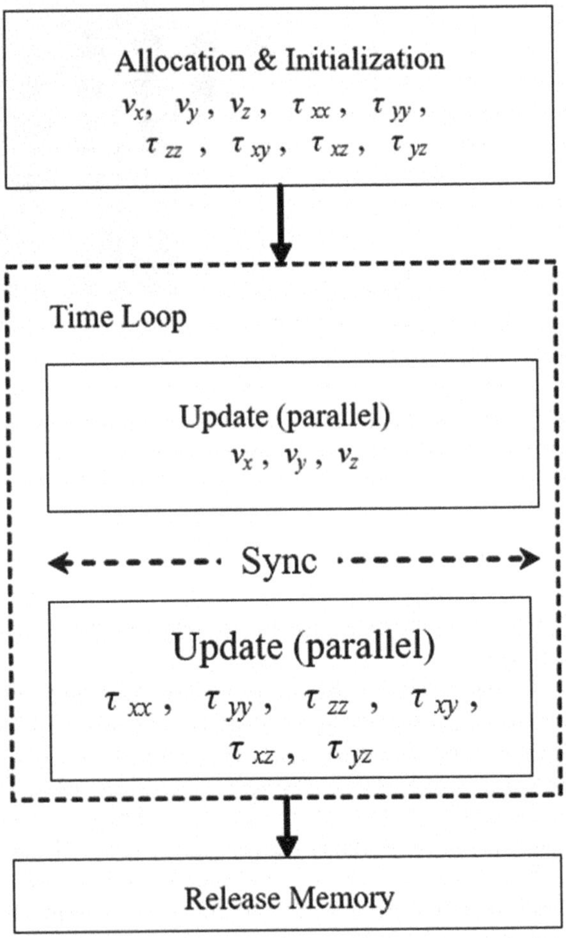

Fig. B.12 The structure of the parallel FDTD program based on the OpenMP (modified from Fig. 2.14 in He [2005])

the previous time step is completed. Memory is not released until all the time steps are computed.

The application programming interface of OpenMP includes a set of compiler directives, library routines, and environment variables, which can be directly embedded into FORTRAN and C/C++ for further usage. The PARALLEL compiler directive divides the code into serial and parallel regions and the parallel region starts with a compiler directive !OMP PARALLEL and ending with !OMP END PARALLEL. In the parallel region, the codes are implemented by multiple threads, while in serial region the codes are implemented via only one thread. The specific markups are responsible for assigning the task into each thread. The local variable inside the thread is assigned by a compiler directive !$OMP DO PRIVATE (name list of local variables). This expression establishes the reproduction for all the local variables at each thread. The values of the local variables are taken from the previous time step but modified during the current time step and are thus not known.

B.6 Validation of the Finite Difference Code for Borehole Acoustics

To check the validity of the FDTD method in borehole environments, the simulation results from the FDTD and DWM (see Appendix A) in wireline and LWD models are compared. The comparisons of the results from 3DFD and DWM for the fluid-filled borehole surrounded by formation F1 listed in Table 2.1, are shown in Figs. B.13 and B.14. The simulation results in Figs. B.13 and B.14 are at 10 kHz for monopole cases in wireline and ALWD, respectively. The geometries of the wireline and LWD models are listed in Tables 2.1 and 5.2 (collar C32 used), respectively.

The solid and dashed lines are the results from the DWM and FDM, respectively. We can clearly see all the modes. For the wireline model shown in Fig. B.13, it is clear to find the P, S and pR, ST, and pR Airy waves in time sequence. For the waveforms in the ALWD model shown in Fig. B.14, we see the collar, P, S, pseudo-Rayleigh (pR.), and ST waves. The amplitude of the collar wave is very small and therefore amplified for display. In the simulation, we did not add any attenuation and thus the amplitude of ST is very large. The results obtained using the two methods are almost identical. These results demonstrate the applicability of the FDTD method. In this book, the wavefields in complex models that cannot be obtained using DWM, such as the eccentered drilling pipe in the ALWD case in Chap. 6 and the borehole acoustic remote reflection image in Chap. 7, are simulated by the finite difference method.

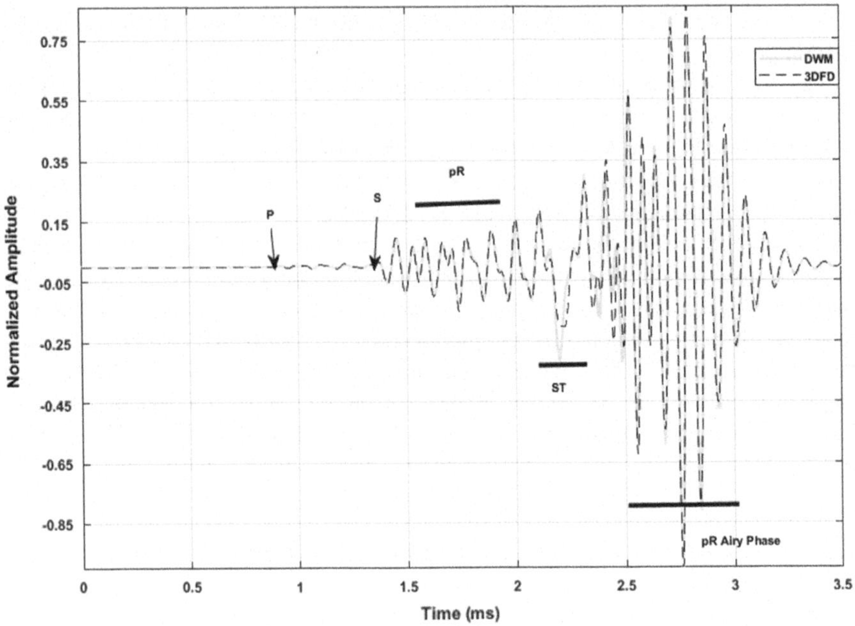

Fig. B.13 Comparisons of FDTD and DWM simulations of the wireline acoustic logging waveforms in a fluid-filled borehole surrounded with formation F1. The waveforms are at 3 m offset from a 10 kHz monopole source

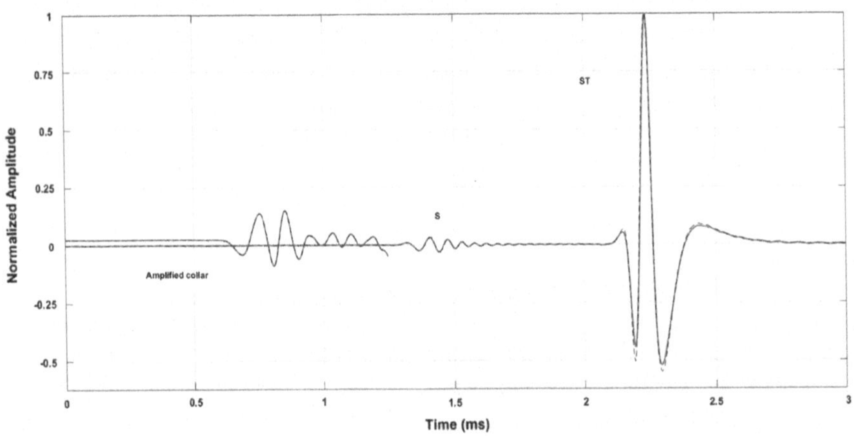

Fig. B.14 Comparisons for FDTD and DWM of the ALWD waveforms in the fluid-filled borehole surrounded with formation F1. The waveforms are at 3 m offset from a 10 kHz monopole source. Collar C32 in Table 5.2 is used

References

Aki K, Richards PG (1980) Quantitative seismology, theory and methods. W.H. Freeman and Co., San Francisco

Allaud LA, Martin MH (1978) Schlumberger: the history of a technique. Schlumberger Ltd. (Translated by Schwob M.)

Al Rougha HAB, Borland WH, Holderson J, Sultan AA, Chakravorty S, Al Raisi M (2005) Integration of microelectrical and sonic reflection imaging around the borehole-offshore UAE. International Petroleum Technology Conference, Doha, Qatar, 2005, Nov. 21–23

Archie GE (1942) The electrical resistivity log as an aid in determining some reservoir characteristics. Petroleum Trans AIME 146:54–62

Arditty PC, Arens G, Staron P (1981) EVA: a long spacing sonic tool for evaluation of velocities and attenuations. In: 51st annual international meeting of the society of exploration geophysicists, Los Angeles, California, United States, October 11–15, 1981

Aron J, Chang SK, Dworak R et al (1994) Sonic compressional measurements while drilling. In: SPWLA 35th annual logging symposium, paper SS

Asquith G, Krygowski D (2006) Basic well log analysis, 2nd edn. AAPG

Bakku SK, Fehler M, Burns D (2013) Fracture compliance estimation using borehole tube waves. Geophysics 78(4):D249–D260

Bancroft JC, Geiger HD, Margrave GF (1998) The equivalent offset method of prestack time migration. Geophysics 63:2042–2053

Baysal E, Kosloff DD, Sherwood JWC (1983) Reverse time migration. Geophysics 48(11):1514–1524

Berenger J (1994) A perfectly matched layer for the absorption of electromagnetic waves. J Comput Phys 114(2):185–200

Berenger JP (1997) Improved PML for the FDTD solution of wave-structure interaction problems. Antennas Propag, IEEE Trans Antennas Propag 45(3):466–473

Beylkin G (1985) Imaging of discontinuities in the inverse scattering problem by inversion of a causal generalized Radon transform. J Math Phys 26:99–108

Biot MA (1952) Propagation of elastic waves in a cylindrical bore containing a fluid. J Appl Phys 223:997–1005

Bleistein N, Cohen JK, Stockwell JW (2001) Mathematics of multidimensional seismic imaging, migration, and inversion. Springer

Boucher FG, Hildebrandt AB, Hagen HB (1951) New diplogging method. In: Second symposium on subsurface geological techniques, School of Geology, University of Oklahoma, March 14–15, pp 101–110

Bouchon M, Aki K (1977) Discrete wave-number representation of seismic-source wave fields. Bull Seismolog Soc Am 67(2):259–277

Brekhovskikh LM (1960) Waves in layered media. Academic press, USA, New York, pp 1–200

H. Wang et al., *Borehole Acoustic Logging—Theory and Methods*,
Petroleum Engineering, https://doi.org/10.1007/978-3-030-51423-5

Brown HD, Grijalva VE, Raymer LL (1970) New developments in sonic wave train display and analysis in cased holes. In: Society of petrophysicists and well-log analysts, 11th annual logging symposium, 3–6 May, Los Angeles, California, paper F

Byun J, Toksöz MN (2006) Effects of an off-centered tool on dipole and quadrupole logging. Geophysics 71(4):F91–F100

Brytik V, de Hoop MV, van der Hilst RD (2012) Elastic-wave inverse scattering based on reverse time migration with active and passive source reflection data. In: Uhlmann G (ed) Inside out—inverse problems. MSRI Publications, Cambridge University Press, pp 411–453

Cameron I (2013) SPE "Back to basics" bond log theory and interpretation. https://zh.scribd.com/document/238612532/Cement-Bond-Log-SPE. Accessed 25 Apr 2018

Capon J (1969) High resolution frequency wave number-spectrum analysis. Process. IEEE 57:1408–1418

Castagna JP, Batzle ML, Eastwood RL (1985) Relationships between compressional-wave and shear-wave velocities in clastic silicate rocks. Geophysics 50(4):571–581

Cawley P, Lowe MJS, Simonetti F, Chevalier C, Roosenbrand AG (2002) The variation of the reflection coefficient of extensional guided waves in pipes from defects as a function of defect depth, axial extent, circumferential extent and frequency. Pro. Inst. Mech. Eng. C. 216(11):1131–1143

Cerjan C, Kosloff D, Kosloff R, Reshef M (1985) A nonreflecting boundary condition for discrete acoustic and elastic wave equations. Geophysics 50(4):705–708

Chabot L, Henley DC, Brown RJ, Brown RJ (2001) Single-well imaging using the full waveform of an acoustic sonic. In: SEG 71th annual meeting, expanded abstracts, pp 420–423

Chabot L, Henley DC, Brown RJ, Bancroft J C (2002) Single-well seismic imaging using full waveform sonic data: An update. In: SEG 72th annual meeting, 2002, Oct. 6–11

Chai X, Zhang W, Wang G, Liu J, Xu M, Liu D, Song C (2009) Application of remote exploration acoustic reflection imaging logging technique in fractured reservoir. Well Logg Technol (In Chinese with English Abstract) 33:539–543

Chang SK, Everhart A (1983) Acoustic waves along a fluid filled borehole with a concentric solid layer. J Acoust Soc Am 74(Supplement 88)

Charara M, Vershinin A, Deger E et al (2011) 3D spectral element method simulation of sonic logging in anisotropic viscoelastic media. SEG Tech Prog Expand Abstra 30(1):432–437

Chen YH, Chew WZ, Liu QH (1998) A three dimensional finite difference code for the modeling of sonic logging tools. J Acoust Soc Am 103(2):702–712

Chen T, Wang B, Zhu Z, Burns D (2010) Asymmetric source acoustic LWD for improved formation shear velocity estimation. SEG Tech Prog Expand Abstr 548–552

Cheng CH, Toksöz MN (1982) Generation, propagation and analysis of tube waves in a borehole. In: SPWLA 23th annual logging symposium, paper P

Cheng CH, Toksöz MN (1980) Modelling of full wave acoustic logs. In: Society of professional well log analysts annual logging symposium, 21st, paper J

Cheng CH, Toksöz MN (1981) Elastic wave propagation in a fluid-filled borehole and synthetic acoustic logs. Geophysics 46(7):1042–1053

Cheng CH, Toksöz MN, Willis ME (1982) Determination of in situ attenuation from full waveform acoustic logs. J Geophys Res 87:5477–5484

Cheng NY, Zhu Z, Cheng CH, Toksöz MN (1992) Experimental and finite difference modeling of borehole Mach waves. Earth Resources Laboratory Industry Consortia Annual Report, 1992–10

Cheng N (1994) Borehole wave propagation in isotropic and anisotropic media: three-dimensional finite difference approach. PhD dissertation. MIT, Cambridge, MA, USA

Chew WC, Weedon WH (1994) A 3D perfectly matched medium from modified Maxwell's equations with stretched coordinates: microwave Optical Tech. Letters 7:599–604

Chew WC, Liu QH (1996) Perfectly matched layers for elastodynamics: a new absorbing boundary condition. J Comp Acoust 4:341–359

Chin WC (2014) Wave propagation in drilling, well logging and reservoir environments. Wiley

Chin WC, Zhou Y, Feng Y, Yu Q (2015) Formation testing: low mobility pressure transient analysis. Scrivener Publishing of Wiley, Beverly, MA

Claerbout JF (1971) Toward a unified theory of reflector mapping. Geophysics 36:467–481

Claerbout JF (1976) Fundamentals of geophysical data processing with applications to petroleum prospecting. McGraw-Hill Book Co., Inc., New York

Clayton R, Engquist B (1977) Absorbing boundary conditions for acoustic and elastic wave equations. Bull Seismol Soc 67:1529–1540

Close D, Cho D, Horn F, Edmondson H (2009) The sound of sonic: a historical perspective and introduction to acoustic logging. Schlumberger Oil Field Rev 30:34–43

Coates R, Kane M, Chang C, Esmersoy C, Fukuhara M, Yamamoto H (2000) Single-well sonic imaging: high-definition reservoir cross-sections from horizontal wells. Paper SPE 65457 presented at the 2000 SPE/petroleum society of CIM international conference on horizontal well technology, Calgary, Alberta, 2000, Nov. 6–8

Cotes G, Xiao L, Prammer M (2000) NMR logging principles and applications. Gulf Publishing Company, USA

Collino F, Tsogka C (2001) Application of the perfectly matched absorbing layer model to the linear elastodynamic problem in anisotropic heterogeneous media. Geophysics 66(1):294–307

Courant R, Friedrichs O, Lewy H (1967) On the partial difference equations of mathematical physics. IBM J (translated from Courant et al 1928) 11:215–234

Coutant O, Virieux J, Zollo A (1995) Numerical source implementation in a 2D finite difference scheme for wave propagation. Bull Seismol Soc Am 85(5):1507–1512

Cowles CS, Leveille JP, Hatchell PJ (1994) Acoustic multi-mode wideband logging device. U.S. Patent No. 5, 289, 433

de Hoop M, Smith H, Uhlamann G, Van der Hilst R (2009) Seismic imaging with the generalized Radon transform: a curvelet transform perspective. Inver Prob 25:025005

Deepwater Horizon Study Group (2011) Final report on the investigation of the Macondo well blowout, p 45. http://ccrm.berkeley.edu/pdfspapers/beapdfs/dhsgfinalreport-march2011-tag.pdf. Last accessed Apr 2019

Dmitriev M, Lisitsa V (2011) Application of M-PML reflectionless boundary conditions to the numerical simulation of wave propagation in anisotropic media. Part I: reflectivity. Numeri Anal Appl 4:271–280

Doll HG (1949) Introduction to induction logging. Petrol Technol 1(6):148–162

Dunham W (2005) The calculus gallery: materpieces from Newton to Lebesgue. Princeton University Press, p 197

Edwards GR, Gan T (2007) Detection of corrosion in offshore risers using guided ultrasonic waves. In: Proceedings of the 26th international conference on offshore mechanics and arctic engineering, June, 2007, San Diego, California, USA. OMAE 2007-29407

Ekstrom MP (1995) Dispersion estimation from borehole acoustic arrays using a modified matrix pencil algorithm. IEEE Sig Syst Comput Conf Asil 29(1):449–453

Ellefsen KJ, Daniel RB, Cheng CH (1993) Homomorphic processing of the tube wave generated during acoustic logging. Geophysics 58(2):1400–1406

Ellis DV, Singer JM (2007) Well logging for earth scientists. Springer

Embree P, Burg JP, Buckus MM (1963) Wide-band velocity filtering—the pie-slice process. Geophysics 28(6):948–974

Esmersoy C, Koster K, Williams M, Boyd A, Kane M (1994) Dipole shear anisotropy logging. In: 64th SEG annual meeting expanded abstracts. Los Angeles

Esmersoy C, Chang C, Tichelaar BW, Kane M, Coates RT, Quint E (1998) Acoustic imaging of reservoir structure from a horizontal well. Leading Edge 17(7):940–946

Fletcher RP, Flower PJ, Kitchenside P (2006) Suppressing unwanted internal reflections in prestack reverse-time migration. Geophysics 71(6):E79–E82

Fortin JP, Rehbinder N, Staron P (1991) Reflection imaging around a well with the eva full-waveform tool. Log Analyst 32:271–278

Frisch GJ, Graham WL, Griffith J (1999) Assessment of foamed-cement slurries using conventional evaluation logs and improved interpretation methods. SPE Rocky Mountain Regional Meeting, Wyoming, America, 1999, SPE 55649

Frisch GJ, Graham WL, Griffith J (2000) A novel and economical processing technique using conventional bond logs and ultrasonic tools for enhanced cement evaluation. In: SPWLA 41st annual logging symposium, paper EE

Frisch G, Fox P, Hospedales D, Lutchman K (2015) Using radial bond segmented waveforms to evaluate cement sheath at varying depths of investigation. Society of Petroleum Engineers, paper 174829

Froelich B, Pittman D, Seeman B (1983) Cement evaluation tool—a new approach to cement evaluation. Society of Petroleum Engineers, paper 1027

Froelich B (2008) Multimode evaluation of cement behind steel pipe. J Acoust Soc Am 123:3648

Geerits T, Mandal B, Schmitt D (2006) Acoustic logging-while-drilling tools having a hexapole source configuration and associated logging methods. US Patent Published No. US20060198242 A1

Gollwitzer LH, Masson JP (1982) The cement bond tool. In: SPWLA, 23rd Annual Symposium

Grosmangin M, Kokesh FP, Majani P (1961) A sonic method for analyzing the quality of cementation of borehole casings. J Petrol Technol 13(2):165–171

Haldorsen JBU, Voskamp A, Thorsen R, Vissapragada B, Williams S, Fejerskov M (2006) Borehole acoustic reflection survey for high resolution imaging. In: SEG 76th annual meeting, New Orleans, Expanded Abstracts, pp 314–318

Hardage BA (1981) An examination of tube wave noise in vertical seismic profiling data. Geophysics 46:892–903

Harness PE, Sbins FL, Griffith JE (1992) New technique provides better low-density-cement evaluation. SPE 24050. Western Regional Meeting, Bakersfield, California, USA, March 1992

Havira RM (1979) Ultrasonic bond evaluation in multilayered media. J Acoust Soc Am 66:S41

Hayden R, Russell C, Vereide A, Babasick P, Shaposhnikov P, May D (2011) Case studies in evaluation of cement with wireline logs in a deep water environment. In: Society of petrophysicists and well log analysts 52nd annual symposium

Hayman AJ, Hutin R, Wright PV (1991) High-resolution cementation and corrosion imaging by ultrasound. In: Society of petrophysicists and well log analysts 32nd annual symposium

He X, Wang X, Chen H (2017) Theoretical simulations of wave field variation excited by a monopole within collar for acoustic logging while drilling. Wave Motion 72:287–302

He F (2005) The study on the simulation of the borehole acoustic reflection imaging logging tool and it's waveform processing method. PhD thesis (In Chinese with English Abstract), China University of Petroleum, Beijing, China

Herold B, Marketz F, Froelich B, Edwards JE, Kuijk R, Welling R, Leuranguer C (2006) Evaluating expandable tubular zonal and swelling elastomer isolation using wireline ultrasonic measurements. In: IADC/SPE Asia Pacific drilling technology conference and exhibition, paper 103893

Higdon RL (1990) Radiation boundary conditions for elastic wave propagation. SIAM J Numer Anal 27(4):831–869

Hilchie DW (1978) Applied openhole well log interpretation (for geologists and engineers). Douglas W. Hilchie Inc., Golden Colorado

Hirabayashi N, Torii K, Yamamoto H, Haldorsen J, Voskamp A (2010) Fracture detection using borehole acoustic reflection survey data. SEG Tech Prog Expand Abstr 29:523–527

Hirabayashi N, Martinez GA, Wielemaker E (2016) Case Studies of borehole acoustic reflection survey (BARS). In: 22nd formation evaluation symposium of Japan, September 29–30

Hornby BE, Pasternark E (2000) Analysis of full-waveform sonic data acquired in unconsolidated gas sands. Petrophysics 41:363–374

Hornby BE (1989) Imaging of near-borehole structure using full-waveform sonic data. Geophysics 54(6):747–757

Huang C, Hunter JA (1980) The correlation of "Tube wave" events with open fractures in fluid-filled boreholes, Atomic Energy of Canada Seismic downhole survey progress report—1979

Huang X (2003) Effects of tool positions on borehole acoustic measurements: a stretched grid finite difference approach. PhD dissertation. Massachusetts Institute of Technology, Cambridge, MA, USA

Huang X, Yin H (2005) A data-driven approach to extract shear and compressional slowness from dispersive waveform data. SEG Tech Prog Expand Abstr 2005:384–387

Ikelle LT, Amundsen L (2018) Introduction to petroleum seismology. Society of Exploration Geophysicists

Jarrot A, Gelman A, Kusuma J (2018) Wireless digital communication technologies for drilling: communication in the bits/s regime. IEEE Signal Process Mag 35(2):112–120

Johnson HM (1962) A history of well logging. Geophysics 27(4):507–527

Johnson SG (2007) Notes on perfectly matched layers (PMLs). MIT course 18.369

Joyce B, Patterson D, Leggett J, Dubinsky V (2001) Introduction of a new omni-directional acoustic system for improved real-time LWD sonic logging-tool design and field test results. In: SPWLA 42nd annual logging symposium, Paper SS

Jutten J, Corrigall E (1989) Studies with narrow cement thickness lead to improved CBL in concentric casings. J Pet Tech 1158–1192

Jutten J, Hayman A (1993) Microannulus effect on cementation logs: experiments and case histories. Society of Petroleum Engineers, paper 25377

Jutten J, Parcevaux P (1987) Relationship between cement bond log output and borehole geometrical parameters. Society of Petroleum Engineers, paper 16139

Kabir MN, Verschuur D (1995) Restoration of missing offsets by parabolic Radon transform. Geophys Prospect 43(3):347–368

Kay SM (1988) Modern spectral estimation: theory and application. Prentice Hall, p 225

Kelly KR, Ward RW, Treitel S, Alford RM (1976) Synthetic seismograms: a finite-difference approach. Geophysics 41:2–27

Kimball CV, Marzetta TL (1984) Semblance processing of borehole acoustic array data. Geophysics 49:274–281

Kimball CV (1998) Shear slowness measurement by dispersive processing of borehole flexural mode. Geophysics 63(2):337–344

Kingsbury NG (2001) Complex wavelets for shift invariant analysis and filtering of signals. J Appl Computat Harmonic Anal 10(3):234–253

Kinoshita T, Dumont A, Hori H, Sakiyama N, Morley J, Garcia-Osuna F (2010) LWD sonic tool design for high-quality logs. Soc. Exp. Geophys., Tech Prog Expand Abstr 29(1):513–517

Kitsunezaki C (1980) A new method for shear wave logging. Geophysics 45:1489–1506

Komatitsch D, Martin R (2007) An unsplit convolutional perfectly matched layer improved at grazing incidence for the seismic wave equation. Geophysics 72(5):M155–M167

Kurkjian AL (1985) Numerical computation of individual far-field arrivals excited by an acoustic source in a borehole. Geophysics 50:852–866

Kurkjian A, Chang SK (1986) Acoustic multipole sources in fluid filled boreholes. Geophysics 51(1):148–163

Kuzuoglu M, Mittra R (1996) Frequency dependence of the constitutive parameters of causal perfectly matched anisotropic absorbers. IEEE Microwave Guid Wave Lett 6(12):447–449

Lamb H (1917) On waves in an elastic plate. Proc R Soc London (Series A), 93(648):114–128

Lang SW, Kurjian AL, McClellan JH, Morris CF, Parks TW (1987) Estimating slowness dispersion from arrays of sonic logging waveforms. Geophysics 52(4):530–544

Lay T, Wallace TC (1995) Modern global seismology. Academic Press

Lebedev NN (1972) Special functions and their applications. Dover, reprint (Translated from Russian)

Lebourg M, Fields RQ, Doh CA (1956) A method of formation testing on logging cable. TP No. 701-G, Fall Meeting, Society of Petroleum Engineers of A.I.M.E., Los Angeles, California, United States

Lecampion B, Quesada D, Loizzo M et al (2011) Interface debonding as a controlling mechanism for loss of well integrity: importance for CO2 injector wells. Energy Procedia 4:5219–5226

Leggett JV, Dubinsky V, Patterson D, Bolshakov A (2001) Field test results demonstrating improved real-time data quality in an advanced LWD Acoustic system, SPE71732

Le Calvez J, Brill TM (2018) Separation of flexural and extensional modes in multi modal acoustic signals. European Patent Application, 16306114.6

Li C, Yue W (2015) High-resolution adaptive beamforming for borehole acoustic reflection imaging. Geophysics 80(6):D565–D574

Li C, Yue W (2017) High-resolution Radon transforms for improved dipole acoustic imaging. Geophys Prospect 65(2):467–484

Li Y, Zhou R, Tang X, Jackson JC, Patterson DJ (2002) Single-well imaging with acoustic reflection survey at Mounds, Oklahoma, USA. In: EAGE 64th conference and exhibition, paper 141, Florence, Italy, 2002, May 27–30

Li J, Tao G, Zhang K et al (2014) An effective data processing flow for the acoustic reflection image logging. Geophys Prospect 62(3):530–539

Li M, Tao G, Wang H et al (2016) An improved multiscale and leaky P-wave removal analysis for shear-wave anisotropy inversion with crossed-dipole logs. Petrophysics 57(3):270–293

Li W, Tao G, Torres-Verdín C (2015) Forward and backward amplitude and phase estimation method for dispersion analysis of borehole sonic measurements. Geophysics 80(3):D295–D308

Li Y, Wang H, Fehler M, Fu Y (2017) Wavefield characterization of perforation shot signals in a shale gas reservoir. Phys Earth Planet Interi 267:31–40

Lisitsa V, Vishnevskiy D (2010) Lebedev scheme for the numerical simulation of wave propagation in 3D anisotropic elasticity. Geophys Prospect 58:619–635

Liu D, Hu W, Chen Z (2008) SVD-TLS extending Prony algorithm for extracting UWB radar target feature. J Syst Eng Electron 19(2):286–291

Love AEH (1952) A treatise on the mathematical theory of elasticity, 4th edn. Dover Publications

Luthi (2000) Geological well logs: their use in reservoir modeling. Springer

Market J, Bilby C (2011) Introducing the first LWD cross-dipole sonic imaging service. In: SPWLA 52nd annual logging symposium, paper DDD

Marple SL (1987) Digital spectral analysis: with applications. Prentice Hall Press

Martin RD, Komatitsch SD, Gedney et al (2010) A high-order time and space formulation of the unsplit perfectly matched layer for the seismic wave equation using auxiliary differential equations (ADE-PML). CMES 56:17–40

Matuszyk PJ, Torres-Verdin C (2011) HP-adaptive multi-physics finite-element simulation of wireline borehole sonic waveforms. SEG Tech Prog Expand Abstr 30(1):444–448

McMechan GA (1983) Migration by extrapolation of time-dependent boundary values. Geophys Prospect 31(3):413–420

Meza-Fajardo KC, Papageorgiou AS (2008) A nonconvolutional, split-field, perfectly matched layer for wave propagation in isotropic and anisotropic elastic media: stability analysis: Bull Seismol Soc Am 98:1811–1836

Miller U (1977) Symmetry and separation of variables. Addison-Wesley

Miller D, Oristaglio M, Beylkin G (1987) A new slant on seismic imaging: migration and integral geometry. Geophysics 52(7):943–964

Miller D, Stanka FE (1999) Method of analyzing waveforms. US Patent 5859811

Minear J, Birchak R, Robbins CA et al (1995) Compressional slowness measurements while drilling. In: SPWLA 36th annual logging symposium, paper VV

Morris C, Sabbagh L, Wydrinski R, Hupp J, van Kuijk R, Froelich B (2007) Application of enhanced ultrasonic measurements for cement and casing evaluation. In: SPE/IADC drilling conference, paper 105648

Nakken EI, Mjaaland S, Solstad A (1995) A new MWD concept for geological positioning of horizontal wells, SPE, 30454

Nolte B, Rao R, Huang X (1997) Dispersion analysis of split flexural waves. In: Borhole acoustic and logging/Reservoir delineation consortia annual report, MIT

Op't Root T, Stolk CC, de Hoop MV (2012) Linearized inverse scattering based on seismic reverse time migration. J Math Pures Appl 98(2):211–238

Paillet FL, Cheng CH (1986) A numerical investigation of head waves and leaky modes in fluid-filled boreholes. Geophysics 51(7):1438–1449

Paillet F (1981) Predicating the frequency content of acoustic waves in boreholes. In: Society of professional well log analysts annual logging symposium, 22nd, paper SS

Paillet F, Cheng C (1991) Acoustic waves in boreholes. CRC Press

Pardue GH, Morris RL, Gollwitzer LH, Moran JH (1963) Cement bond log—a study of cement and casing variables. J Pet Tech 5:545–555

Patterson D, Bolshakov A, Matuszyk P (2015) Utilization of electromagnetic acoustic transducers in downhole cement evaluation. In: SPWLA 56th annual logging symposium

Peng C, Toksöz MN (1992) Tube wave generation at a layer boundary for an incident compressional plane wave. MIT Earth Resources Lab Consortium Annual meeting report

Pistre V, Kinoshita T, Endo T et al (2005) A modular wireline sonic tool for measurements of 3D (azimuthal, radial, and axial) formation acoustic properties. In: SPWLA 46th annual logging symposium, New Orleans, Louisiana, United States, June 26–29

Plona B, Sinha S, Kostek, Chang SK (1992) Axisymmetric wave propagation in fluid-loaded cylindrical shells. II: theory versus experiment. J Acoust Soc Am 92(2):1144–1155

Poletto F, Miranda F (2004) Seismic while drilling: fundamentals of drill-bit seismic for exploration. Elsevier Publishing

Prony R (1795) Essai experimental et analytique: L'ecole Polytech 1:24–76

Qleibo M (2012) SonicScope the next generation of sonic while drilling. https://www.fesaus.org/webcast/2012/09/NTF/2_SonicScope_Schlumberger.pdf. Access 25 April 2018

Rao R, Toksöz MN (2005) Dispersive wave analysis - method and applications. Earth Resources Lab Consortium Annual report

Reiter E (1991) Imaging of large offset ocean bottom seismic data. Ph.D. Thesis. Massachusetts Institute of Technology, Cambridge, MA, United States

Ripley HE, Harms WW, Sutton DL, Watters LT (1981) Ultra-low density cementing compositions. J Canad Petrol 1(6):112–118

Roden JA, Gedney SD (2000) Convolutional PML (CPML): an efficient FDTD implementation of the CFS-PML for arbitrary media. Microwave Opti Technol Lett 27:334–338

Rose JL (1999) Ultrasonic waves in solid media. Cambridge University Press, Cambridge, United Kingdom, pp 1–100

Rust WM Jr (1938) A historical review of electrical prospecting methods. Geophysics 3(1):1–6

Sacchi MD, Ulrych TJ (1995) High-resolution velocity gathers and offset space reconstruction. Geophysics 60(4):1169–1177

Sacchi MD, Porsani M (1999) Fast high-resolution parabolic Radon transform. 69th SEG Annual Meeting, Expanded Abstracts 1477–1480

Saenger EH, Bohlen T (2004) Finite-difference modeling of viscoelastic and anisotropic wave propagation using the rotated staggered grid. Geophysics 69:583–591

Sanyal SK, Meidav HT (1976) We11 Logging in the Geothermal Industry, Paper presented at the 17th Annual Logging Symp. of the SPWLA, Denver, Colorado, June 1976

Schlumberge (1987) Log interpretation principles/applications. Schlumberger educational services, Houston, Texas, United States

Schmitt DP, Bouchon M (1985) Full wave acoustic logging: synthetic microseismograms and frequency-wavenumber analysis. Geophysics 50:1756–1778

Schmitt DP (1988) Shear-wave logging in elastic formations. J Acoust Soc Am 84(6):2215–2229

Selesnick IW, Baraniuk RG, Kingsbury NG (2005) The dual-tree complex wavelet transform. IEEE Trans Signal Process 22(6):123–151

Shahvali A, Azin R, Zamani A (2014) Cement design for underground gas storage well completion. J Nat Gas Sci Eng 18:149–154

Shang XF (2014) Inverse scattering: theory and application to the imaging of the earth's seismic discontinuities. PhD Thesis. Massachusetts Institute of Technology, Cambridge, MA, United States

Sinha BK, Zeroug S (1999) Applications of sonics and ultrasonics in geophysical prospecting. IEEE Ultraso Symp 521–532

Skelton EA, Adams S, Craster R (2007) Guided elastic waves and perfectly matched layers. Wave Motion 44:573–592

Smith WD (1974) A nonreflecting plane boundary for wave propagation problems. J Comput Phys 15(4):492–503

Smolen J (1996) Cased hole and production log evaluation. Pennwell Pub 161–191

Song FX, Toksöz MN (2009) Model-guided geosteering for horizontal drilling. MIT, Earth Resources Laboratory Industry Consortia Annual Report, 2009, 05

Song R, Liu J, Hou C, Wang K (2012) Numerical simulation of Sector Bond log and improved cement bond image. Geophysics 77(4):95–104

Stephen RA, Cardo-Casas F, Cheng CH (1985) Finite-difference synthetic acoustic logs. Geophysics 50:1588–1609

Stoica P, Li HB, Li J (1999) A new derivation of the APES filter. IEEE Signal Process. Lett. 6(8):205–206

Stolt RH (1978) Migration by Fourier transform. Geophysics 63(1):23–48

Su Y, Tang X, Xu S, Zhuang C (2015) Acoustic isolation of a monopole logging while drilling tool by combining natural stopbands of pipe extensional waves. Geophys J Int 202:439–445

Summers GC, Broding RA (1952) Continuous velocity logging. Geophysics 17(3):598–614

Tang XM (1997) Predictive processing of array acoustic waveform data. Geophysics 62(6):1710–1714

Tang XM, Wang T, Patterson D (2002a) Multipole acoustic logging-while-drilling. SEG Tech Prog Expand Abstr 21(1):364–367

Tang XM, Dubinsky V, Wang T, Bolshakov A, Patterson D (2002b) Shear-velocity measurement in the logging-while drilling environment: modeling and field evaluations. In: SPWLA 43rd annual logging symposium, paper RR

Tang XM, Patterson D, Dubinsky V, Harrison CW, Bolshakov A (2003) Logging-while-drilling shear and compressional measurements in varying environments. In: SPWLA 44th annual logging symposium, paper II

Tang XM, Cheng CH (2004) Quantitative borehole acoustic methods. Elsevier

Tang X (2004) Imaging near-borehole structure using directional acoustic-wave measurement. Geophysics 69:1378–1386

Tang XM, Patterson D (2005) Analyzing and processing acoustic logging data for poorly bonded cased boreholes. In: Society of petrophysicists and well log analysts 46th annual symposium

Tang XM, Zheng Y, Vladimir D (2005) Logging while drilling acoustic measurement in unconsolidated slow formations. In: SPWLA 46th, paper R

Tang XM, Zheng Y, Patterson D (2007) Processing array acoustic-logging data to image near-borehole geologic structures. Geophysics 72(2):E87–E97

Tang XM, Li C, Patterson D (2009) A curve-fitting method for analyzing dispersion characteristics of guided elastic waves. SEG Tech Prog Expand Abstr 461–465

Tang XM, Patterson D (2009) Single-well S-wave imaging using multicomponent dipole acoustic-log data. Geophysics 74(6):WCA211–WCA223

Tang XM, Wei Z, Su Y, Zhuang C (2013) A review on the progress and application of dipole acoustic reflection imaging technology. Well Logg Technol (In Chinese with English Abstract) 37(4):330–340

Tang X (2016) Reflection imaging away from borehole: a paradigm shift. SEG Tech Prog Expand Abstr 4873–4877

Tang XM (2017) Acoustic characterization and imaging of shale gas fractures in horizontal wells. SEG 87th Annual Meeting, Expanded Abstracts 959–963

Tang XM, Li S, Xu S, Su Y, Zhuang C (2017) Acoustic characterization and imaging of shale gas fractures in horizontal wells: field case study in the Sichuan Basin of Southwest China. Well Logg Technol (In Chinese with English Abstract) 41(5):501–505

Tao G, Cheng CH Toksöz MN (1997) Measurements of shear-wave azimuthal anisotropy with cross-dipole logs. In: Industry consortia annual report of Earth Resources Laboratory at Massachusetts Institute of Technology, 1997–10

Tao G (2006) Logging information processing and application. China University of Petroleum Press, Dongying, China

Tao G, He F, Yue W, Chen P (2008a) Processing of array sonic logging data with multi-scale STC technique. Petrol Sci 5(3):238–241

Tao G, He F, Wang B, Wang H, Chen P (2008b) Study on 3D simulation of wave fields in acoustic reflection image logging. Sci China Earth Sci 51:186–194

Trad DO, Ulrych TJ, Sacchi MD (2002) Accurate interpolation with high resolution time-variant Radon transforms. Geophysics 67(2):644–656

Trad D, Ulrych T, Sacchi M (2003) Latest views of the sparse Radon transform. Geophysics 68(1):386–399

Thierry S, Klieber C, Lemarenko M, Le Calvez J, Brill T, Barrou T, Hayman A, Mege F, Van Os R (2016) New-generation ultrasonic measurements for quantitative cement evaluation in heavy muds and thick-wall casings. In: Society of petroleum engineers annual technical conference and exhibition, SPE-181450-MS

Thierry S, Klieber C, Lemarenko M, Brill T, Le Calvez J, Mege F, Barrou T, Lavrentiev M, Constable K (2017) Ultrasonic cement logging: expanding the operating envelope and efficiency. In: SPWLA 58th annual logging symposium, Paper number: SPWLA-2017-WWWW

Tichelaar BW, van Luik KW (1995) Sonic logging of compressional-wave velocities in a very slow formation. Geophysics 60(6):1627–1633

Tixier MP, Martin M (1954) History of logging and future development. Canadian Oil Gas Indust 7(8)

Toksöz MN, Johnston DH (1981) Seismic wave attenuation. Society of Exploration Geophysicists

Toksöz MN, Cheng CH, Garcia G (1984a) Full waveform acoustic logging—where we are, where we are going. MIT Earth Resources Lab Consortium Annual meeting report

Toksöz MN, Wilkens RH, Cheng CH (1984b) Determination of shear wave velocity and attenuation from waveforms in low velocity formations. MIT Earth Resources Lab Consortium Annual meeting report

Tsang L, Rader D (1979) Numerical evaluation of transient acoustic waveform due to a point source in a fluid-filled borehole. Geophysics 44(10):1706–1720

Tsang L (1978) Tim-harmonic solution of the elastic head wave problem incorporating the influence of Rayleigh poles. J Acoust Soc Am 63:1302–1309

Tubman KM (1984) Full waveform acoustic logs in radially layered boreholes. PhD Thesis. Massachusetts Institute of Technology, Cambridge, MA, United States

Tubman KM, Cole SP, Cheng CH, Toksöz MN (1984) Dispersion curves and synthetic microseismograms in unbonded cased boreholes. Earth Resources Laboratory Industry Consortia Annual Report, 1984–02

Van Kuijk R, Zeroug S, Froelich B, Allouche M, Bose S, Miller D, Calvez J, Schoepf V (2005) A novel ultrasonic cased-hole imager for enhanced cement evaluation. In: International petroleum technology conference, paper 10546

van Kuijk R, Calvez JL, Froelich B (2006) Determination of the impedance of a material behind a casing combination two sets of ultrasonic measurements. United States Patent, NO. 7149146 B2

Varsamis GL, Wisniewski L, Arian A et al (1999) A new MWD full wave dual mode sonic tool design and case histories. In: SPWLA 40th annual logging symposium, paper F

Virieux J (1986) P-SV wave propagation in heterogeneous media: velocity-stress finite-difference method. Geophysics 51:889–901

Vogel CB (1952) A seismic velocity logging method. Geophysics 17:586–597

Walker T (1968) A full-wave display of acoustic signal in cased holes. J Pet Tech 818–824

Waters KH (1978) Reflection seismology. Wiley
Wang B, Tao G, Wang H, Li W, Tan B (2011) Extraction method of P and S reflection waves from array acoustic logging data. J China Univ Petrol 35(2):57–63 (in Chinese with English abstract)
Wang T, Tang X (2003) Finite-difference modeling of elastic wave propagation: a nonsplitting perfectly matched layer approach. Geophysics 68:1749–1755
Wang P, de Hoop M, van der Hilst R, Ma P, Tenorio L (2006) Imaging of structure at and near the core mantle boundary using a generalized radon transform: 1. Construction of image gathers. J Geophys Res Solid Earth 111:B12304
Wang H, Tao G, Wang B, Li W, Zhang X (2009) Wave field simulation and data acquisition scheme analysis for LWD acoustic tool: Chinese. J Geophys (In Chinese) 52:2402–2409
Wang H, Tao G (2011) Wavefield simulation and data-acquisition-scheme analysis for LWD acoustic tools in very slow formations. Geophysics 76(3):E59–E68
Wang H, Tao G, Shang X, Fang X, Burns D (2013) Stability of finite difference numerical simulations of acoustic logging-while-drilling with different perfectly matched layer schemes. Appl Geophys 10(4):384–396
Wang H, Tao G, Zhang K (2013) Wavefield simulation and analysis with the finite-element method for acoustic logging while drilling in horizontal and deviated wells. Geophysics 78(6):D525–D543
Wang H, Tao G, Fehler M (2015) Investigation of the high-frequency wavefield of an off-center monopole acoustic logging-while-drilling tool. Geophysics 80(4):D329–D341
Wang H, Tao G, Shang X (2015) A method to determine the strike of interface outside of borehole by monopole borehole acoustic reflections. J Petrol Sci Eng 133:304–312
Wang H, Tao G, Shang X (2016a) Understanding acoustic methods for cement bond logging. J Acoust Soc Am 139(5):2407–2416
Wang X, He X, Zhang X (2016) Generalized collar waves in acoustic logging while drilling. Chin Phys B 25:124316
Wang H, Fehler M, Tao G, Wei Z (2016) Investigation of collar properties on data-acquisition scheme for acoustic logging-while-drilling. Geophysics 81(6):611–624
Wang P, Bose S, Sinha BK, Blyth M (2016d) Dipole shear anisotropy using logging-while-drilling sonic tools. In: SPWLA 57th annual logging symposium, paper PPPP
Wang H, Fehler MC, Miller D (2017) Reliability of velocity measurements made by monopole acoustic logging-while-drilling tools in fast formations. Geophysics 82(4):D225–D233
Wang H, Fehler MC (2018) The wavefield of acoustic logging in a cased-hole with a single casing—part I: a monopole tool. Geophys J Int 212(1):612–626
Wang H, Fehler M (2018) The wavefield of acoustic logging in a cased hole with a single casing—part II: a dipole tool. Geophys J Int 212(2):1412–1428
Wang H, Fehler M, Fournier A (2018) Assessing CO_2 leak paths by analysis of borehole monopole wavefield modes. SEG Techn Prog Expand Abstr
Wang B, Zhang K (2019) Investigation on acoustic reflection logging of LWD using a flexible finite difference technique. Explorat Geophys 50(2):172–180
Wei Z, Wang H (2016) Numerical simulation and field data analysis of an off-center quadrupole acoustic wireline tool. SEG Tech Prog Expand Abstr 820–824
White JE (1965) Seismic waves, radiation, transmission and attenuation, ch. 4. McGraw-Hill, New York
White JE (1967) The hula log: a proposed acoustic tool. In: Transform 8th Annual Logging Symposium Society Professor Well-Log Analysis, paper I
Whitmore ND (1983) Iterative depth imaging by backward time propagation. SEG 53rd Annual Meeting, Expanded Abstracts, pp 382–384
Winbow GA, Chen ST, Rice JA (1991) Acoustic quadrupole shear wave logging device. US Patent: 5027331
Wu PT, Darling HL, Scheibner D (1995) Low-frequency P-wave logging for improved compressional velocity in slow formation gas zones. SEG Tech Prog Expand Abstr 14(1):9–12
Yilmaz Ö (2001) Seismic data analysis. Society of Exploration Geophysicists, Tulsa, OK, USA

Yu S (1984) The resonance modes of elastic wave in a fluid-filled borehole in acoustic well logging. Chinese J Geophys 27(1):94–102 (In Chinese with English Abstract)
Zemanek J, Angona FA, Williams DM, Caldwell RL (1984) Continuous shear wave logging, paper U. In: 25th annual logging symposium transactions. Society of Professional Well Log Analysts
Zemanek J, Caldwell RL, Glenn E, Holcomb SV, Norton LJ, Straus A (1969) The Borehole televiewer—a new concept for fracture location and other types of borehole inspection. J Pet Tech 25:762–774
Zeroug S, Froelich B (2003) Ultrasonic leaky-Lamb wave imaging through a highly contrasting layer. In: IEEE ultrasonic symposium, pp 794–798
Zeroug S, Bose S, Sinha B, Liu Y, D'Angelo R (2016) Sonic and ultrasonic measurement applications for cased oil wells. Insight 58(8):423–430
Zhang XD (1996) Modern signal processing. Tinghua University Press, Beijing
Zhang T, Tao G, Li J, Wang B, Wang H (2009) Application of the equivalent offset migration method in acoustic log reflection imaging. Appl Geophys 6(4):303–310
Zhan X, Zhu Z, Chi S, Toksöz M (2010) Elimination of LWD (Logging While Drilling) tool modes using seismoelectric data: Commun. Comput Phys 7:47–63
Zhang W, Shen Y (2010) Unsplit complex frequency-shifted PML implementation using auxiliary differential equations for seismic wave modeling. Geophysics 75:T141–T154
Zhang H, Xie D, Shang Z, Yang X, Wang H, Tao H (2011) Simulated various characteristic waves in acoustic full waveform relating to cement bond on the secondary interface. J Appl Geophys 73(2):139–154
Zhang K (2016) Discussion on borehole wave modes excited by two types of acoustic logging while drilling sources. Mathemat Comput Sci 1(3):48–55
Zhang X, Wang X, Zhang H (2016) Leaky modes and the first arrivals in cased boreholes with poorly bonded conditions. Sci China-Phys Mech Astron 59(2):624301
Zhang J, Xu X, Yu Q, Zhai Y, Chao Y, Ji Z (2018) Development and application of victory multicomponent reflected shear wave imager. J Appl Acoust (In Chinese with English Abstract) 37(2):268–272
Zheng H (2000) Interpretating Respon ses of сгдт casedhole gamma-density-thickness tool using numerical modeling techniques. Well Logg Tech 24(4):243–252
Zheng Y, Tang X (2005) Imaging near-borehole structure using acoustic logging data with pre-stack F-K migration. In: SEG 75th annual meeting, expanded abstracts, pp 360–363
Zheng Y, Hu H (2017a) Nonlinear signal comparison and high-resolution measurement of surface-wave dispersion. Bull Seismol Soc Am 107(3). https://doi.org/10.1785/0120160242
Zheng X, Hu H (2017) A theoretical investigation of acoustic monopole logging-while-drilling individual waves with emphasis on the collar wave and its dependence on formation. Geophysics 82:D1–D11
Zhu J (1999) A transparent boundary technique for numerical modeling of elastic waves. Geophysics 64(3):963–966

Index

H. Wang et al., *Borehole Acoustic Logging—Theory and Methods*, Petroleum Engineering, https://doi.org/10.1007/978-3-030-51423-5

D

E

F

G

S

The manufacturer's authorised representative in the EU is Springer Nature Customer Service Centre GmbH, Europaplatz 3, 69115 Heidelberg, Germany. If you have any concerns regarding our products, please contact ProductSafety@springernature.com

Printed and bound by CPI Group (UK) Ltd, Croydon, CR0 4YY
22/09/2026
02221060-0015